Desert V

SAND VERBENA (*Abronia*) and DUNE EVENING PRIMROSE (*Oenothera*) of the sand dunes.

Desert Wild Flowers

Edmund C. Jaeger

Author of
The North American Deserts
The California Deserts
Denizens of the Deserts
Desert Wildlife

Revised Edition

Stanford University Press
Stanford, California

Stanford University Press
Stanford, California

Printed in the United States of America
Cloth SBN 8047-0364-7
Paper SBN 8047-0365-5
First published 1940
Revised edition 1941
Last figure below indicates year of this printing:
88 87

Dedicated to

Samuel Bonsall Parish

and

Harvey Monroe Hall

California botanists, representatives of the highest traditions of their science, eminent scholars of the desert flora.

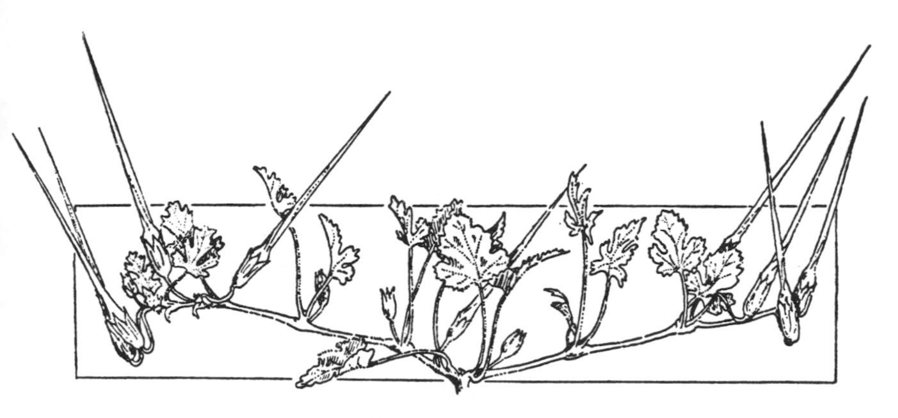

Erodium texanum
(No. 264)

Preface

TWENTY-FIVE years ago the author, with a pack burro, began trekking through the deserts of the Southwest. With pencil and sketch pad, he recorded the forms and detailed structures of the strange plants he saw. Continuing that practice through the intervening years, he has now collected his sketches in a book which may help those traveling similar roads and byways to identify the flowers they encounter. The line drawings show better than any photographs (even colored ones) the details of structure which make identification possible, and their aid will be appreciated by those who would really know the species they find.

With few exceptions all of the sketches were made from living material in the field. Those which were sketched from herbarium material are marked by an asterisk (*). To draw so many plants in their native habitat involved traveling over thirty thousand miles of desert roads and trails. Every effort has been made to show each plant in its natural relations as to stem angle, position of leaves, and flower detail; all of these are very characteristic for each species, and are easily lost by those who draw from pressed herbarium specimens. Color is a changing factor in a herbarium specimen; hence special pains were taken to record colors from living specimens. Those who would use this book most profitably would do well not only to familiarize themselves thoroughly with the illustrations but also to carry with them on their travels a box of colored indelible pencils so that they may fill in the natural colors directly from the flower. This if carefully done will not only greatly increase the attractiveness of one's copy of the book but will also impress the plant more firmly in one's memory.

It has not always been possible, owing to many problems facing the illustrator in the field, to have the genera and species in the exact order in which they occur in floral manuals. But plants are grouped into families according to the accepted order.

Practically all known plants of the deserts of California and related portions of Nevada (including the Lake Mead area at Hoover Dam) are included. A considerable proportion of these have never before been illustrated in popular or scientific literature. Here, then, the desert traveler has a veritable "pocket picture-book herbarium" which he can easily carry with him, confident that he may find illustrated nearly every species occurring in the desert area. Unless otherwise indicated, all the drawings show the plants one-half natural size. Allowance must be made for variations in size due to differences in habitat.

Special attention has been given to the natural history of the desert plants rather than to their description. Since there are some details of structure not evident from the illustrations alone, brief general floral descriptions are sometimes appended. Features perhaps new to a popular flower handbook are the explanations of the meanings of scientific names and the short accounts of the more than one hundred fifty desert explorers commemorated in botanical names. Thus, an opportunity is given to glean a good knowledge of desert exploration and botanical history. Up to this time a majority of the desert plants have never had English or so-called common names. In supplying these here every effort has been made to select simple, dignified, and appropriate names, based upon some diagnostic character, a natural history feature, or a prominent botanist's or collector's connection with the plant. The scientific names are for the most part those used by Munz in his *Manual of Southern California Botany*. Where trinomials are employed in botanical names, the third term refers to a varietal, not a subspecific, name.*

Contrary to the usual opinion, there is not a month in the year when no plants are in flower on the desert; we may truly say that the flower season of the desert lasts all through the year. Of course the great spectacular show of color comes in the spring—February through March on the Colorado Desert, and April through May on the Mohave. In the arid piñon-juniper areas the floral season begins early in May and lasts through June. Some of the shrubby composites, such as the rabbit brushes, do not come into flower until autumn.

* If specific and varietal names are apparently nouns treated as adjectives, they are defined as adjectives; such are names ending in *-theca*, *-folia*, and *antha*.

The California deserts are part of the great Sonoran Biotic Province, and comprise an enormous area of some thirty-five thousand square miles, extending from the high mountains north of Death Valley south along the state's eastern borders to the Mexican boundary. It is an area of varied topography, with broad basins segregated wholly or partly by numerous mountain groups, which vary in altitude from a few hundred to over ten thousand feet. Some of the basins are far below sea level, while contiguous areas may be so high as to support a vegetation almost montane in character.

The plant life of the desert is as varied as its topography. Many of the species have been immigrants: some coming into the Mohave Desert from the higher mountainous areas to the north and east and creeping southward even to the Colorado Desert; others slipping over the edge of the Sierra chain to the west and adding a coastal element to the floral assemblage; and another host migrating northward from the Mexican plateau region of Sonora. The present California desert flora are thus a complex mixture of at least four elements. That is why the plant list is so extensive and varied in its make-up.

The deserts of California, described in detail in the author's *The California Deserts*, are two in number. The larger, more elevated, northern one, lying north of the Chuckawalla Mountains and containing both the Death Valley and the Joshua Tree national monuments, is called the Mohave Desert (often spelled Mojave). The smaller, southern one, including the region immediately contiguous to the lower Colorado River and the rather low-lying areas which drain into the Salton Sea, is called the Colorado Desert. The latter lies almost wholly in southern California; its Baja California extensions are not treated here.

This book will prove of interest not only to schooled professional and amateur botanists but also to travelers and out-of-doorists. Cattlemen, ornithologists, entomologists, zoölogists, ethnologists, beekeepers, agriculturists, and allergists will all find their special interests amply regarded throughout the pages. Text and drawings are designed to facilitate quick identification of the plants, no matter who the user of the book may be or where his interests may lie.

Inasmuch as this book is intended primarily as a field reference book and not one for continuous reading, brevity has been stressed, somewhat altering the ordinary style of literary composition and making the introduction of a few abbreviations of often-used terms desirable. The amount of text varies much from plant to plant. If a sketch of some plant seems unusually brief, it is because there is really little to say concerning that plant's natural history.

Grateful acknowledgments of critical suggestions and generous loans of herbarium specimens for study and comparison are due to Dr. Philip Munz of Pomona College; Dr. LeRoy Abrams, Mrs. Roxana Ferris, and Dr. Ira Wiggins of the Dudley Herbarium at Stanford University; Dr. David Keck of the Carnegie Institution of Washington; Dr. Lincoln Constance of the University of California at Berkeley; Dr. Ivan Johnston of the Gray Herbarium; and Dr. S. Stillman Berry of Redlands, California. Special thanks are also in order for the many notes on natural history and flower colors supplied by Mr. M. French Gilman, able naturalist of the Death Valley National Monument; and to Mr. Lloyd Smith, Mr. Dean Hollingsworth, and Mr. Ernest Gifford for numerous suggestions and for much highly exacting clerical work. Loans of photographs are specifically acknowledged by placing the names of contributors beneath the illustrations, except the end-sheet photograph, which is by Mr. Hollingsworth.

E. C. J.

RIVERSIDE, CALIFORNIA
January 1, 1940

Carl Eytel

List of Families

Glossary

(Citations are to plant numbers, not to pages)

Alternate leaves. Only one from each place on a stem where a leaf is borne. Illustrations: 158, 271.

Axillary. Borne or occurring in the angle between a leaf and the stem.

Bract. Modified leaf reduced in size, subtending a flower or flower cluster.

Caespitose. Growing in matted turf-like tufts.

Compound leaf. A leaf in which the blade is made up of several separate leaflets. Illustrations: 198, 218.

Head. Said of flowers in a globose cluster, all without separate flower stalks, and gathered on a common receptacle surrounded by bracts. Illustrations: Sunflower Family 608–764.

Inflorescence. Arrangement of flowers in a cluster.

Linear. Very narrow, with parallel sides, several times as long as wide.

Opposite leaves. Two from each leaf-bearing region on the stem and proceeding from opposite sides of the stem. Illustrations: 78, 84, 283.

Palmately. With the leaflets diverging radiately from the apex of a common leaf-stem. Illustrations: 214–223.

Parasite. A plant which may be recognized from its lack of green coloring.

Pinnately. With the leaflets in pairs along a common leaf-stem. Illustrations: 195–200.

Raceme. A flower cluster in which the flowers are borne along the stem on flower stalks of nearly equal length, the oldest being toward the base of the cluster. Illustrations: 159, 167.

Revolute. Rolled backward.

Scale leaf. Minute structures usually closely appressed to the stem. Illustrations: 7, 34.

Semi-parasite. A plant like mistletoe, which grows on and penetrates the tissues of its host plant, but is green colored.

Simple leaf. The blade composed of one piece. Illustrations: 277, 306, 473.

Spatulate. Rounded apically and contracted below to a narrow and slender base. Illustration: 527.

Spike of flowers. A flower cluster in which the flowers are without individual flower stalks and more or less densely arranged along a common stem. Illustrations: 240, 246.

Umbel. A flower cluster in which the individual flower stalks are of approximately equal length all originating at the same point on a common stem. Illustrations: 382–391.

Whorled leaves or flowers. With the parts borne in a circle around the stem. Illustrations: 525, 593–596.

Introductory Remarks Concerning the Key

The chief concern of the key is the quick identification of unknown plants. To accomplish this, there have been listed all the species described in this book and these have been reassembled into groups based on simple but distinctive non-technical characteristics, such as habit of growth, gross anatomical features, arrangement of leaves, color of flowers, etc. This grouping is achieved in the key by arranging contrasting characteristics in an indented outline form. In some cases there are only two alternatives to choose between; in other cases there are numerous choices or divisions. Since some confusion may result because the user may not know how many divisions to look for in case they appear on different pages, the divisions have been designated with symbols to aid in recognizing them. For example, on page xv the first choice is between two items: *1A Woody plants* and *1B Non-woody plants.* If the plant in question is woody, attention is focussed on the first indentation under group *1A Woody plants.* Thus the second choice to be made is between *2A Semi-parasitic plants* and *2B Non-parasitic plants.* If the plant in question is determined to be a woody parasite growing on a tree, then one immediately turns to numbers 33 to 37 in the text, and with the aid of the discussion and drawings, the species can be identified. On the other hand, if the plant in question is not a parasite, the three headings under group *2B Non-parasitic plants* are scrutinized. If the plant is a tree, one proceeds as directed to examine the group *3A Trees* further down on page xv. Continue in the same manner through the various subsections until you reach one that ends with the name of a plant. On comparison of the plant with the drawings, you may decide that some error has been made, either by yourself or in the construction of the key. If it is your own error or possibly due to some peculiar variations of your specimen, you may discover it by carefully comparing with other specimens and then retracing your steps; if by any chance it is the fault of the key, you may try, at any point of doubtful choice, first one and then another alternative. In case you can reach no satisfactory identification, you may send the plant to the author, and she may be able to advise you on the identification and at the same time make all necessary corrections in the key.

RUTH COOPER, Botanist
Riverside City College
Riverside, California

Key

1A. Woody plants: trees and shrubs (and woody vines)	
2A. Semi-parasitic plants growing on other shrubs or trees........	MISTLETOE FAMILY, 33–37
2B. Non-parasitic plants	
3A. Trees (usually developing a main trunk at the ground level)	See this page, below
3B. Shrubs (usually developing several main branches at the ground level). Plants woody at least at the base, sometimes with herbaceous tops..........................	See pp. xv–xxii
3C. Vines	MILKWEED FAMILY (in part), 402, 403
1B. Non-woody plants	See pp. xxii–xxx

(Be sure to follow the numerical order of indentation of symbols, regardless of distance from edge of page. For mechanical reasons of space-saving, it is frequently necessary to drop the space indentation allotted to the lower numbers in sections where those numbers do not again appear.)

3A. TREES

4A. Leaves apparently absent	
5A. Plants thorny	Palo Verde and Smoke Tree, 195, 196, 236
5B. Plants not thorny..	Elephant Tree, 303
4B. Leaves needle-like or scale-like, evergreen......................	PINE FAMILY, 1, 2
4C. Leaves fan-shaped, plaited, parallel-veined.....................	Fan Palm, 9
4D. Leaves bayonet-shaped, tipped with a strong spine...............	Joshua Tree, 22
4E. Leaves not as above	
5A. Leaves oppositely arranged on stem..........................	Ash, 399, 400
5B. Leaves alternately arranged on stem	
6A. Leaves simple	
7A. Flowers showy	
8A. Flowers pink, in terminal clusters..................	Desert Willow, 586
8B. Flowers yellow; leaves like small fig leaves..........	California Slippery Elm, 315
7B. Flowers inconspicuous	
8A. Fruit an acorn	Scrub Oak, 30
8B. Fruit not an acorn	
9A. Leaves smooth, lustrous, bright green............	Cottonwood, 28
9B. Leaves rough, strongly veined..................	Hackberry, 31
6B. Leaves compound	
7A. Plants with extremely stout trunk and branches.........	Elephant Tree, 303
7B. Plants with trunk and branches of ordinary thickness....	PEA FAMILY (in part), 195–199, 236, 237

3B. SHRUBS

4A. Plants apparently without leaves or almost leafless (scale leaves present or leaves early falling because of drought)	
5A. Stems with milky juice..................................	MILKWEED FAMILY (in part), 404, 408
5B. Stems thick green succulent slabs or cylinders, bearing clusters of spines	CACTUS FAMILY, 338–359
5C. Stems thick, cane-like, several from the base, spiny..........	Ocotillo, 318
5D. Stems not as above	
6A. Flowers clustered in dense heads, the head surrounded by bracts and often resembling a single flower	
7A. Plants with scale leaves..............................	Scale-broom, 729
7B. Plants almost leafless, those leaves present linear and ½ in. or more long	
8A. Plants of restricted range	
9A. Alkaline flats, Death Valley....................	Oxytenia, 663
9B. Salt Creek wash and near Gulliday Well, Colorado Desert	Broom Baccharis, 653

8B. Plants of more general range
    9A. Plants blooming in Autumn; flower heads narrow, less than ¼ in. wide, clustered, bracts arranged in 5 vertical columns........................ Mohave Rubberbrush, 634
    9B. Plants blooming in Spring; flower heads broader than ¼ in., bracts irregularly arranged....... Sweetbush, 690

6B. Flowers not clustered in dense heads
  7A. Stems apparently jointed
    8A. Stems fleshy; leaves reduced to triangular scales alternately arranged on stem........................ Picklebush, 86
    8B. Stems not fleshy; leaves reduced to scales arranged at the joints in pairs or whorls..................... JOINT FIR FAMILY, 3–8
  7B. Stems not jointed
    8A. Plants thorny or spiny
      9A. Flowers conspicuous
        10A. Flowers golden, sweet-scented, April–May..... Desert Cassia, 212
        10B. Flowers bluish-violet, June.................. Smoke Tree, 236
      9B. Flowers inconspicuous, less than ¼ in. across
        10A. Flowers minute, greenish; fruits enclosed by 2 disk-shaped bracts, usually not exceeding ½-in. diameter .... Saltbush, Lenscale, Shadscale, 90, 91, 96, 97, 100
        10B. Flowers minute, yellowish-green; fruits reddish to dark blue, thin-skinned, ⅜-in. diameter.. Gray-leaved Abrojo, 295
        10C. Flowers small, yellow, about 3/16 in. across; fruits like small dry olives in clusters....... Crucifixion Thorn, 266
    8B. Plants not thorny or spiny
      9A. Flowers conspicuous
        10A. Flowers dull red .... Chuparosa, 585
        10B. Flowers purple .... Thamnosma, 292
      9B. Flowers inconspicuous, minute, greenish; fruits enclosed by 2 disk-shaped bracts............... Cattle Spinach, Allscale, 92

4B. Plants with leaves
  5A. Plants with stiff dagger-shaped leaves, sometimes very succulent and thick, forming a basal tuft or rosette
    6A. Leaves narrow, not fleshy
      7A. Leaves very stiff; flowers 1¼ to 3½ ins. long .... Yucca, 20–23
      7B. Leaves more pliant and grass-like; flowers 1/16 to ¼ in. long .... Nolina, 24
    6B. Leaves thick, fleshy... Agave, 25–27
  5B. Plants with leaves not as above
    6A. Leaves compound
      7A. Leaves oppositely arranged on stem
        8A. Leaves divided into 2 leaflets....................... Creosote Bush, 294
        8B. Leaves divided into 3 leaflets....................... Fagonia, 293
        8C. Leaves divided into 5 to 7 filiform spine-tipped divisions .... Thurber Dyssodia, 724
      7B. Leaves alternately arranged on stem
        8A. Plants with thorny, spiny, or prickly stems or leaves, or with spiny, bur-like fruits
          9A. Fruits spiny, bur-like .... Burrobush, 673
          9B. Leaves with spiny teeth
            10A. Leaves divided into 3 to 5 narrow leaflets, each tipped with a sharp point................. San Felipe Dyssodia, 726
            10B. Leaves divided into oval, holly-like leaflets with spiny-toothed margins .... Desert Barberry, 136
          9C. Stems with prickles .... Mohave Rose, 182
          9D. Stems spiny
            10A. Small compact shrubs, 4 to 8 ins. high....... Bud Sagebrush, 730
            10B. Large shrubs, at least 2 ft. high or more ...... PEA FAMILY (in part), 197, 199, 212, 230

8B. Plants without thorns, spines or prickles
9A. Plants woody only at the base
10A. Flowers with 5 light yellow petals, which are all approximately equal in size and shape.............. Rock Five-finger, 180
10B. Flowers with 5 very irregularly shaped petals
11A. Flowers yellow .... Species of Lotus and Deerweed, 205, 207–210
11B. Flowers purple .... Species of Dalea, 227, 232
9B. Plants woody throughout
10A. Leaves divided into 3 leaflets
11A. Leaflets with smooth margins.................... Bladder-pod, 144
11B. Leaflets with scalloped margins.................. Desert Squaw-bush, 301
10B. Leaves divided into more than 3 leaflets
11A. Flowers purple, pea-shaped .... Species of Dalea, 225, 231, 233–235
11B. Flowers white .... Desert Sweet, 194
11C. Flowers yellow to orange
12A. Leaves divided into 2 to 8 oval leaflets......... Coues' Cassia, 211
12B. Leaves divided into many very small leaflets.... Small-leaved Hoffmannseggia, 213
11D. Flowers crimson, several grouped into a soft cluster Calliandra, 200
6B. Leaves simple (often toothed, scalloped, lobed, or dissected into linear divisions)
7A. Plants woody throughout (cf. 7B, p. xxi)
8A. Leaves opposite (at least the upper leaves opposite; note that flower stems in Nos. 228 and 229 have 1 to 3 pairs of opposite leaf-like bracts)
9A. Plants with thorny branches or spiny leaves
10A. Plants with thorny branches
11A. Flowers inconspicuous, in small axillary clusters appearing before the leaves; leaves $\frac{1}{2}$ to $1\frac{3}{4}$ ins. long, $\frac{1}{4}$ to $\frac{1}{2}$ in. wide................. Tanglebrush, 398
11B. Flowers conspicuous
12A. Flowers yellowish, 4-parted; fruit small....... Blackbrush, 186
12B. Flowers purple and white, in clusters 2 to 4 ins. long; fruits inflated, papery pods.......... Paper-bag Bush, 523
10B. Plants with spiny leaves
11A. Flowers conspicuous, lavender to violet.......... Sage, 513, 516
11B. Flowers inconspicuous, small reddish-green....... Holly-leaf Spurge, 288
9B. Plants without thorns or spines
10A. Leaf margins with teeth or scallops
11A. Wart-like appendages present at base of each leaf.. Mohave Buckbrush, 299
11B. Wart-like appendages not present
12A. Flowers inconspicuous, in small crowded clusters; fruits winged .... Single-leaved Ash, 400
12B. Flowers more conspicuous
13A. Flowers in terminal spikes, white.......... Wright Lippia, 509
13B. Flowers in separated whorls, stems 4-sided.. Species of Salvia, Desert Lavender, 514, 515, 522, 524, 525
10B. Leaf margins smooth, without teeth
11A. Wart-like appendages present at base of each leaf.. Mohave Buckbrush, 299
11B. Not as above
12A. Flowers without petals, greenish
13A. Flowers in compact clusters in axils of leaves Goat-nut, 302
13B. Flowers in long, slender, pendant clusters.. Quinine-bush, 392
12B. Flowers with petals
13A. Petals not united; flowers white, very small Whipplea, 177
13B. Petals united
14A. Flowers pink .... Long-flwrd. Snow-berry, 607
14B. Flowers dull red...................... Chuparosa, 585
14C. Flowers yellow .... Bush Penstemon, 552
Bush Monkey Flower, 576
14D. Flowers blue or purple and blue....... Species of Salvia, 518, 524

8B. Leaves alternate or clustered on short lateral stems

9A. Plants with thorny branches, or spiny leaves, or bur-like fruits (cf. p. xix)

10A. Plants with bur-like fruits (sometimes also with spiny-toothed leaves) .... Species of Burbush, 672–674

10B. Plants with spiny leaves

11A. Leaves tipped with a sharp spine

12A. Caespitose shrubs less than 6 ins. high; branchlets often thorny; flowers white .... Spiny-tipped Tongue-flower, 291

12B. More erect shrubs, taller than 1 ft.; flowers yellow... Cotton-thorn, Felt-thorn, 735–739

11B. Leaves with spiny-toothed margins

12A. Flowers greenish

13A. Flowers in nodding heads in terminal spikes; fruits, burs with slender hooked prickles...... Holly-leaved Burbush, 674

13B. Flowers in close axillary clusters; fruits oval, ¼ in. long, red .... Evergreen Buckthorn, 298

12B. Flowers yellow, plants of Death Valley region........ Holly-leaved Goldenbush, 625

12C. Flowers dull white; plants in mountains of Death Valley Prickle-leaf, 670

10C. Plants with thorny or spiny branches

11A. Leaves roundish in outline, deeply cleft into 3 to 5 shallowly toothed lobes; flowers yellow; berries black.... Oak-belt Gooseberry, 176

11B. Leaves spatulate-shaped, margins several-toothed; flowers small, white .... Desert Range Almond, 193

11C. Leaves with smooth margins (sometimes minutely toothed)

12A. Leaves clustered on stem

13A. Flowers very small and inconspicuous

14A. Flowers minute, greenish, without petals....... Purple-bush, 275

14B. Flowers small, usually about 3⁄16-in. diameter, with 5 small ladle-shaped petals alternating with the sepals .... Gray-leaved Abrojo, 295
Parry Abrojo, 296
Calif. Snake-bush, 300

13B. Flowers more conspicuous

14A. Flowers white or pink, with 5 small separate petals .... Wild Peach, 191
Desert Range Almond, 193

14B. Flowers scarlet, tubular, ¾ to 1 in. long....... Ocotillo, 318

14C. Flowers greenish or lavender, trumpet-shaped.. Species of Thornbush, 527–534

14D. Flowers yellow; plants with white woolly stems Cotton-thorn, Felt-thorn, 735, 737, 739

12B. Leaves not clustered on stem, attached singly

13A. Flowers very small and inconspicuous, petals absent (cf. 13B, "flowers larger," p. xix)

14A. Leaves fleshy

15A. Leaves slender .... Greasewood, 82

15B. Leaves broader, flat .... Spiny Hop-sage, 102

14B. Leaves not fleshy

15A. Fruits enclosed by 2 flat disk-shaped membranous bracts, not more than ½ in. across; leaves often with mealy, granular surface Species of Saltbush, Lenscale, Shadscale, 90, 91, 96, 97, 100

15B. Fruits like miniature plums, black or purplish; leaves prominently veined beneath Spiny Abrojo, 297

15C. Fruits very small, tipped with feathery tail, 1 to 1¾ ins. across........ Little-leaved Mahogany, 189

13B. Flowers larger and more showy, petals present
14A. Flowers white to rose
15A. Plants less than 2 ft. high; flowers white
16A. Plants found in western Colorado Desert.... Shrubby Brickellia, 612
16B. Plants found in eastern Mohave Desert..... **Mortonia, 289**
Tongue-flower, 290, 291
15B. Plants 3 to 16 ft. high; leaves to 1¼ ins. long; flowers white to rose.................... Desert Apricot, 192
14B. Flowers deep Indigo blue, pea-shaped........... Indigo-bush, 226
14C. Flowers red-purple; flower stems with 1 to 3 pairs of opposite leaf-like bracts.................. Ratany, 228, 229
14D. Flowers yellowish or a combination of pink, purple, and yellow .... Spiny Milkwort, 267
14E. Flowers brownish-purple outside, white inside.... Spiny Menodora, 397
9B. Plants not spiny or thorny
10A. Low shrubs, less than 2 ft. high (see bottom this page and p. xx for 10B and 10C)
11A. Leaves or stems conspicuously white
12A. Flowers minute, greenish, inconspicuous
13A. Leaves with toothed margins, holly-like.............. Desert Holly, 101
13B. Leaves not toothed, margins smooth and revolute..... Winter Fat, 103
12B. Flowers larger, yellow Paper-flower, 703
13A. Leaves on the average less than ½ in. long..........
Eytelia, 619
13B. Leaves usually much longer than ½ in. .... Felt-thorn, 736, 738
14A. Leaves very slender ....
14B. Leaves broader, oval .... Viguiera, 680
Piñon Wormwood, 732
11B. Leaves or stems not conspicuously white
12A. Plants very rough, like sandpaper..................... Sandpaper-plant, 319–322
12B. Plants with stinging hairs............................ Sting Bush, 323
12C. Plants forming turf-like mats over surfaces of rocks..... Turf Spiraea, 181
12D. Plants not as above
13A. Flowers pinkish .... Shrubby Cream-bush, 183
13B. Flowers white .... Arrow-leaf, 608
Species of Brickellia, 610, 612, 615, 617
13C. Flowers yellow
14A. Leaves ordinarily less than ¾ in. long
15A. Leaves wedge-shaped, glossy green........... Goldenbush, 626
15B. Leaves linear .... Goldenhead, 621
Rabbit Brush, 633
14B. Leaves ordinarily more than ¾ in. long
15A. Leaves linear, reflexed .... Matchweed, 618
15B. Leaves broader .... Alkali Goldenbush, 627, 628
Parish Viguiera, 681
Trixis, 747
13D. Flowers purple .... Thamnosma, 292
10B. Medium-sized shrubs, 2 to 5 ft. high in typical growth
11A. Plants slender, willow-like .... Willow, 29
Knapp Brickellia, 613
Arrow-weed, 656
11B. Plants slender to broad, not willow-like (see also 10C, p. xx)
12A. Leaves slender (cf. with "leaves broad," 12B, p. xx)
13A. Flowers inconspicuous, greenish
14A. Fruits in terminal clusters, with 4 thin, roundish, vertical wings surrounding the seed, not more than ¾-in. diameter .... Wingscale, 99
14B. Fruits in terminal clusters, surrounded by 5 to 12 shining, horizontal, wing-like scales......... Cheese-bush, 678

13B. Flowers more conspicuous

14A. Leaves sparse

15A. Flowers light pink to purplish .... Alkali Aster, 640

15B. Flowers whitish or pale yellow .... Broom Baccharis, 653; Oxytenia, 663

15C. Flowers yellow .... Rubberbrush, 634; Sweetbush, 690

14B. Leaves more thickly clothing the stems

15A. Leaves narrowly wedge-shaped, 3-toothed at apex .. Sagebrush, 731

15B. Leaves 3- to 5-parted into filiform divisions, or the upper ones entire; flowers pale buffish-yellow or yellow .... Oxytenia, 663; Sand-wash Groundsel, 745

15C. Leaves needle-shaped, round in cross-section, with a balsamic odor .... Black-banded Rabbit Brush, 632; Rubberbrush, 635; Desert-fir, 734

15D. Leaves not as above

16A. Flower heads resembling small sunflowers ..... Goldenbush, 629, 630

16B. Flower heads not resembling small sunflowers .. Rabbit Brush, 636

12B. Leaves broad

13A. Flowers inconspicuous, greenish

14A. Fruits enclosed in flat disk-shaped bracts, ⅛- to ⅓-in. diameter .... Allscale, 92; Wingscale, 98, 99

14B. Fruits with feathery tails 1 to 1¼ ins. long .... Little-leaved Mahogany, 189

14C. Fruits globular, ¼- to ½-in. diameter, usually 3-lobed Shrub Euphorbia, 286; Mouse-eye, 287

13B. Flowers more conspicuous

14A. Flowers white (sometimes pale lilac) .... Ragged Rock-flower, 178; Woolly Yerba Santa, 471

14B. Flowers yellow

15A. Leaves about ¾ in. long .... Desert Buckwheat, 45

15B. Leaves longer than ¾ in. .... Viguiera, 680; Encelia, 686–688

14C. Flowers cream-colored

15A. Leaves deeply lobed .... Cliff-rose, 184; Antelope-brush, 185; Apache-plume, 187

15B. Leaves with smooth margins, or shallowly toothed .. Desert Brickellia, 616

14D. Flowers maroon .... Woolly Brickellia, 614

10C. Tall shrubs, usually over 5 ft. high, sometimes shorter

11A. Plants slender, willow-like

12A. Flowers showy, pink .... Desert Willow, 586

12B. Flowers less showy

13A. Flowers whitish .... Knapp Brickellia, 613

13B. Flowers pale roseate-purple .... Arrow-weed, 656

13C. Flowers greenish, in catkins .... Slender Willow, 29

11B. Plants not willow-like

12A. Flowers greenish, without petals .... Mountain Mahogany, 188; Curl-leaf, 190

12B. Flowers conspicuous, of various colors other than green

13A. Flowers pink .... Pink Felt-plant, 314; Desert Manzanita, 395

13B. Flowers yellowish (pale yellow) .... Emory Baccharis, 655; Desert-fir, 734

13C. Flowers yellow .... Yellow Felt-plant, 313; Slippery Elm, 315

13D. Flowers lavender .... Felt-leaved Yerba Santa, 472

13E. Flowers white .... Coville Serviceberry, 179; Narrow-leaved Yerba Santa, 473; Gum-leaved Brickellia, 611

7B. Plants woody only at the base (cf. "plants woody throughout," 7A, p. xvii)

8A. Leaves well distributed on plant; i.e., not restricted to the base of plant

9A. Leaves apparently whorled on stem........................ Species of Bedstraw, 593–596

9B. Leaves oppositely arranged on stem

10A. Leaves broad

11A. Plants with milky juice............................ Fendler Spurge, 282

11B. Plants without milky juice

12A. Flowers white (sometimes pale pink)............ Yerba de la Rabia, 108
White Four-o'clock, 120

12B. Flowers flesh-colored, scarlet, or rose-purple...... Species of Penstemon, 551, 553, 555, 557–560, 565

10B. Leaves (or divisions of leaves) slender

11A. Leaves usually shorter than ½ in.

12A. Flowers small, greenish-membranous .... Rixford Rockwort, 129

12B. Flowers pale blue or pale lavender in dense terminal heads .... Narrow-leaved Monardella, 517
Rock Pennyroyal, 520

12C. Flowers pinkish, scattered singly on branchlets.... Alkali-flat Frankenia, 317

11B. Leaves usually longer than ½ in.

12A. Flowers greenish-white .... Mountain Catch-fly, 130

12B. Flowers yellow .... Thurber Dyssodia, 724

12C. Flowers pinkish-red

13A. Flowers less than ½ in. long................. Stansbury Phlox, 414

13B. Flowers longer than ½ in., sometimes purplish Species of Penstemon, 550, 556, 562

9C. Leaves alternately arranged on stem

10A. Plants with stinging hairs.......................... Tragia, 269
Sting Bush, 323

10B. Plants without stinging hairs

11A. Stems and leaves very noticeably sandpapery......... Species of Sandpaper-plant, 319, 322

11B. Stems and leaves not as above

12A. Leaves spiny-tipped or spiny-toothed

13A. Flowers whitish to blue .... Blue Mantle, 416
Prickly Gilia, 419
Mohave Gilia, 437

13B. Flowers yellow .... Spiny Goldenbush, 622

13C. Flowers orange .... Cooper Dyssodia, 725

13D. Flower heads lavender with yellow center..... Mecca Aster, 637
Mohave Aster, 639

12B. Leaves not spiny

13A. Leaves or leaf divisions slender; i.e., several to many times longer than broad

14A. Flowers very small and inconspicuous

15A. Flowers greenish to wine color.......... Inkweed, 83
Red Molly, 85
Tooth-leaf, 268

15B. Flowers white .... Woody Forget-me-not, 498

14B. Flowers larger and more conspicuous

15A. Flowers in showy vermilion tufts........ Woolly Paintbrush, 569

15B. Flower heads orange.................. San Felipe Dyssodia, 726

15C. Flower heads whitish.................. Short-leaved Baccharis, 652

15D. Flower heads deep pink............... Spanish Needle, 704

15E. Flower heads purplish, herbage ill-scented Odora, 728

15F. Flower heads aster-like, white, pink, or lavender with yellow centers........ White Aster, 638
Species of Fleabane, 645–650

15G. Flower heads sunflower-like, yellow..... Sand-wash Groundsel, 745
Long-stemmed Eriophyllum, 707

15H. Flowers white, fragrant, with 4 petals.. Desert Alyssum, 165

13B. Leaves or leaf divisions broad

14A. Flower heads aster-like, whitish to pink, with yellow center ... Pygmy Fleabane, 646

14B. Flowers white or pinkish, 5 petals, each with red or purple base ... Rock Hibiscus, 311

14C. Flowers yellow

15A. Flowers irregular in shape, pea-shaped ... Species of Deer-weed and Lotus, 205, 207–210

15B. Flowers regular in shape

16A. Petals 4 ... Desert Plume, 149; Primrose, 378, 379

16B. Petals 5 ... Rock Five-finger, 180

14D. Flowers brownish, about ⅛ in. long ... Desert Ayenia, 316

14E. Flowers apricot, pink to peach-red to scarlet ... Species of Mallow, 304, 305, 307, 308

14F. Flowers light blue to lavender ... Species of Coldenia, 474, 475, 477

8B. Leaves restricted largely to base of plant

9A. Flowers very small and numerous in a much-branched inflorescence, white, pink, or yellow ... Species of Buckwheat, 48–52, 56–59, 61, 62

9B. Flowers not as above

10A. Flowers greenish to white, leaves opposite ... Desert Sandwort, 131

10B. Flowers yellow ... Broom Twinberry, 396; Wild Marigold, 699; Arizona Actinea, 723

10C. Flowers purple, 4-petaled ... Species of Rock-cress, 167, 170

10D. Flowers tubular, blue or deep pink, leaves opposite ... Species of Penstemon, 548, 554

10E. Flower heads resembling asters, white, pink, and lavender with yellow centers ... Species of Fleabane, 646, 647, 650

10F. Flowers pink to reddish; leaves alternate or clustered ... Desert-straw, 751; Thorny Skeleton-plant, 760

1B. Non-woody plants

2A. Parasitic, non-green plants ... Scaly-stemmed Sand Plant, 393; Sand-food, 394; Dodder, 412, 413; Piñon Strangleroot, 588; Burro-weed Strangler, 589

2B. Non-parasitic, green plants

3A. Leaves parallel-veined, grass-like; flower parts in 3's or 6's ... Iris Family, 10; Lily Family (in part), 11–19

3B. Plants not as above

4A. Plants climbing through and over other plants ... See this page, below

4B. Plants prostrate ... See p. xxiii

4C. Miniature plants—less than 4 ins. ... See pp. xxiii–xxiv

4D. Plants taller ... See pp. xxv–xxx

4A. Plants climbing through and over other plants

5A. Stems milky ... Purple Climbing Milkweed, 406; Rambling Milkweed, 407

5B. Stems not milky

6A. Leaves alternate, 3- to 5-parted ... Brandegea, 602

6B. Leaves opposite, entire ... Species of Snapdragon, 542, 543, 544; Blue Maurandya, 545

4B. Plants prostrate or trailing on ground	
5A. Plants with spiny-toothed leaves	Mohave Spiny-herb, 74 Yellow Spiny-cape, 77
5B. Plants not as above	
6A. Leaves in whorls of 3	Golden Carpet, 79
6B. Leaves opposite	
7A. Plants with milky juice	Species of Sand-mat, 273, 279–281, 283–285
7B. Plants without milky juice	
8A. Flowers very small in axillary clusters, bright silvery-white	Frost-mat, 132
8B. Flowers blue	Thompson Penstemon, 563
8C. Flowers greenish-white or reddish, fruits pinkish with 3 wings	Small-flowered Abronia, 117
8D. Flowers rose, lavender, red, to purple	West Indian Boerhaavia, 110 Hairy Sand-verbena, 118 Mohave Sand-verbena, 119 Windmills, 123 Desert Heron's-bill, 264
6C. Leaves alternate, clustered, or basal	
7A. Leaves lobed or divided into leaflets	
8A. Flowers white or pinkish	White Mallow, 309
8B. Flowers yellow	
9A. Fruits gourds	Palmate-leaved Gourd, 599 Finger-leaved Gourd, 600
9B. Fruits small pods	Yellow Pepper-grass, 166 Hairy Lotus, 203 Hill Lotus, 206
7B. Leaves entire, not lobed	
8A. Flowers in small clustered woolly heads	Dwarf Filago, 661
8B. Flowers daisy-like, white to pinkish with yellow center	Mohave Desert-star, 643
8C. Flowers small, tubular, white with 5 yellow spots and purplish throat	Chinese Pusley, 479
8D. Flowers yellow, large	Calabazilla, 601
8E. Flowers deep pink to purple	Purple Mat, 461
8F. Flowers white	
9A. Leaves clustered along stems	Small-leaved Nama, 460 Narrow-leaved Nama, 463 Nuttall Coldenia, 476 Parishella, 603
9B. Leaves not clustered	Sand-cress, 127 Species of Comb-bur, 480–483
4C. Miniature plants—generally less than 4 ins. high (cf. 4D, p. xxv)	
5A. Leaves opposite (those uppermost on the branches)	
6A. Leaves divided into 3 to 7 linear radiating segments	Golden Gilia, 415 Parry Gilia, 420
6B. Leaves undivided, occasionally slightly toothed	
7A. Flowers yellow	Alkali Gold-fields, 700 Small-rayed Baeria, 701 Chinch-weed, 733
7B. Flowers tubular, yellow, purple-dotted	Spotted Mimulus, 577
7C. Flowers rose to red to reddish-purple, sometimes with yellow shadings	
8A. Foliage white-woolly	Lemmon Xerasid, 708
8B. Foliage not white-woolly	Species of Mimulus, 572–574, 578, 579
5B. Leaves basal or mostly basal	
6A. Plants with fleshy leaves	Bitterroot, 125 Desert Pot-herb, 126 Narrow-leaved Miner's Lettuce, 128 Pygmy Poppy, 142

6B. Plants with spiny leaves or bracts	Xantus Spiny-herb, 68
	Brittle Spine-flower, 69
	Corrugata, 73
6C. Plants not as above	Spiny Plantain, 591
7A. Leaves entire, undivided; flowers whitish	
8A. Plants white-woolly	Whisk Broom, 65
	Pursh Plantain, 590
	Woolly Plantain, 592
8B. Plants not white-woolly	Mouse-tail, 135
	Wedge-leaved Draba, 171
	Species of Thread Plant, 604–606
7B. Leaves lobed, divided into leaflets, or finely dissected into small segments	
8A. Flowers yellowish with blue wings	Short-stemmed Blue Lupine, 220
8B. Flowers white or yellowish-white tipped with purple	Death Valley Locoweed, 243
	Cliff Locoweed, 262
8C. Flowers maroon	Gilman Parsley, 390
8D. Flowers rose-lavender, violet, or purple	Gray Locoweed, 242
	Keel-beak, 256
	Pursh Locoweed, 258
	Utah Cymopterus, 384
8E. Flowers creamy or yellowish; plants with milky juice	Keyesia, 762
5C. Leaves alternate (at least upper ones) or clustered	
6A. Leaves spiny-toothed	
7A. Flowers greenish in axillary spikes	Broad-leaved Stillingia, 276
7B. Flowers white with maroon spots	Humble Gilia, 422
7C. Flowers yellow, very small	Rigid Spiny-herb, 70
	Watson Spiny-herb, 71
7D. Flowers pale pink or tan or purple, usually dotted with purple spots	Spotted Gilia, 426
	Schott Gilia, 427
	Desert Calico, 429
6B. Leaves not spiny-toothed	Bristly Gilia, 430
7A. Herbage (at least main stems) white-woolly	
8A. Flowers minute, surrounded by papery bracts like an Everlasting Flower, often in woolly clusters	Nest-straw, 657, 658
	Filago, 661, 662
8B. Flowers white, tubular, sometimes with purplish tube	Pearl-o'rock, 445, 446
8C. Flowers yellow or yellow and white	
9A. Heads with rays like a daisy	Wallace Eriophyllum, 706
	Frémont Xerasid, 709
	Woolly Eriophyllum, 711
9B. Heads without rays	Pringle Eriophyllum, 705
	Mohave Eriophyllum, 712
	Yellow-head, 713
7B. Herbage not white-woolly	
8A. Leaves deeply lobed, dissected, or divided into leaflets	Little Gold-poppy, 141
	Notch-leaved Locoweed, 259
	Ives Phacelia, 449
	Mohave Owl Clover, 567
8B. Leaves entire, sometimes shallowly toothed	
9A. Flowers purple and yellow, aster-like	Ground-daisy, 642
9B. Flowers bluish-purple	Thick-leaved Phacelia, 452
9C. Flowers yellow	Species of Primrose, 361, 362, 364, 380
9D. Flowers white	
10A. Petals united to form a tubular or bell-shaped corolla	Gilia, 424, 428
	Weasel Phacelia, 441
	Lemmonia, 464
	Forget-me-not, 489, 497
10B. Petals separate, not fused to each other	Saw-toothed Ditaxis, 270
	Pygmy Primrose, 360

4D. Plants generally taller than 4 ins., especially if upright in habit

5A. Leaves basal .... See this page, below
5B. Leaves whorled or opposite .... See p. xxvi
5C. Leaves alternate .... See p. xxvii

5A. Leaves basal

6A. Plants with milky juice

7A. Flowers white (sometimes marked with rose or purple) .... Parachute Plant, 743
White Tack-stem, 759
7B. Flowers yellow — Egbertia, 761
8A. Heads nodding in the bud .... Cleveland Yellow-saucers, 754
Snake's-head, 755
Yellow-saucers, 756
Desert Dandelion, 757
8B. Heads not nodding in the bud .... Scale Bud, 748
Silver-puffs, 749
Yellow Tack-stem, 758

6B. Plants without milky juice — Western Crepis, 763

7A. Plants with very succulent fleshy leaves .... Rock Echeveria, 175
7B. Plants with spiny bracts
8A. Flowers small, white .... Thurber Spiny-herb, 72
Punctured Bract, 75
Trilobia, 76
8B. Flowers lavender, borne in dense whorls .... Thistle Sage, 512
7C. Plants not as above
8A. Leaves white-woolly or silvery-hairy
9A. Flowers very small, yellow, white, or pink, usually borne on much-branched forking stems .... Species of Buckwheat, 39–41, 43, 44, 46, 54, 60, 63, 64, 80
9B. Flowers with 4 yellow petals each about ¼ in. long; fruits globose pods .... King Bean-pod, 169
9C. Flowers pea-shaped, scarlet .... Scarlet Locoweed, 263
9D. Flowers yellow, sunflower-like .... El Caparossa, 666
Large-flowered Sunray, 684
8B. Leaves not conspicuously white
9A. Leaves entire or shallowly toothed
10A. Flowers yellow, white, or rose, very small, borne on much-branched forking stems .... Species of Buckwheat, 38, 4 47, 53
Wild Rhubarb, 81
10B. Flowers white, petals 6, stamens numerous .... Cream-cups, 139
10C. Flowers with 4 separate petals, purplish or white .... Inflated-stem, 152
Spectacle-pod, 163
Woody Bottle-washer, 365
10D. Flowers bluish, purplish, or red, with the petals fused to form a bell-shaped or tubular corolla .... Parish Phacelia, 443
Three-hearts, 459
Utah Firecracker, 549
Beard-tongue, 561
10E. Flowers sunflower-like, yellow or yellow and purple .... Great Hulsea, 665
Naked-stemmed Sunray, 685
9B. Leaves deeply toothed or divided into leaflets or dissected into small segments
10A. Leaves deeply toothed
11A. Flowers rose, violet to pink .... Desert Windflower, 134
Species of Gilia, 421, 425, 433
11B. Flowers white, sometimes aging to pink
12A. Petals 4 .... Fringe-pod, 172
Species of Primrose, 368, 371
12B. Petals 5 or 6 .... Bear Poppy, 137
Species of Gilia, 423, 434
11C. Flowers yellow .... Species of Primrose, 366, 367

10B. Leaves divided into rather coarse leaflets
11A. Flowers white .... Brown-eyed Primrose, 375
11B. Flowers white tipped with purple.......... Layne Locoweed, 248
11C. Flowers deep purple .... Inyo Locoweed, 257
11D. Flowers yellow .... Frost-stemmed Primrose, 370

10C. Leaves dissected into fine segments
11A. Flowers very small, borne in umbels.......... Species of Wild Parsley, 386–389, 391
11B. Flowers white to rose-tinted, in heads.......... Mohave Pincushion, 718
11C. Flowers light blue to cream.......... Lesser Gilia, 436
11D. Flowers purplish .... Purple-bell Phacelia, 448
11E. Flowers deep blue .... Chia, 525
11F. Flowers yellow .... Desert Gold-poppy, 140; Species of Coreopsis, 691, 692; Yellow Cut-leaf, 764

5B. Leaves whorled or opposite
6A. Leaves in whorls of 2 to 8.......... Species of Bedstraw, 597, 598
6B. Leaves opposite
7A. Plants diffuse, weak-stemmed .... Pterostegia, 78; Species of Baby Blue-eyes, 469 470
7B. Plants stouter-stemmed
8A. Leaves spiny-toothed .... Rock Maurandya, 546
8B. Leaves not spiny-toothed
9A. Leaves succulent, fleshy .... Alkali-weed, 84; Sticky-ring, 109; Ring-stem, 112; Lowland Purslane, 124
9B. Leaves not succulent
10A. Plants grayish or whitish
11A. Stems with milky juice.......... Species of Milkweed, 405, 410
11B. Stems without milky juice.......... Angle-stemmed Buckwheat, 66; Honey-sweet, 105; Turkey Mullein, 278
10B. Plants not as above
11A. Leaves entire, not toothed
12A. Flowers greenish, tinged with either white or pink .... Desert Wing-fruit, 107; Green Saucers, 122
12B. Flowers white .... Wishbone Bush, 114; Four-o'clock, 116; Mohave Pennyroyal, 519
12C. Flowers red or scarlet.......... Creeping Sticky-stem, 111; Eaton Firecracker, 547
12D. Flowers pale pink to bright purple.......... Pink Three-flower, 121; Frankenia, 317; White-margined Penstemon, 564
12E. Flowers purple .... Boerhaavia, 113; Giant Four-o'clock, 115; Thyme Pennyroyal, 511; Showy Penstemon, 566
12F. Flowers yellow .... False Mimulus, 580; Tailed Pericome, 671; Nevada Viguiera, 679
11B. Leaves slightly scalloped or toothed
12A. Flowers pinkish to white.......... Horsemint, 510; Parish Mimulus, 575
12B. Flowers purple .... Caltha-leaved Phacelia, 453
12C. Flowers pale blue .... Low Germander, 521
12D. Flowers yellow; fruits woody with 2 curved prongs .... Unicorn Plant, 587

11C. Leaves more deeply toothed
12A. Flowers white, brownish-streaked outside ... Evening Snow, 417
12B. Flowers white to pink or bluish-pink ... Long-tubed Gilia, 418
12C. Flowers light to dark blue ... Species of Baby Blue-eyes, 469, 470
12D. Flowers purplish ... Goodding Verbena, 508

5C. Leaves alternate
6A. Stems stout, inflated; flowers purple to white, 4-petaled ... Squaw Cabbage, 164
6B. Stems brittle ... Lance-leaved Ditaxis, 271
Flor de la Piedra, 337
Emory Rock Daisy, 702
6C. Stems very weak ... Pellitory, 32
Slender Keel-fruit, 162
Death Valley Phacelia, 439
White Fiesta-flower, 465
6D. Stems with milky juice
7A. Flowers rather inconspicuous, greenish ... Cut-lobed Spurge, 272
Desert Poinsettia, 274
7B. Flowers larger, about ½ in. long, greenish-white, or with purplish hoods ... Four-o'clock Milkweed, 409
Antelope Horns, 411
7C. Flowers sky blue ... Amsonia, 401
7D. Flowers white, heads ¾ to 1½ ins. broad ... California Chicory, 750
7E. Flowers pink to reddish ... Annual Mitra, 752
Parry Rock-pink, 753
6E. Stems not as above
7A. Leaves fleshy ... Patata, 87
7B. Leaves rough with barbed hairs, adhering very tightly to clothing ... Species of Blazing Star, 324–333
7C. Leaves not as above
8A. Plants with spiny bur-like fruits ... Burweed, 675, 676
8B. Plants with spiny stems or leaves
9A. Flowers inconspicuous, greenish ... Russian Thistle, 104
9B. Flowers pale lavender, trumpet-shaped ... Desert Gilia, 431
9C. Flowers white, sometimes changing to pink or brown ... Prickly Poppy, 138
Mexican Devil-weed, 641
Mohave Thistle, 746
9D. Flowers orange ... San Felipe Dyssodia, 726
9E. Flowers yellow, later pink or purplish ... Hole-in-the-sand Plant, 727
9F. Flowers violet to purple with yellow centers ... Piñon Aster, 651
8C. Plants without spines

9A. Leaves with entire margins, not toothed ... See this page, below
9B. Leaves shallowly toothed or scalloped ... See p. xxviii
9C. Leaves deeply toothed or dissected ... See p. xxix
9D. Leaves divided into entirely separate rather coarse leaflets ... See p. xxx

---

9A. Leaves with entire margins, not toothed
10A. Leaves rather narrow
11A. Flowers greenish-yellowish in numerous papery dense clusters; herbage woolly ... Small-flowered Cudweed, 664
11B. Flowers deep pink ... Spanish Needle, 704
11C. Flowers rose or lavender to purplish-red ... Fringed Amaranthus, 106
Hispid Nama, 462
11D. Flowers scarlet ... Long-leaved Paintbrush, 571
11E. Flowers blue ... Blue Flax, 265
11F. Flowers brownish ... Brown Turbans, 609

11G. Flowers yellow

- 12A. Flowers in loose terminal clusters — Yellow Flax, 265; Thread-stemmed Gilia, 432
- 12B. Flowers in close axillary clusters — Golden Forget-me-not, 499; Sulphur-throated Forget-me-not, 503
- 12C. Flowers in heads — Rock Goldenrod, 620; Hispid Golden Aster, 631

11H. Flowers white

- 12A. Fruit very small, 4-lobed capsules with numerous black seeds — Linear-leaved Cambess, 174
- 12B. Fruit of 4 flat spreading nutlets — Species of Comb-bur, 480–483
- 12C. Fruit inconspicuous, of 4 very small oval nutlets — Species of Forget-me-not and Popcorn Flower, 486–488, 490–496, 500–502, 504–507

10B. Leaves broader

- 11A. Flowers very inconspicuous, greenish; fruits enclosed by 2 disk-shaped bracts 1/8 to 5/16 in. long — Wheelscale, 89; Arrow Scale, 93; Wedge Scale, 94
- 11B. Flowers white — Long-capsuled Primrose, 376; False Morning-glory, 478; Flexuous Forget-me-not, 491; Desert Tobacco, 541
- 11C. Flowers greenish to greenish-white or greenish-yellow — Big-leaved Caulanthus, 155; Desert Croton, 277; Western Poverty Weed, 668
- 11D. Flowers yellow
  - 12A. Flowers in terminal clusters — Palmer Bead-pod, 168
  - 12B. Flowers in coiled spikes — Shiny-seeded Fiddleneck, 484; Checker Fiddleneck, 485
  - 12C. Flowers in leafy spikes, large — Ghost-flower, 583; Lesser Mohavea, 584
  - 12D. Flowers in heads — Exalted Laphamia, 669
- 11E. Flowers yellow and maroon — Heart-leaved Twist-flower, 150

9B. Leaves shallowly toothed or scalloped

10A. Leaves rather narrow

- 11A. Flowers white — Narrow-leaved Primrose, 372; White Tidy-tips, 693
- 11B. Flowers yellowish, petals yellowish, sepals green to purplish — Long-beaked Twist-flower, 157
- 11C. Flowers yellow — Desert Wallflower, 173; Spencer Primrose, 381; Autumn Vinegar-weed, 694; Peirson Vinegar-weed, 695; Kellogg Tarweed, 696; Yellow-frocks, 710

10B. Leaves broader

- 11A. Herbage white-woolly or white-hairy — Hairy-leaved Sunflower, 682; Silver-leaved Sunflower, 683; Greater Monolopia, 714; Velvet Rosette, 741; Fan-leaf, 742
- 11B. Herbage not white-woolly
  - 12A. Flowers very inconspicuous, petals none; fruits enclosed by 2 disk-shaped bracts 1/16 to 1/4 in. long — Bractscale, 88; Redscale, 95

12B. Flowers with the petals separate from each other, not fused

13A. Petals 4 .... Hall Caulanthus, 154
Cooper Caulanthus, 156
Booth Primrose, 363
Purple Primrose, 373
Paiute Primrose, 377

13B. Petals 5 .... Mohave Desert Mallow, 304
Desert Five-spot, 306
Alkali Pink, 310
Small-leaved Abutilon, 312

12C. Flowers with the petals fused to form a bell-shaped, trumpet-shaped, or saucer-shaped corolla

13A. Flowers borne in coiled spikes.................. Species of Phacelia, 438, 442, 444, 447, 455

13B. Flowers not in coiled spikes

14A. Flowers trumpet-shaped, large .... Western Jimson, 535
Small Datura, 536

14B. Flowers saucer-shaped .... Species of Ground-cherry, 537–540

12D. Flowers clustered together in dense heads, the head surrounded by bracts and often resembling a single flower

13A. Flowers greenish, heads small, numerous, nodding in fruit .... Desert Dicoria, 659
Single-fruited Dicoria, 660

13B. Flowers yellow .... Smooth-stemmed Goldenbush, 624
Desert Sunflower, 689
Mohave Groundsel, 740

9C. Leaves deeply toothed or dissected into fine segments

10A. Plants white-woolly .... Yellow Paintbrush, 570
Lax-flower, 698
Long-stemmed Eriophyllum, 707
Douglas Pincushion, 720

10B. Plants not white-woolly

11A. Flowers with separate petals, these all approximately the same size and shape

12A. Petals 4

13A. Petals white .... Alkali Crucifer, 151
Hairy-stemmed Caulanthus, 153
Rock Mustard, 158
California Primrose, 369

13B. Petals yellow .... Yellow Tansy Mustard, 159
Hairy-leaved Caulanthus, 160
Coulter Lyre-fruit, 161
Pale Yellow Primrose, 374

12B. Petals 5; flowers very small, in umbels............. Parish Wild Parsley, 382
Parry Wild Parsley, 383
Desert Umbel, 385

11B. Flowers with petals fused together to form a trumpet-shaped or bell-shaped corolla

12A. Flowers pale yellow, persistent, becoming papery.... Whispering-bells, 468

12B. Flowers white or bluish, small, sometimes with yellow tube .... Species of Eucrypta, 466, 467

12C. Flowers rose-lavender or purple with yellow tube.... Rock Gilia, 435
Two-colored Phacelia, 457
Frémont Phacelia, 458

12D. Flowers pale lavender or light purple to deep violet or bluish-purple .... Species of Phacelia, 440, 450, 451, 454, 456

11C. Flowers of separate or fused petals, but these very irregular in shape and size

12A. Flowers light but lively sky-blue .... Parish Larkspur, 133

12B. Flowers purplish or purple and yellowish-white ...... Bird's-beak, 581, 582

12C. Flowers red, in dense spikes .... Desert Paintbrush, 568

11D. Flowers clustered together in dense heads, the head surrounded by bracts and often resembling a single flower

12A. Heads with white to light lavender rays, and yellow centers .... Silver Lake Daisy, 644

12B. Heads greenish, small .... Nevada Poverty Weed, 667
Western Ragweed, 677

12C. Heads with yellow flowers .... Slender Goldenbush, 623
Biennial Goldflower, 721
Groundsel, 744

12D. Heads with white flowers .... Species of Pincushion, 715–717, 719

9D. Leaves divided into entirely separate rather coarse leaflets

10A. Leaflets pinnately arranged

11A. Herbage gland-dotted; flowers rose .... Prairie Clover, 240

11B. Herbage not gland-dotted

12A. Leaves equally pinnate; i.e., terminal leaflet not present .... Colorado River Hemp, 239

12B. Leaves odd-pinnate; i.e., terminal leaflet present

13A. Flowers in umbels or solitary, yellow .... Species of Lotus, 204, 205, 207–210

13B. Flowers in spikes or racemes

14A. Pods inflated, papery .... Species of Locoweed, 244, 246, 247, 249, 254, 260, 261

14B. Pods not inflated .... Species of Locoweed, 241, 245, 250–253, 255

10B. Leaflets palmately arranged

11A. Leaves with 3 leaflets

12A. Flowers with 4 petals, of the same size and shape, or nearly so .... Species of Stinkweed, 143, 146, 147
False Clover, 148

12B. Flowers with 5 petals, very irregular in shape, pea-shaped

13A. Flowers yellow .... Species of Lotus, 205, 207, 208

13B. Flowers purple to pink .... Species of Clover, 201, 202

11B. Leaves with more than 3 leaflets

12A. Flowers with 4 petals .... Yellow Spiderwort, 145

12B. Flowers with 5 petals, pea-shaped

13A. Leaves bearing glandular dots .... Beaver-dam Breadroot, 238

13B. Leaves without glandular dots .... Species of Lupine, 214–219, 221–224

Desert Wild Flowers

PINACEAE. Pine Family

1. SINGLE-LEAVED PIÑON (see p. 3*). *Pinus monophylla* (L., pine; Gr., one-leaved). The single-leaved piñon is common on all the high desert ranges. Its black-barked trunks, much-branched crown, and solitary needles distinguish it from other pines. The large nuts, borne in small cones, have long been a staple food of the desert Indians. Weathered beating-poles, heaps of old cones, and rings of rocks outlining former camps of nut-gathering Indians may still be found in many of the desert mountains. The nuts, ripe in early September after the second season's growth, are eaten raw, or roasted, or ground into a meal. Unfortunately, abundant crops do not occur every year. In autumn, flocks of piñon jays gather to share the feast of nuts, and their strange corvine cries are among the few bird notes breaking the silence of the quiet, sun-drenched, desert woodlands. The piñon is the host of a special mistletoe (see **37**).

The similar, 2-needle piñon, *Pinus edulis* (L., edible), first collected in California and reported to the Southern California Academy of Sciences by the author in 1928, is found on the slopes of the New York Mountains together with the single-leaved species.

Parry's 4-needle piñon, *Pinus Parryana* (after Dr. C. C. Parry; see **24**), occurs in the upper reaches of desert canyons and slopes of the Santa Rosa Mts. and southward into Lower California.

2. CALIFORNIA JUNIPER (see p. 4). *Juniperus californica* (L., juniper; of California. The ending of the specific name is feminine because we follow the Latin rule that all trees are feminine). This juniper is abundant on slopes of mountains on the west side of the Colorado Desert, and also on the desert bases of the mountains bordering the Mohave Desert on the west. Limited moisture and shallow soils seem to be the chief factors governing its size. The average tree is from 10 to 15 ft. spread; but the author has recently measured one in the Joshua Tree National Monument with a spread

* Photographic illustrations used in lieu of or in addition to drawings are cited by page number.

of 42 ft. The small, fleshy, berry-like cones are a very handsome silver-blue or green, turning reddish when mature at the end of the second year. Indians ate the fruit fresh or ground the dry berries to make cakes of the meal.

The closely related Utah juniper, *J. utahensis*, either alone or in company with the nut-pine, clothes the summits of most of the high ranges of the eastern Mohave Desert and southern Nevada. Magnificent specimens have recently been discovered by the author on the desert slope of the San Bernardino Mountains near Baldwin Lake. It is a tree more erect in habit than the California juniper; and the leaves, instead of being in whorls of three as in that species, are usually opposite. Especially fine forests of this tree are to be found east of Panaca in Nevada, where there is an almost pure stand nearly twenty miles across. The pulp of the hard-seeded fruits is dry and sweet and was a staple article of Indian diet. A red dye was made from the ashes. During June phainopeplas are seen in numbers in the juniper thickets of the mountains of the eastern Mohave Desert.

GNETACEAE. Joint firs

The ephedras or joint firs are low, straggling, seemingly leafless shrubs inhabiting the warm, arid regions of the Mediterranean Basin, China, and western North and South America. They are closely related to the pines, the cycads, and the ginkgos but show in the structure of their stems strong affinities to the higher flowering plants. The long-jointed, fluted stems have at the nodes small, scale-like, papery leaves arranged in two's or three's, which dry up soon after appearing on the young tender branches.

The male and female blossoms are, in our species, found on separate plants. The small but attractive staminate (male) flowers, occurring in catkins or cones, have prominent, protruding, yellowish stamens. The female cones are less spectacular, consisting only of greenish, papery or thickened scales; the lower scales are sterile, the fertile flowers being found only at the top.

The small swellings or galls found on many of the stems of joint fir are made by the fusiform gallfly *Lasioptera ephedrae.* (See illustration of *Ephedra viridis*, 7.)

The stems contain a considerable amount of tannin, and for this reason a decoction made by boiling the stems was used among the Pima and other southwestern Indians as a remedy for intestinal ailments. The tea has long been used as a beverage among both Indians and Mexicans and is really quite delicious. The rather large, black seeds were sometimes roasted and eaten. The antelope ground squir-

1. SINGLE-LEAVED PIÑON or Desert Nut-Pine, a characteristic tree of the higher desert mountains, generally growing in association with California or Utah juniper

2. CALIFORNIA JUNIPER in association with agave and Mohave yucca in a zone just above the creosote belt

rels, well aware of the nutritious qualities of the seeds, harvest them in quantity.

3. CALIFORNIA JOINT FIR. *Ephedra californica* (Anc. Gr. name used by Pliny for the horsetail, *Hippuris*, which it resembles; of California). The leaf scales are in whorls of three. Common at low altitudes on both deserts and west to San Diego and south to Baja California. The stems are deep yellowish-green and of greater diameter than those of our other species. Healthy shrubs growing on sand hummocks or in deep gravelly washes often attain a diameter of 5 or 6 ft. and reach a height of a meter. This is the plant most commonly harvested commercially. (The drawing has purposely been enlarged to show details of structure of the cones.)

4. DEATH VALLEY JOINT FIR. *Ephedra funerea* (L., funereal, that is, of the Funeral Mountains). A spinose, intricately branched species with elongate, taper-pointed fruits recently found in the Death Valley area. Says Mr. M. French Gilman: "How odd it is that in the original description no mention was made of the shape of the bush. It is really very symmetrically rounded and does not sprawl out as the other joint firs frequently do. The gray-green branches are in general shorter than in other species, and radiate from definite centers on the main stem. I do not understand why such an outstanding plant so long escaped notice." It may be distinguished from the Nevada joint fir, which it somewhat resembles, by the ternate arrangement of the scale leaves and the sessile fruits.

5. LONG-LEAVED JOINT FIR, POPOTILLA, TEPOSOTE, CANATILLA. *Ephedra trifurca* (L., three-forked). A small bush, about 2 ft. high, distinguished by its long leaves, large, papery fruits, and spinose branches. The leaf scales occur in whorls of three. Rare in the California deserts: Los Algodones Dunes, near Yuma; Superstition Mt.; to Tex. and Mex.

6. NEVADA JOINT FIR. *Ephedra nevadensis* (of Nevada). The gray-based leaf scales and bracts are arranged in two's, as are also the long-stemmed fruits and the smooth seeds within them. An erect shrub, 2–4 ft. high, with small, glaucous-green, divergent branches. It grows mostly in rocky soils of the foothills. Its range includes both the Mohave and Colorado deserts in California and similar arid regions of Nevada, Utah, and Arizona. Its bluish-green stems help to distinguish it from all of our other species except the Death Valley joint fir. Desert folk use a tea made from it as a cure for mouth-canker. In certain areas of Nevada this joint fir is much grazed by livestock.

7. MOUNTAIN JOINT FIR. *Ephedra viridis* (L., green). This ephedra is easily distinguished by the exceptionally bright green color of its numerous, slender, erect, broom-like branches. The leaf scales occur in two's. It is a species of the higher altitudes and is generally plentiful among the junipers and piñons of the Mohave D.; to Utah and N.Mex.

8. CLOKEY JOINT FIR. *Ephedra Clokeyi* (Ira W. Clokey, botanist, of South Pasadena). A much-branched, gray-green joint fir, with leaves early deciduous and arranged in pairs. The seeds are solitary in the fruit and are furrowed.

PALMACEAE. Palm Family

9. DESERT PALM, CALIFORNIA FAN PALM (see p. 8). *Washingtonia filifera* (George Washington; L., "thread-bearing," referring to the leaf edges). There is a reserved majesty and beauty in the form of the desert palm which has aroused great admiration and has led to its introduction into gardens throughout the world. It is definitely known to be a native of the Colorado Desert and not an introduction by the early padres, as some would have us believe. The finest and most extensive palm groves are to be found in and near Thousand Palms Canyon to the north of Indio and in Palm Canyon on the south side of the Cahuilla Basin. Scores of smaller groups, numbering from 1 to 20 trees, known only to prospectors and stockmen, are scattered throughout the alkaline, rocky canyons near the base of mountains rimming the Salton Sink both north and south. It is to such small, unfrequented palm groups that one must go to see the trees in their natural beauty. There the fire vandals have not been and there are no naked, leaf-denuded trunks, but instead beautiful, soft, fawn-gray, leaf-thatched stems crowned with graceful leaves of living green. Under the leaf mantle of the trunks, canyon wrens find retreat; and from the copious loose threads hanging from the ends of the leaves, Arizona hooded and Scott orioles construct their unique, pendent nests. Under the leafy fronds of palms bordering streams, numerous little hylas or tree frogs live and all through the night make their distinctive music.

Some of the old palms living in deep, secluded gorges reach a height of 60–70 ft. Since the trunks have no annual rings there is no way to determine their exact age; some of the largest are probably as much as 200 years old. Dead palms are generally found to be infested by the larvae of the great bostrychid beetle, *Dinapate wrighti* (named after W. G. Wright, veteran entomologist, of San Bernardino). The robust grubs eat extensive galleries throughout

3. *Ephedra californica*
4. *Ephedra funerea*
5. *Ephedra trifurca*
6. *Ephedra nevadensis*
7. *Ephedra viridis*
8. *Ephedra Clokeyi*

9. The native DESERT PALM generally growing about alkaline seeps and small streams in gorges of the southern desert mountains

Photo by Randall Henderson

the trunks and after several years finally emerge as adults. The large openings where the beetles have gnawed themselves out are readily seen on the surfaces of fallen trees.

Both the thin, sweet, outside pulp and the hard, horny, pea-sized seeds of the palm fruits were eaten by the Cahuilla Indians. The seeds were pounded into a flour.

IRIDACEAE. Iris Family

10. BLUE-EYED GRASS. *Sisyrinchium bellum* (Gr., name for a bulbous plant of the Iris family; L., beautiful). **Fl.**: bluish-violet. This handsome irid with grass-like leaves is common to moist slopes throughout most of California but is confined to alkaline seeps on the desert.

LILIACEAE. Lily Family

11. DESERT MARIPOSA. *Calochortus Kennedyi* (Gr., "beautiful grass," in allusion to the flowers and grass-like leaves; Wm. L. Kennedy, who sent specimens from Kern Co. to T. C. Porter, who described the species in the *Botanical Gazette*). **Fl.**: flame-color or brilliant vermilion. This short-stemmed mariposa is most often found singly but may occur in large colonies, where it then makes a brilliant display of color which may be seen from considerable distances as patches of bright flame on the desert floor. When the flower is viewed closely, the contrast between the clear vermilion of the petals, the blackish-red patches below, and the purplish anthers provides an added charm. The black, hairy gland on each petal is ringed about by vermilion. If the plant is growing near a bush, the stem elongates and goes right up through its branches, displaying the flowers at the top. In the higher mountains of the eastern Mohave Desert, and growing in a zone above, is the somewhat similar *C. aureus*, with pale yellow flowers. It too climbs up into shrubs whenever opportunity presents itself.

12. DESERT LILY. *Hesperocallis undulata* (Gr., "evening beauty"; L., "wavy," referring to the ruffle-edged leaves). **Fl.**: white, with bluish-green band down the middle of the back of each petal. Distinctively a sand-flat and dune species, and therefore most abundant in the southeastern Colorado and southern Mohave deserts. Flowering begins in February or March, and in some localities blossoms may be found as late as the first of May. Propagation is probably mostly by seeds. The flowers are pleasantly fragrant. The bulbs of this day-blooming lily are deep-seated, often occurring from 18 in. to 2 ft. beneath the surface of the sands in which they grow. The

10. *Sisyrinchium bellum*
11. *Calochortus Kennedyi*
12. *Hesperocallis undulata*
13. *Allium fimbriatum*
14. *Zygadenus brevibracteatus*

Indians found their onion flavor agreeable, as did the early Spanish explorers, who called them *ajo,* garlic.

13. FRINGED ONION. *Allium fimbriatum* (L., garlic; L., "fringed," because of the prominent fringed crests of the ovaries). **Fl.**: purple to pale rose, with deeper midvein. The flowers are borne on a stem but 2–3 in. high, and the single leaf is generally curled close to the ground. Dry slopes and flats of both deserts. The var. *mohavense,* with pale pink flowers, is confined to the western Mohave D.

14. DESERT ZYGADENE. *Zygadenus brevibracteatus* (Gr., "yoke-like gland," the glands being arranged in pairs like an inverted U at the base of the perianth segments; L., short-bracted). **Fl.**: greenish-white or yellowish. Though belonging to a genus commonly referred to as Death Camas, because of a deadly toxin that affects both stock and man, desert zygadene is not known to share the poisonous properties of its near relations. The bulbs have coats like an onion but do not possess the distinctive odor of that plant. Long after blooming and withering the flower parts remain in place. The plants generally grow singly and are usually about 2 ft. tall. Almost entirely confined to the Mohave D.

15. CRESTED ONION. *Allium cristatum* (L., "crested," in reference to the prominent triangular crests of the ovary). **Fl.**: papery-white, striped with a reddish vein. This single-leaved onion is found in the Little San Bernardino Mts. and the eastern Mohave D.; to Utah.

16. PITTED ONION. *Allium lacunosum* (L., "full of holes or pits," because of the remarkable squarish pits seen when the light brown onion-coats are examined with a hand lens). **Fl.**: pinkish, with darker midvein. On the rocky benches near Black Canyon large fields of this tall-stemmed onion are in bloom in late May. Open places, western Mohave D.; to central Calif., and Santa Rosa I.

17. STRAGGLING MARIPOSA. *Calochortus flexuosus* (L., full of turns). **Fl.**: deep purple, rarely white. Stem slender to stout, commonly wandering over the ground or running up among bushes. The oblong glands of the flower bear densely tufted hairs. On slopes and benches up to 4,000 ft., Chuckawalla Mts. of Colorado D. and mountains of the Death Valley area; to Utah. The bulbs, like those of other mariposas, are eaten by the Indians.

The rather inconspicuous straight-stemmed Palmer Mariposa (*C. Palmeri*), with pallid lavender flowers, blooms in June along the western borders of the Mohave D., as at Cushenbury Springs.

15. *Allium cristatum*
16. *Allium lacunosum*
17.* *Calochortus flexuosus*
18.* *Muilla serotina*
19. *Brodiaea capitata pauciflora*

18. LONG-STEMMED MUILLA. *Muilla serotina* (the word "Muilla" was coined by spelling the name *Allium* backwards; L., "late," in the season). **Fl.**: greenish-white. This is the tallest of the California muillas, its stems often ranging in length from 14 to 20 in. Though common in interior southern California, it is seen only along the western edges of the deserts. A species, *M. coronata,* the crowned muilla, with flower stems but 1½–3 in. tall and bearing only a few flowers with broad strap-like filaments, is found in Antelope Valley and El Paso Mts.

19. DESERT HYACINTH. *Brodiaea capitata pauciflora* (James J. Brodie, Scotch botanist, specializing in cryptogams; L., having a head; L., few-flowered). **Fl.**: pale blue. This few-flowered brodiaea is widely scattered on our deserts, even reaching Arizona and New Mexico. The small bulbs, being edible, served as a source of food both for the Indians and for the early white settlers, who called them "grass-nuts."

20. WHIPPLE YUCCA (see p. 14). *Yucca Whipplei* (the Haitian name; Lt. A. W. Whipple—see **177**). **Fl.**: whitish. Whipple yucca, so common on the coastal side of the high mountains bordering the desert, is found but sparingly on their arid eastern slopes. When we reach the low, rocky ranges of the Mohave Desert farther eastward, we come in contact with the small-statured variety, *caespitosa,* which generally grows in clumps among rocks.

21. FLESHY-FRUITED YUCCA (see p. 16). *Yucca baccata* (L., berried). **Fl.**: whitish. On the far eastern Mohave Desert, often consorting with tree yuccas and the long-leaved Mohave yucca; a low, stemless species with enormous, smooth-skinned, fleshy fruits and saber-shaped leaves 18–30 in. long. The green pods, often 6 in. long, were an important food among the Indians, especially the Navahos, who ate them raw or roasted. Some they cooked and dried for winter use; others they made into a conserve used to sweeten beverages. The fruits chosen to be stored were first dried before the fire, then ground and kneaded into small cakes, which were afterward sun-dried. The roots yielded a soap, called *amole,* used especially in ceremonial wedding hair-washings, whereas the leaves provided strips for basket making. Many specimens of cloth and sandals made of the fiber of this yucca have been found in prehistoric ruins. This species can be distinguished immediately from all other yuccas of its range by the grayish blue-green color of its long leaves. Rocky slopes of piñon-juniper belt of limestone mountains of the eastern Mohave D.; to Colo., Tex., and Mex.

20. WHIPPLE YUCCA occurs along the desert's western edge

22. JOSHUA TREE, TREE YUCCA (see pp. 17 and 18). *Yucca brevifolia* (L., short-leaved). **Fl.**: greenish-white. Great forests of this sturdy tree are spread over widely scattered areas of the Mohave Desert and we believe its distribution marks the limit of that desert's true extent better than any other plant. In ancient times, when more humid conditions prevailed, the tree yucca was much more widespread and occurred at lower altitudes. Evidence of this is seen in the fossil remains and dung of the extinct giant yucca-feeding ground sloth, *Nothrotherium*, in areas of southern Nevada where the yucca is no longer found. At present, tree yuccas are found only upon and about the bases of those high desert ranges which receive a rainfall of 8 to 10 in. a year. Their southernmost station is near Piñon Wells in the Joshua Tree National Monument, and from here they range northward to central Nevada and east to extreme southwestern Utah and northern Arizona. Along the slopes of the mountains rimming the desert's western edge, they attain an average height of 20–30 ft. and sometimes a spread of 20 ft.

In the mountains of eastern San Bernardino County and southern Nevada there are vast areas densely populated with the var. *Jaegeriana* (named after Edmund C. Jaeger, of Riverside Junior College), a tree characterized by shorter leaves and different manner of branching. This tree is best seen in the broad valleys about the base of the New York Mountains, as at Cima on the route of the Union Pacific Railroad. Auto travelers pass through fine forests of it between Baker and Valley Wells en route to Las Vegas.

Joshua trees do not bloom every year, the length of the interval between times of flowering probably being determined by both rainfall and temperature. From the spreading, disk-like bases of the trunks great numbers of small, fibrous roots penetrate the soil both downward and horizontally, securing needed moisture and most effective anchorage against the strong gales which often sweep across the desert. The smallest of the roots are red, and the Indians used these for making patterns in their baskets. Propagation is effected both by rodents and by wind, and also by peculiar subterranean stems which push outward from the main trunk like bamboo runners. It is the sprouting of these runners that often accounts for small plants growing about the parent tree. These young plants are singled out by a butterfly called the Navaho yucca borer, *Megathymus yuccae navaho*, as the proper place to lay eggs, and the larvae which hatch not only eat the plant fibers but also the underground stem leading to the parent. Branching is caused either by the activity of the larvae of the yucca-boring beetle, *Scyphophorus yuccae*, or by flowering. In both cases the terminal bud is killed. Wood rats some-

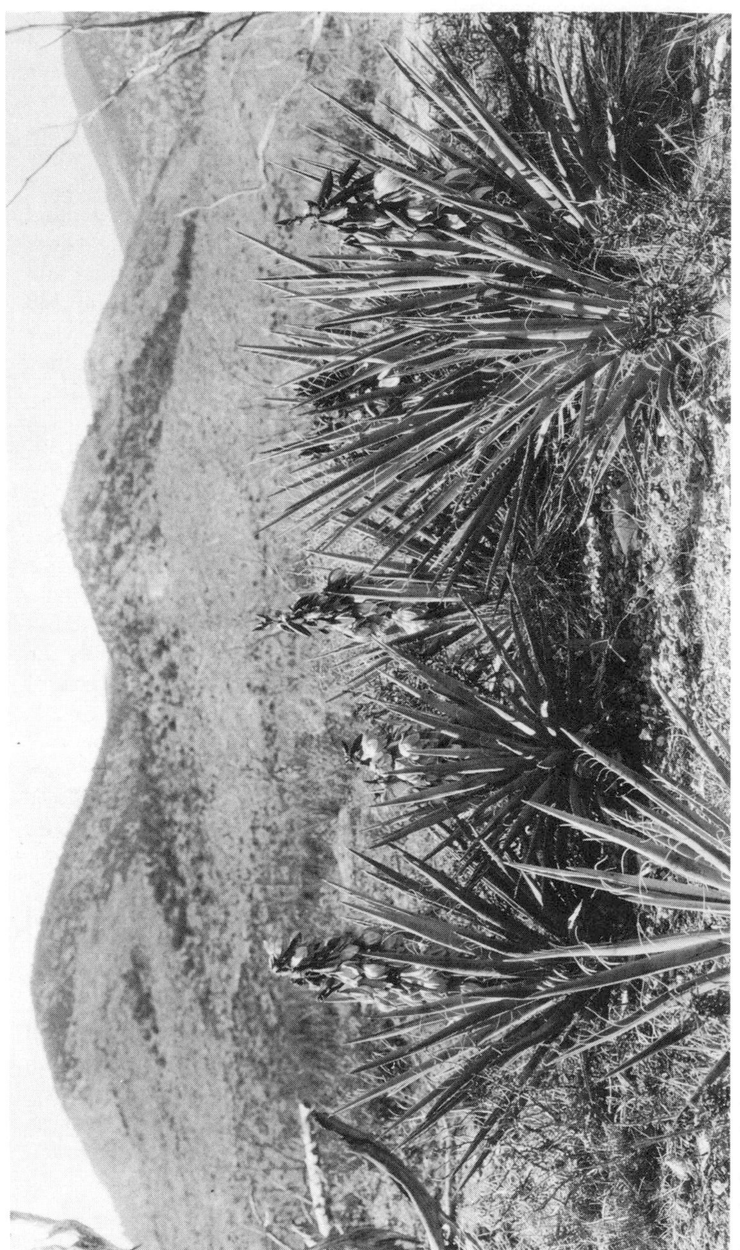

21. The FLESHY-FRUITED YUCCA (*Yucca baccata*)

Photo by Susan D. McKelvey

22. JAEGER TREE YUCCA (*Yucca brevifolia Jaegeriana*)
as found in southern Nevada

22. TREE YUCCAS are found in the Mohave Desert in a belt at the upper limit of the creosote bush or among junipers and Desert Nut-Pine

times climb the living trees and gnaw off the spiny leaves. These they incorporate into the nests which they build about prostrate trunks. While there are vertebrate animals closely associated with the tree yucca, there is only one, the night lizard, *Xantusia vigilis,* which is entirely dependent upon the Joshua tree and could not survive without it. This little saurian lives under the bark and in cavities of rotting, decumbent trunks. It subsists largely on termites, ants, and insect larvae.

At least twenty-five species of desert birds are known to utilize the tree yuccas as nest sites. The red-shafted flicker and the little speckle-check or cactus woodpecker dig holes in the fibrous trunks and branches; and these are later occupied by the ash-throated flycatcher, the Baird wren, the plain titmouse, the Western bluebird, and the Pasadena screech owl. The Scott oriole hangs its pendant nest among the needles and builds most of the structure from yucca fibers.

23. MOHAVE YUCCA, SPANISH DAGGER (see p. 20 and frontispiece). *Yucca schidigera* (L., "bearing a splinter of wood," "applicable to the coarse marginal fibers of the leaf-blade"). **Fl.**: whitish or with a purplish tinge. This yucca, commonly called *Yucca mohavensis* but recently renamed *Y. schidigera,* is scattered widely in the semiarid coastal counties of southern California, attaining perfection only on the broad, gravelly benches of the southern Mohave and eastern Colorado deserts. Where it is associated with ironwood trees in the washes of the Chocolate Mountains, it often reaches a height of 10 ft. This is the commonest yucca of the desert area. It occurs in many situations entirely unsuitable for the other species and often dominates wide areas. The distinctive field-mark is the long, yellow-green, trough-like leaf. The fruits, which ripen in midsummer, were utilized widely by the Indians. Pack rats often gnaw off the bitter outer covering of the fruit, which is rich in sugar.

24. PARRY NOLINA (see p. 21). *Nolina Parryi* (P. C. Nolin, French agricultural writer of the eighteenth century; Dr. C. C. Parry, 1823–1890, American botanist, who many times visited the California deserts and who is commemorated in the names of more than a score of native plants. Many of his interesting desert observations were described in the pages of the *San Francisco Bulletin*). **Fl.**: whitish. Visitors to the Joshua Tree National Monument or to the rocky canyons of the Palm Springs area frequently confuse this plant with certain yuccas. It differs radically from the yucca in its narrow, more pliant, grass-like leaves, much smaller flowers, and papery, dry-winged fruits containing small, round, hard seeds. Nolinas are con-

23. The MOHAVE YUCCA (*Yucca schidigera*), a short-statured species widespread on both California deserts

Photo by Susan D. McKelvey

24. PARRY NOLINA occupies the northern slopes of the higher mountains bordering the Coachella Valley. It flowers in late May and June and is mistaken by many for a yucca.

fined to rocky mountain sides and never occur on flat mesas, as do many yuccas. Old trunks, clothed only in ragged leaf bases, appear like giant barrels 2–3 ft. in diameter. The handsome, plumy flower panicles appear in May and may persist, laden with dry fruits, until late autumn. Some extraordinarily fine specimens of what appears to be this same nolina are found in the Kingston Mountains far to the north. The plants have very long leaves and make a spread of 12 ft. The flower stalks are a foot in diameter at the base and fully 12 ft. high. What magnificent plants and what a sight by moonlight! Parry nolina was first collected by Dr. Parry south of the Whitewater River on the dry, rocky ridges of Mount San Jacinto. On this same trip he collected *Phacelia campanularia* (see **455**).

In the desert mountains of San Diego County to Baja California and Arizona will be found another nolina, *N. Bigelovii* (after Dr. J. M. Bigelow; see **338**), which differs in having narrower flatter leaves with entire or scarcely serrulate margins which shred into brown fibers. It is on the whole a smaller plant and occupies lower more arid areas. The author has recently discovered several hundred plants widely scattered in a number of rocky canyons of the Orocopia, Chukawalla, Eagle, and Old Woman Mountains east of the Salton Sea.

AMARYLLIDACEAE. Amaryllis Family

25. DESERT AGAVE. *Agave deserti* (Gr., noble; L., of the desert). **Fl.**: yellow. Desert agaves are easily distinguished from both yuccas and nolinas by their thick, succulent leaves and by the fact that they commonly grow in large colonies. They blossom in late May. As soon as the fruits mature, the plants die and their place is taken by young offsets which spring up in a circle about the parent. In the fast-sprouting flower stems the Cahuilla Indians found a copious supply of food; the plump young buds they pried from the rosette of leaves with long juniper poles, and baked the white, solid butts, rich in saccharine, in great stone-lined pits. The well-browned portions they cut in slices, dried, and stored away for future use. The product is soft, sweet, and very nutritious and resembles in flavor a well-roasted yam. Eaten too freely, it has the unwholesome effect of producing bowel complaints. From the fibrous leaves the resourceful Indians made cloth, rope, and sandals. The adult plants serve as hosts to a most peculiar butterfly known as the Stephens skipper (*Megathymus Stephensi*), the larvae of which both feed and pupate in the bases of the leaves. The plant in attempting to protect itself from injury builds up a hard wall about the growing larva and thus unwittingly gives the developing insect unusual protection. The dried stems of this agave are used as nesting sites by the cactus woodpecker. With the exception of a few small colonies in the Providence and Granite mountains north of Amboy, the desert agave is confined to the dry desert mountains bordering the Salton Sink on the west; thence to Baja Calif.

26. UTAH AGAVE. *Agave utahensis* (of Utah). **Fl.**: yellow. This species is figured because it is listed in most of the botanical manuals dealing with California plants. Recent investigations made by the auther lead him to believe that this species does not occur in the state. Plants labeled *A. utahensis* from the Granite Mountains north of Amboy are now definitely known to be *Agave deserti.* The Utah agave was first collected near St. George, Utah, and is probably confined to that vicinity. The narrow flower panicle is like that of its variety *nevadensis.*

27. PYGMY AGAVE. *Agave utahensis nevadensis* (of Nevada). **Fl.**: yellowish. On rocky slopes and ridges in the higher limestone mountains of the eastern Mohave Desert grows the pygmy agave, with its small, spine-tipped leaves in rosettes shaped like an artichoke head not more than 10 in. in diameter. The uniquely slender flower-bearing stalks, which reach a height of 6–8 ft., seem quite out of proportion to the dwarf plant which bears them. A large woodcutting bee, *Xylocopa arizonensis,* builds nesting cells in the tall stems and provisions them with pollen from the flowers.

SALICACEAE. Willow Family

28. FRÉMONT COTTONWOOD. *Populus Fremontii* (L., "people," from the number and continued motion of its leaves like a populace; Capt. John C. Frémont—see **315**). A broad, open-crowned tree, often with gracefully drooping branches. It is abundant in many of the canyons on the desert slopes of the coastal mountains and frequently is the only sizable tree offering shade about desert springs. It ventures eastward along the Mohave River to Afton Canyon. Where it occurs about alkaline springs the lower trunks of Frémont's cottonwood are often encrusted with white salts and stained brownish by exuding sap. In autumn, huge numbers of buzzards congregate in the thickets of cottonwood trees along the Mohave River. The birds remain together as if gathered for some sort of convention in preparation for their long southward migration. They roost in the trees nightly for a week or more. In the morning as soon as the sun is up, they spread their wings to fly aloft and soar in marvelous spirals. Occasionally they go through the same antics in the spring. Along

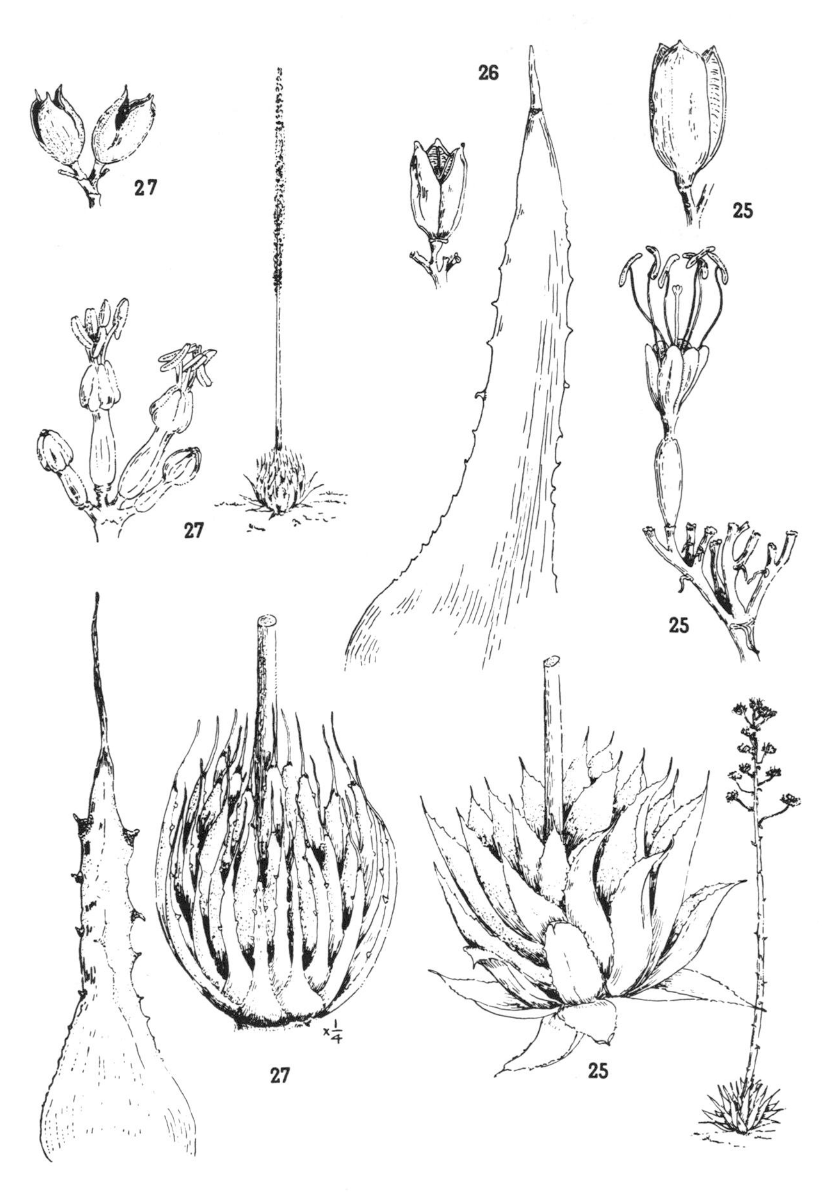

25. *Agave deserti* 26. *Agave utahensis*
27. *Agave utahensis nevadensis*

the Colorado River and in the Salton Sink the Frémont cottonwood is largely replaced by the MacDougal cottonwood, *P. MacDougalii,* a species with bluish-green leaves, whiter bark, and more upright habit. This tree was named after Dr. D. T. MacDougal (1865–), of the Carnegie Institution, authority on the vegetation of African and American deserts.

29. SLENDER WILLOW. *Salix exigua* (L., "to leap" or spring, from the quickness of its growth; L., small, scanty). A dwarf willow, 3–6 ft. tall, with grayish leaves, sometimes found about desert springs and streamlets.

Along the Mohave and Colorado rivers grow other species of willows common to California and Arizona stream banks. Their exact identification is often difficult, even under the critical study of experts, and it has been deemed inexpedient to deal with them here.

FAGACEAE. Beech Family

30. TURBINELLA OAK. *Quercus dumosa turbinella* (L., name of the oak; shrub-like; "a little top," referring to the top-shaped acorn cups). This scrub oak is confined to the piñon-juniper belt of the higher desert ranges and is very variable in the forms of its leaves and acorns. It is a member of the white-oak group, in which the acorns mature the first season. On its small twigs we often see dozens of "oak apples" made by gallflies of the genus *Andricus.* On the same oak tiny cynipoid wasps sometimes produce on the leaf edges beautiful thin-walled galls, which appear for all the world like little yellow marbles mottled with red. These remarkable galls sometimes occur by thousands on a single tree. Other cynipoid wasps produce conspicuous, spindle-shaped stem swellings. But the most interesting insect parasite is the "oak scale," *Cerococcus quercus,* colonies of which appear on the small branchlets, covering them with bright yellow, warty kernels of wax. Though it is dreadfully bitter to the taste, the Indians used this animal wax for chewing gum.

Occasionally in the higher desert mountains the broad-crowned, gold-cup live oak, *Q. chrysolepis,* grows in sheltered canyons.

ULMACEAE. Elm Family

31. DOUGLAS HACKBERRY. *Celtis Douglasii.* (L., for an African species of Lotus; David Douglas, Scotch botanist, 1798–1834, after whom the Douglas fir was named. Douglas traveled under the auspices of the Horticultural Society of London and visited California in 1831–32, getting as far south as Santa Barbara. He collected about 500 species in the state and "added more to the knowledge of California botany than all the botanists who had gone before him." His death, met in 1834 while on a trip to the Sandwich Islands, as the Hawaiian Islands were then called, was most tragic. He fell, while collecting, into a pitfall made to trap wild animals. His body was mangled by a wild bull which had previously or just afterward fallen into the same pit. A passer-by was attracted to the place by seeing Douglas' pet dog standing guard over a bundle of plants left on the ground by his master shortly before he stepped to his death.)

In the California deserts this is indeed a rare tree, being known only from a narrow gorge on the east side of Clark Mountain of the eastern Mohave Desert. Beyond our borders its range extends to Utah and eastern Washington. The sweet berries were eaten by the Indians.

URTICACEAE. Nettle Family

32. PELLITORY. *Parietaria floridana* (L., of walls, upon which it grows; of Florida). Slender, weak-stemmed perennial, found at low elevations growing in damp earth in the shade of rocks. It occurs on both deserts; also found in central Calif., Baja Calif., and Fla. The name "pellitory" is a corruption of the generic name, *Parietaria.*

LORANTHACEAE. Mistletoe Family

33. COLORADO RIVER MISTLETOE. *Phoradendron coloradense* (Gr., tree thief; of Colorado [River]). On mesquites along the Colorado R. Leaves thickish, usually with no veins evident.

34. JUNIPER MISTLETOE. *Phoradendron ligatum* (L., "tied with a band," because the scale-leaves are conspicuously constricted at the base). Occasional on desert junipers, but more often found on the mountain-dwelling Western juniper.

35. DESERT MISTLETOE. *Phoradendron californicum* (of California). Desert mistletoe is common on leguminous trees, particularly the large ones of the Colorado Desert. It frequently hangs in great festoons, generally high in the tree. From a distance its brownish-green mass of stems often resembles a large swarm of bees. In this, as in other mistletoes, the flowers are of separate sexes and are sunken in the joints of the spikes. At evening time the staminate flowers are exceedingly fragrant and their apple-blossom-like scent can be detected at a considerable distance. Most of the seeds are fertile, and with few exceptions those which are favorably placed on the host plants germinate. But the successful survivors are few, for most vigorous host plants show remarkable resistance to the inroads of the young seedlings. In the case of ironwood, many of the seeds are actually pushed away from the limb surface by a gummy substance exuded by the tree at the point of injury. The exudate eventually hardens and drops to the ground, carrying with it the young mistletoe plant. The handsome coral-pink berries of desert mistletoe are eaten in quantity by phainopeplas, and we often see twigs, above which the birds have perched, covered by quantities of seed-filled excreta. Bluebirds, robins, thrashers, and desert quail also feed upon the fruit. The berries are available to the birds as sources of water and food from November to April—a period spanning at least half the year—the time, be it noted, when their young are incubated and reared. Later, in the hotter months, juicy red fruits of the desert thorn (*Lycium*) supply them with the needed water. Knowing this we can account for many of the flocks of quail we find flourishing far from the water holes. The small plumbeous gnatcatcher, as well as the verdin, often builds its nest in the center of dead clumps of this mistletoe. The Pima Indians of Arizona boiled the fruit-bearing stems and then stripped the berries into their mouths. Both Calif. deserts; eastward to Ariz.; most common on mesquite, ironwood, and palo verde; occasional on creosote bush and condalia.

Along the Mohave River many of the cottonwood trees are infested with the large-leaved mistletoe, *Phoradendron flavescens macrophyllum.*

36. TUFTED MISTLETOE. *Phoradendron densum* (L., closely set). Occurs in dense, ball-like, upright tufts on California juniper, sometimes 15–20 to a tree. The leaves and stems are yellowish-green; the berries straw-colored. So. Calif.; north to Ore.

37. PIÑON MISTLETOE. *Arceuthobium divaricatum* (Gr., juniper life, i.e., living on juniper; L., "diverging," "spread asunder" or "forked"). In this species the stems are usually erect. The berries, olive-green in color, contain a sap which is under considerable pressure, and when ripe they burst with the slightest touch, ejecting the sticky seed with sufficient force to throw it several feet. This mistletoe is almost confined to desert nut-pines. (The illustration indicates the differences between branchlets bearing male and female flowers.)

28. *Populus Fremontii*
29. *Salix exigua*
30. *Quercus dumosa turbinella*
31. *Celtis Douglasii*

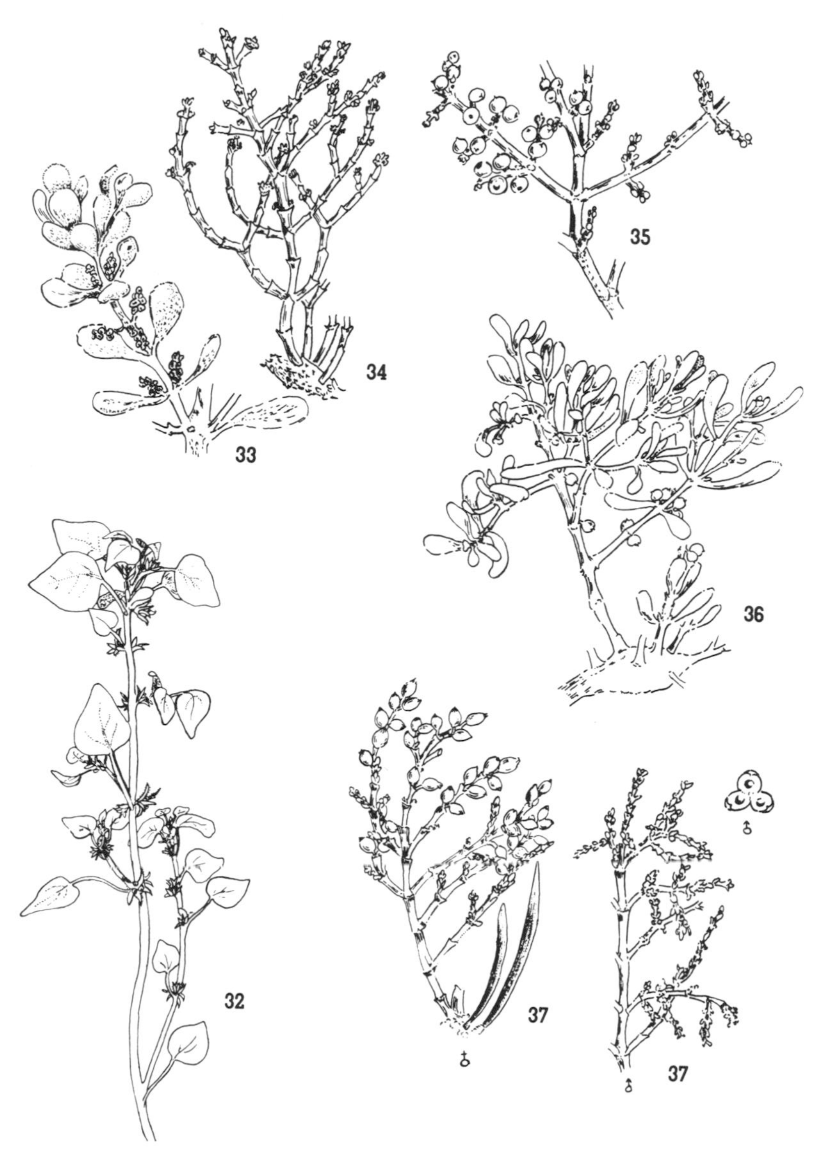

32. *Parietaria floridana*
33. *Phoradendron coloradense*
34. *Phoradendron ligatum*, ×¾
35. *Phoradendron californicum*
36. *Phoradendron densum*
37. *Arceuthobium divaricatum*, ×1

POLYGONACEAE. Buckwheat Family

38. LITTLE TRUMPET. *Eriogonum trichopes* (Gr., "woolly joint or angle," in allusion to the hairy joints of some species; L., hairy-footed. **Fl.**: yellow. Annual, but sometimes perennial, with pronounced bright yellowish-green stems and leaves. Remember that as a rule only the short basal node is much inflated; this makes it easy to distinguish from the darker green *Eriogonum inflatum*, or desert trumpet, which shows large inflations in many of the upper nodes (see **60**). Very common about dry-lake borders and on clay hills impregnated with salts. When in flower and again in age the plants turn reddish, the broad areas occupied by them present a colorful picture, visible miles away. Both deserts; to Utah and Mex.

39. YELLOW TURBAN. *Eriogonum pusillum* (L., small, weak). **Fl.**: bright yellow to reddish. Annual, 4–12 in. high, the leaves somewhat greenish above but white-felty beneath. A common species of the sandy flats and bordering mountains of the Mohave D., also the Colorado D. at Palm Springs. Called yellow turban because of the yellow flowers surmounting the turban-shaped, glandular involucres.

40. KIDNEY-LEAVED BUCKWHEAT. *Eriogonum reniforme* (L., "kidney-shaped," with reference to the leaf). **Fl.**: bright yellow. The kidney-shaped leaves, somewhat ruffled about the edges, are white-woolly beneath but often quite bare of hairs above. It is closely related to the yellow turban, but its involucres are not glandular. The flowers are exceedingly small. Common on sandy flats of the Colorado and Mohave deserts, to Owens Valley; Nev. and Baja Calif.

41. RED-ROOT BUCKWHEAT. *Eriogonum racemosum* (L., "clustered," with reference to the arrangement of the 5-toothed, bell-shaped involucres on the flowering branches). **Fl.**: rose-pink or white. The one or more erect stems of this perennial species rise to a height of 8–32 in. from a reddish taproot. The densely woolly-hairy, basal leaves are long-stalked. Dry plains and slopes, Providence Mts.; to Colo. and Tex.

42. THOMAS ERIOGONUM. *Eriogonum Thomasii* (Major Thomas, stationed at Fort Yuma about 1850). **Fl.**: very small, at first dull yellow, later white to rose. Annual, 4–8 in. high, branching from the base. The exceedingly small, wiry, terminal stems are at first bright green but turn deep red-brown in age; the leaves are white-woolly beneath but hairless above. The flower is peculiar in that the outer calyx segments show a sac-like dilation on each side of the heart-shaped base. Sandy plains and alluvial fans of the Colorado D. north to Inyo Co.; to Utah. The drawing shows but a portion of a small plant.

43. SKELETON WEED. *Eriogonum deflexum* (L., "turned-down," in allusion to the position of the upper branches). **Fl.**: whitish, sometimes pink. Short annual, with a very broad, flat-topped crown, often 2 ft. across. The branches are loaded with myriads of pendant pearly flowers the size of small rice kernels. Very common in late summer (September to November) along the highways where the soil has been disturbed. In winter the dried stalks turn maroon and are very conspicuous. Prominent along highways of the eastern Mohave D. is its more upright variety, *ascendens,* with a height of 2 or 2½ ft.

44. TECOPA SKELETON WEED. *Eriogonum deflexum brachypodum* (Gr., broad-footed). **Fl.**: white to pink. A much more widespreading plant than the species; the flat tops are often up to a yard across. The involucres have quite long stalks. It is seen to wonderful advantage along roads leading into Death Valley. If one at all closely examines the old red stems of this and other buckwheats, one is certain to note the broad-ringed scars where the feeding larvae of the buckwheat butterfly (*Apodemia mormo deserti*) have girdled them.

45. DESERT BUCKWHEAT. *Eriogonum deserticola* (L., desert inhabitant). **Fl.**: yellow. The largest of our desert buckwheats; up to 5 ft. tall and sometimes with woody stalks 2 in. in diameter! The widely branching plants are conspicuous on the barren soils of the south end of the Salton Sink, also on the dunes west of Yuma. The white-woolly, somewhat ruffled leaves are seen only on the young, tender roots. The roots penetrate deeply in the sands.

46. MOHAVE BUCKWHEAT. *Eriogonum mohavense* (of Mohave [Desert]). **Fl.**: calyx bright yellow. 4–12 in. high, the stems bluish-green except at the nodes. The small, white-felty leaves generally lie flat to the earth forming a rosette. Occasional, western Mohave D. to Owens Valley.

47. NODDING BUCKWHEAT. *Eriogonum cernuum* (L., "nodding-pendulous," with reference to the flowers). **Fl.**: white. Annual, 4–12 in. high, the basal leaves felty beneath; leaf stems white-woolly and reddish. A rare plant of the desert ranges; Owens Valley; Ore.; Alberta; Kans.; N.Mex.

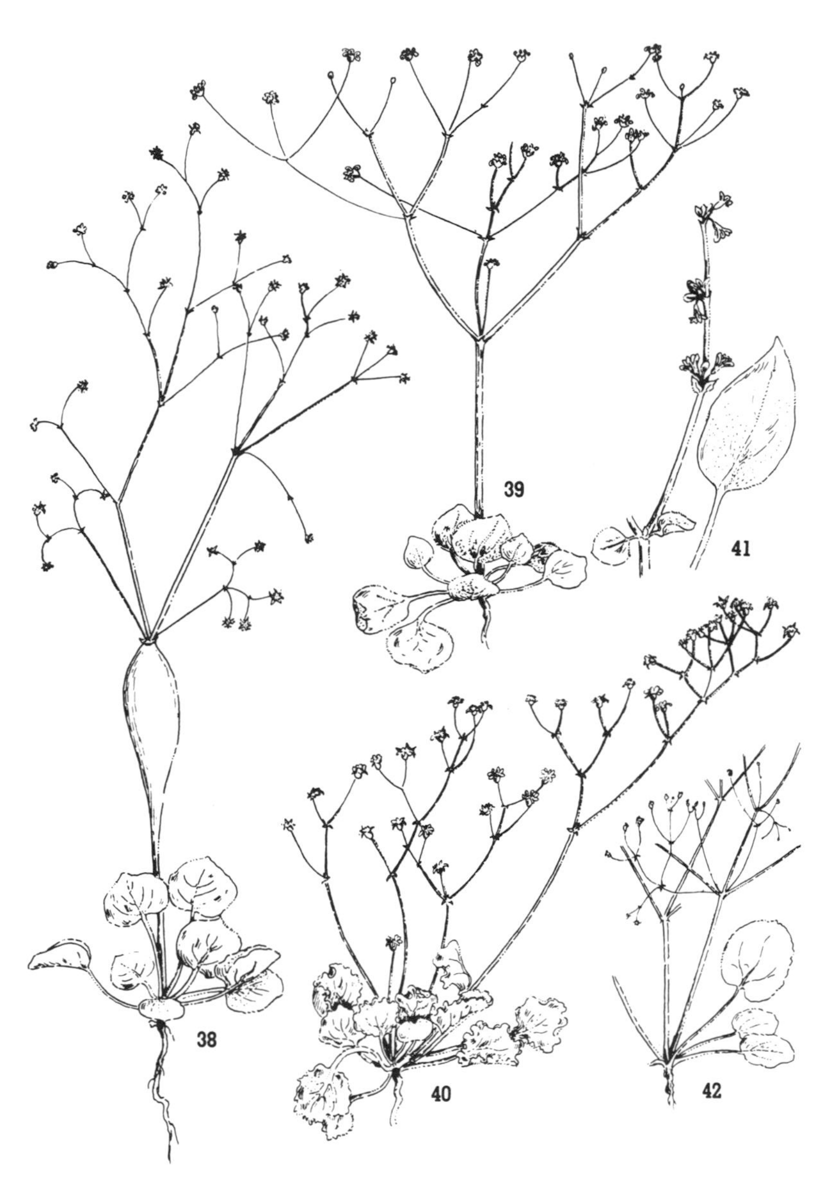

38. *Eriogonum trichopes*
39.* *Eriogonum pusillum*
40. *Eriogonum reniforme*
41. *Eriogonum racemosum*
42. *Eriogonum Thomasii*

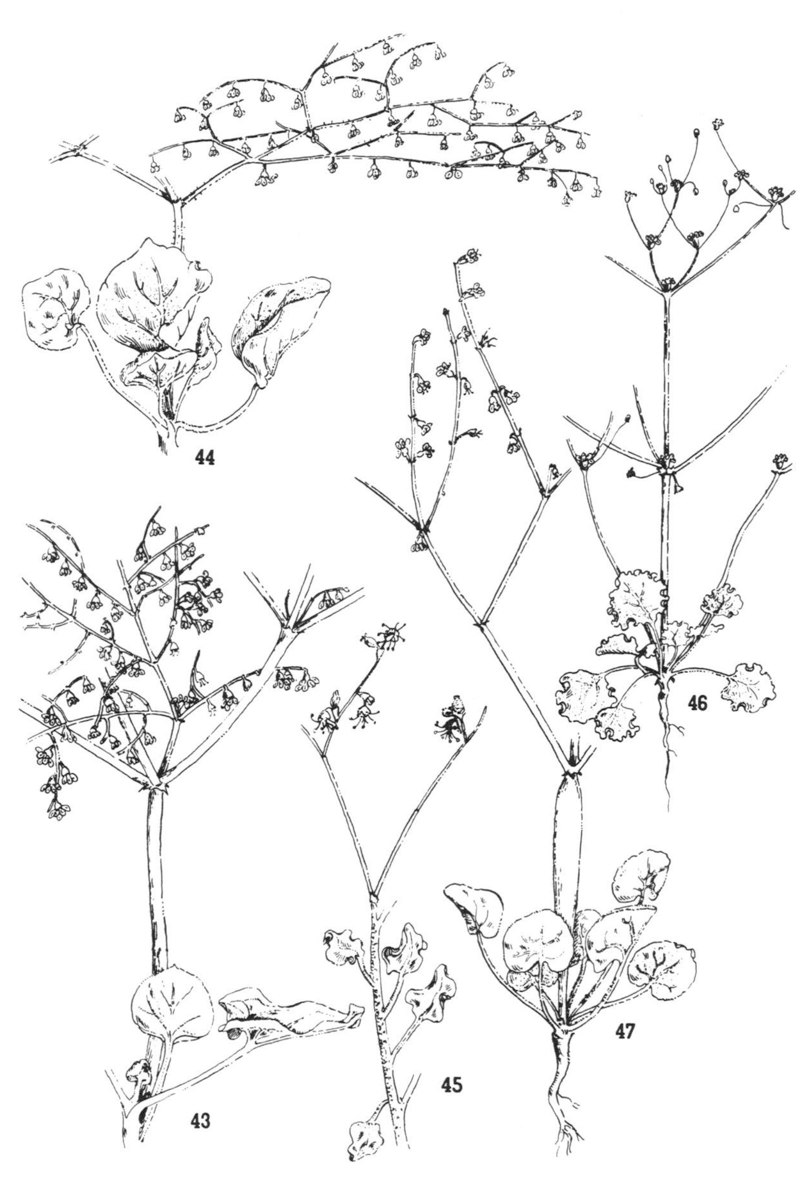

43. *Eriogonum deflexum*
44. *Eriogonum deflexum brachypodum*
45. *Eriogonum deserticola*
46. *Eriogonum mohavense*
47. *Eriogonum cernuum*

48. WRIGHT BUCKWHEAT. *Eriogonum Wrightii* (C. Wright; see **209**). **Fl.**: white or pink. Woody-stemmed perennial, 1–1½ ft. tall; the parts covered with white, matted, wool-like hairs. Common on the desert's edge at Cajon Pass and to be found occasionally at similar altitudes in the desert ranges to the east; to Nev., Utah, and N.Mex.

49. KNOTTY BUCKWHEAT. *Eriogonum nodosum* (L., having knots or nodes). **Fl.**: pink or whitish. A shrubby perennial, 1–2½ ft. high, whitish throughout, with numerous leaves on the basal portion of the slender, mostly upright stems. It should be noted that the flowers are only on the diverging lateral branches and not on the main ascending branches as in the Wright buckwheat. The involucres are white-woolly. Mountain slopes and canyons along the western edge of the Colorado D.

50. FLAT-TOP. *Eriogonum plumatella* (L., small-feathered). **Fl.**: white. A very erect plant, leafy below, and with several closely set, ascending stems 1–2½ ft. high, which are peculiarly forked above and horizontally branched to form flattish "platforms." The whole plant is grayish; but in the var. *Jaegeri*, confined to the close vicinity of the upper Morongo Wash at the west end of the Colorado Desert, the upper stems are devoid of hair and therefore green or (in age) brownish. Dry, rocky places, 2,500–4,000 ft. Little San Bernardino Mts. and desert base of the San Gabriel Mts. to the Providence Mts.

51. MACDOUGAL BUCKWHEAT. *Eriogonum microthecum MacDougalii* (Gr., small-cupped; Dr. D. T. MacDougal—see **28**). **Fl.**: pinkish. Low, much-branched shrub, with leaves white or woolly beneath. On dry, rocky slopes of higher piñon-covered mountains bordering our deserts; to Wash. and the Rockies. The drawing was made from a specimen taken from mountains at the north end of the Death Valley trough.

52. ROCK BUCKWHEAT. *Eriogonum saxatile* (L., "rock dwelling," found among rocks). **Fl.**: white to pale yellow. A perennial, arising from a woody, short-branching stem. The silvery-felty leaves, varying much in size, are crowded close to the soil. Mountain slopes of the piñon-juniper area of the Joshua Tree National Monument, north to the Coast Range and Sierra Nevada.

53. PAGODA BUCKWHEAT. *Eriogonum Rixfordii* (G. P. Rixford; see **129**). **Fl.**: white to pinkish. A very handsome green annual coming into flower in late summer in the Death Valley area. The gracefully curved ascendant stems, its exceedingly intricately divided branchlets, and its numerous platforms or "stories" of bloom one

48. *Eriogonum Wrightii* 49. *Eriogonum nodosum*
50.* *Eriogonum plumatella*

above the other give it a form at once distinctive and suggestive of a Japanese pagoda or some of the African acacias. Sometimes the plants are 2 ft. high.

54. LONG-STEMMED BUCKWHEAT (not illustrated). *Eriogonum elongatum* (L., "lengthened," because of the long stems). **Fl.**: rose or pink. A very erect plant with several almost naked, whitish stems up to 4 ft. high, these arising from a leafy, branching base. Coastal and cismontane southern Calif. reaching the Colorado Desert's edge as at Jacumba and Cabazon.

55. WRIGHT BUCKWHEAT. *Eriogonum Wrightii membranaceum* (Charles Wright [see **209**]; L., membranaceous). **Fl.**: white or pink. A shrubby, much-branched perennial, 1–1½ ft. tall, with the lower stems leafy and whitish. The bases of the leaf stems are dilated into hairless, brownish sheaths which clasp the stem. Little San Bernardino Mts., Cajon Pass area among junipers; to Baja Calif.

56. OVAL-LEAVED BUCKWHEAT. *Eriogonum ovalifolium* (L., oval-leaved). **Fl.**: yellow. This pretty Sierra Nevada perennial occurs in the dry piñon-juniper area of the Inyo and Panamint ranges; thence to Ariz. and N.Mex. The large, pompon-like flower heads are very conspicuous in late June. A form with pink to white flowers has recently been assigned the varietal name *vineum* (L., belonging to wine).

57. HEERMANN BUCKWHEAT. *Eriogonum Heermannii* (Dr. A. L. Heermann, collector on the Williamson Survey in 1853). **Fl.**: whitish. A shrubby species, 1½–2 ft. high, leafy below, with stems repeatedly, often densely, di- or trichotomously branched above to make a tangle of rather rigid branchlets. The involucral teeth are glabrous, i.e., not hairy. Rocky gorges and slopes of the Mohave D. to Inyo Co.; and Nev.

58. BOULDER BUCKWHEAT. *Eriogonum Heermannii floccosum* (L., having loose "flocks" or tufts of wool, with reference to the herbage). **Fl.**: whitish. A compact, more intricately branched shrub of the rocky areas of the Little San Bernardino, New York, and Providence mountains.

59. TUFTED BUCKWHEAT. *Eriogonum caespitosum* (L., "turfy," i.e., having low stems forming a dense turf or sod). **Fl.**: white or yellow, fading reddish. This silver-leaved dwarf grows close to the earth on plateaus and slopes of the arid piñon forests of the Inyo Mts. It is also widespread in similar situations over much of the Great Basin area.

60. DESERT TRUMPET. *Eriogonum inflatum* (L., inflated). **Fl.**: yellow. One of the unique plants discovered by John C. Frémont on his notable journey across the Mohave Desert in 1844. An annual or perennial, from 8 to 32 in. tall, often several "stories" high, with upper parts of the nodes inflated. The leaves, which lie flat to the soil, have a remarkable silver sheen. The tender tips have a sour taste like that of sheep-sorrel and may be used advantageously in salads. Wasps of the genus *Onyerus* sometimes use the hollow stems as a larder. This the wasp accomplishes by drilling a hole near the top of one of the inflations and filling the lower constriction with small pebbles. Then the tiny wasp will pile in a surprisingly large number of insect larvae, pack them tightly down into the cavity, and lay her eggs upon them. The remaining part of the inflated stem is filled up with more grains of sand to make the storehouse secure. The stored larvae insure an adequate food supply for her offspring.

Sometimes the stems of this buckwheat "forget" to inflate, and we then have the variety *deflatum*. The species is common to both deserts, but the variety is confined to the southern Colorado D.; to Baja Calif.

61. LONG TRUMPET. *Eriogonum nudum* (L., naked). **Fl.**: yellowish. Several-stemmed perennial, often 3 ft. high, with blue-green herbage. A distinguishing feature is the great length of the inflated lower joints. The illustration was made from plants collected twelve miles south of Darwin.

62. NAPKIN-RING BUCKWHEAT. *Eriogonum intrafractum* (L., broken within). Perennial, glaucous-green herb, with erect, leafless stems 3–4 ft. long, usually solitary and unbranched except above where the flowers occur. When old, as Cilman states, the bark of the stems falls away "leaving the stems made up of a series of white sections or joints looking like tiny napkin rings. The rings easily unjoint when weathered enough, and on the ground around old plants the fallen stems indicate the vintages of the different years' growth of stalks. Grapevine Mt., Death Valley Nat. Mon.

63. YELLOW BUCKWHEAT. *Eriogonum brachyanthum* (Gr., short-flowered). **Fl.**: lemon-yellow. 3–12 in. high; base of stem and leaves white-woolly. The stems turn deep reddish-brown with age. Sand and mildly alkaline clay flats below 4,000 ft. western Mohave D.; to Owens Valley. This species comes into flower in July along with *E. trichopes*, *E. baileyi*, and *Lessingia germanorum*, the four often closely associated.

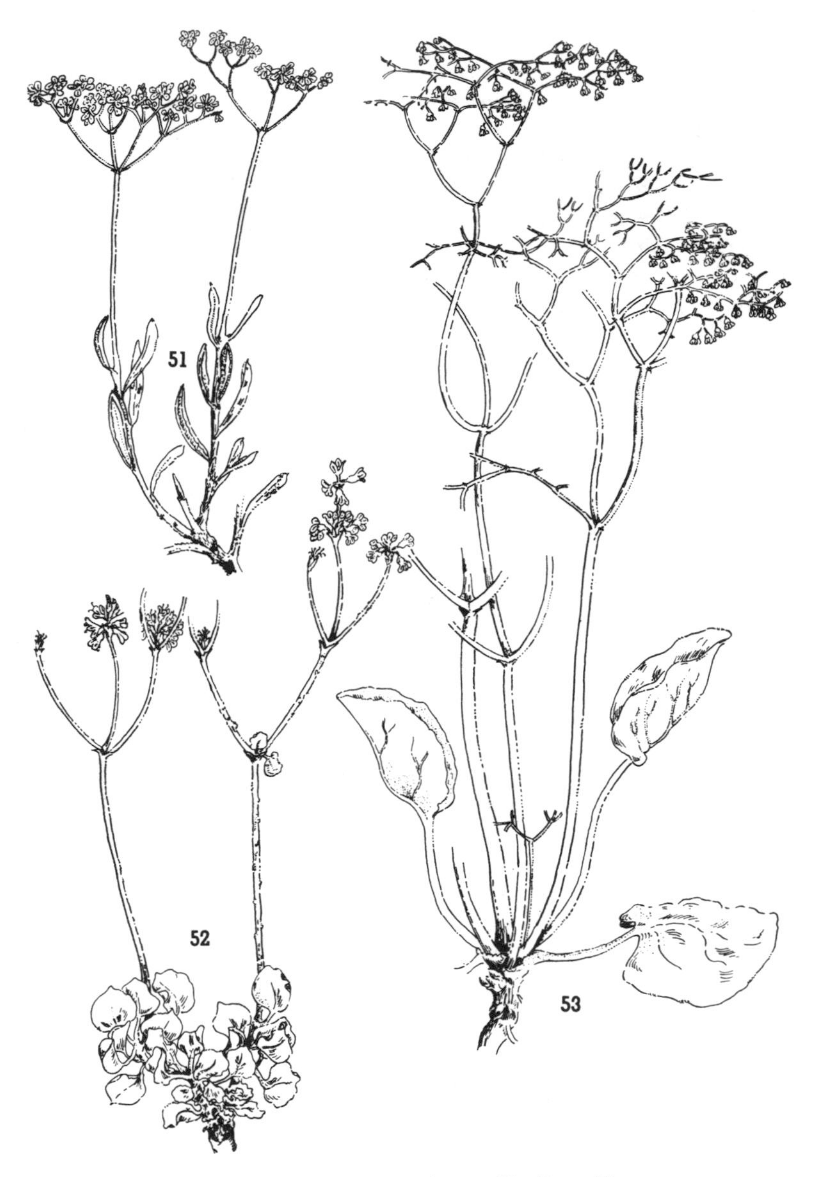

51. *Eriogonum microthecum MacDougalii*
52. *Eriogonum saxatile*
53. *Eriogonum Rixfordii*

55. *Eriogonum Wrightii membranaceum* 57. *Eriogonum Heermannii*.
56. *Eriogonum ovalifolium* 58. *Eriogonum Heermannii floccosum*
59. *Eriogonum caespitosum*

64. BAILEY BUCKWHEAT. *Eriogonum Baileyi* (W. W. Bailey, botanist on the King Expedition to Nevada in 1867). **Fl.**: white. Erect, spreading annual of bluish-green color, 4 in. to 1 ft. high. The leaves are white-woolly and all basal. Common on somewhat alkaline soils of flat areas of the western Mohave D.

65. WHISK BROOM. *Eriogonum nidularium* (L., "nest-like," referring to the form of old plants). **Fl.**: white or yellowish. A pleasing little annual, 3–8 in. high, having parts covered with cobwebby hairs. Dry sands and gravels of the Mohave D.; to Nev.

66. ANGLE-STEMMED BUCKWHEAT. *Eriogonum angulosum* (L., "angled," with reference to the 4–6-angled stems). **Fl.**: white to pink. Western Mohave D.; to Ida., Ariz., and Baja Calif. The rare variety *maculatum*, with yellowish calyx spotted with purple, is known from the western Mohave D. to Inyo Co. The flowers appear late in June and during July.

67. SLENDER-STEMMED BUCKWHEAT. *Eriogonum gracillimum* (L., "most slender," in reference to the thread-like flower stems). **Fl.**: rose-pink. Freely branching annual, 4–10 in. high, with thin, woolly herbage. Sandy plains and mountain valleys of the middle and western Mohave D. to San Joaquin Valley and westward. The flowering season is April and May.

68. XANTUS SPINY-HERB. *Chorizanthe Xanti leucotheca* (Gr., "divided-flowered," on account of the parted involucre; Xantus de Vesey [see **719**]; Gr., "white case," in reference to the white-haired involucres). **Fl.**: pale pink. Northwestern Colo. D. along Cabazon, Whitewater, and Mission creeks. Outstanding characters are the ovate leaves, the whitened involucres, and the foliaceous bracts.

69. BRITTLE SPINE-FLOWER. *Chorizanthe brevicornu* (L., short-horned). **Fl.**: white. Erect, yellowish-green annual, 3–8 in. high, with oblanceolate leaves, curved involucres, and a lax inflorescence. The plants retain their green color even when mature. The dry stems are exceedingly fragile and readily break into many pieces when handled. Very common to both deserts, mostly on dry, rocky hills; to southwestern Utah, Nev., and western Ariz. *C. spathulata*, much like *brevicornu* except for the spatulate leaves, is known from the Panamint Mts. Its stems are reddish at maturity.

70. RIGID SPINY-HERB. *Chorizanthe rigida* (L., stiff). **Fl.**: yellowish. Flourishes in coarse gravels and on black pebble beds of the hottest deserts of Calif., Nev., Ariz., and southwestern Utah. It and the creosote bush are sometimes the only plants able to withstand the intense heat. The dry, woody plants retain their long spines and often remain in place in the soil a year or more after reaching maturity.

71. WATSON SPINY-HERB. *Chorizanthe Watsoni* (Sereno Watson; see **459**). **Fl.**: yellow. 1–4 in. high, often with several greenish or reddish stems. Dry plains, western Mohave D.; to Wash., Ida., Utah, and Nev.

72. THURBER SPINY-HERB. *Chorizanthe Thurberi* (G. Thurber; see **319**). **Fl.**: white. Erect annual, 4–10 in. high, with 7–10 oblong to elliptic leaves in a basal rosette. Each of the involucres bears 3 large, sharp, basal horns. Frequent in dry, sandy places of both deserts; to Nev. and Ariz.

73. CORRUGATA. *Chorizanthe corrugata* (L., wrinkled). **Fl.**: white. A small plant marked by the strong horizontal corrugations of the cylindric, three-lobed involucres. Open places of the eastern Mohave and Colorado deserts; Baja Calif., southern N.Mex., southwestern Ariz.

74. MOHAVE SPINY-HERB. *Chorizanthe spinosa* (L., "thorny," the leaves being spine-tipped). **Fl.**: white. Prostrate annual forming a spiny mat 5–15 in. across. The plants when in full bloom may be mistaken, by the novice, for a low-growing, white-flowered forget-me-not. The basal leaves are oval or broadly oblong, the involucres 4–5-ribbed. Western Mohave D.

75. PUNCTURED BRACT. *Oxytheca perfoliata* (Gr., "sharp-cased," referring to the spiny involucre; L., leaf-piercing). **Fl.**: whitish. A plant up to 1 ft. broad which at once draws attention by its perfoliate bracts, i.e., broad bracts through the centers of which the stems pass. In age these large, almost leafy bracts turn reddish, adding much to the attractiveness of this unusual spring annual. The small blue butterfly, *Philotes speciosa,* lays its eggs on the plants and the tiny larvae upon hatching begin to feed at the center of the saucer-like bracts. When at rest the larvae look like little seeds at the bottoms of the spiny saucers. Plains above 2,400 ft., Mohave D. to northern Calif.; Nev. and northern Ariz.

76. TRILOBIA. *Oxytheca trilobata* (L., three-lobed). **Fl.**: white. An exceedingly attractive, low, maroon-stemmed annual. The small flower, having calyx parts with eroded margins, is much enhanced

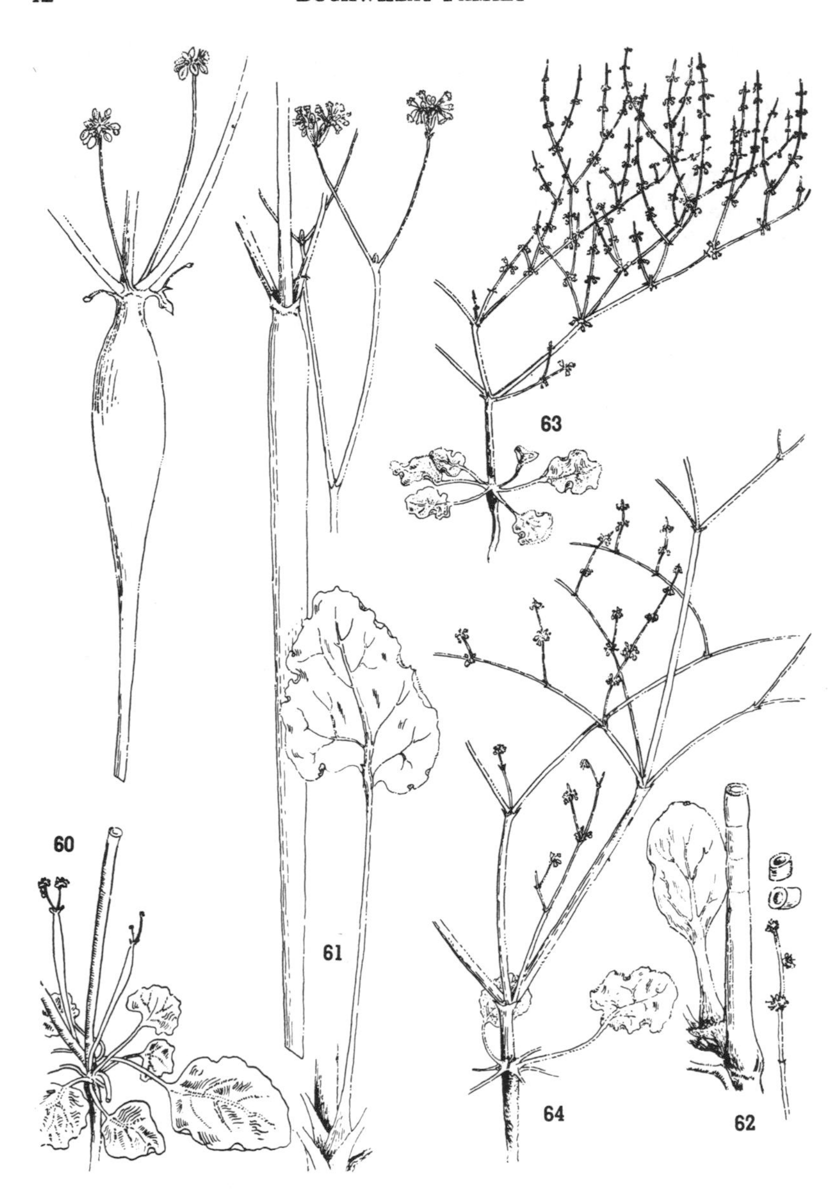

60. *Eriogonum inflatum*
61. *Eriogonum nudum*
62. *Eriogonum intrafractum*
63. *Eriogonum brachyanthum*
64. *Eriogonum Baileyi*

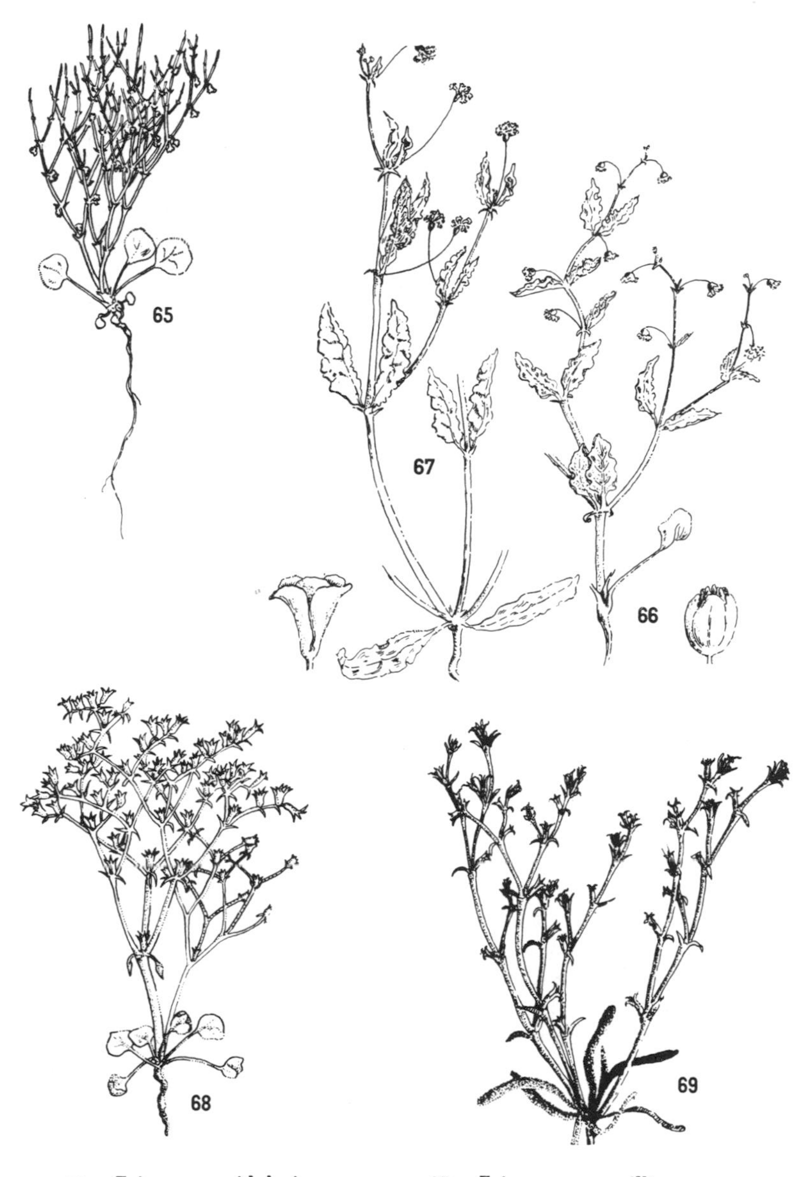

65. *Eriogonum nidularium*
66. *Eriogonum angulosum*
67. *Eriogonum gracillimum*
68. *Chorizanthe Xanti leucotheca*
69. *Chorizanthe brevicornu*

in beauty when viewed under a lens. Mountains of cismontane Calif.; occurring at the western edge of the Mohave D. in the Joshua Tree National Monument.

77. YELLOW SPINY-CAPE. *Oxytheca luteola* (L., yellowish). **Fl.**: yellow. Prostrate annual, with stems 2–5 in. long, and yellowish herbage. It is one of the plants collected by C. C. Parry on the Mohave Desert, "growing in the moist sand formed by seepage from the railway water-tank at Lancaster, and in the neighboring alkaline soil with *Kochia californica* and *Atriplex Parryi*." Dry lakes about far western end of the Mohave D.

78. PTEROSTEGIA. *Pterostegia drymarioides* (Gr., "wing-covering," referring to the bract; L., like the genus *Drymaria*). **Fl.**: reddish. Slender, weak-stemmed annual, common in the shade of rocks and shrubs of the western Colorado and eastern Mohave deserts. Also plentiful in coastal Calif.; to Ore. and Baja Calif.

79. GOLDEN CARPET. *Gilmania luteola* (first described in 1893 as a *Phyllogonum*, the genus was renamed in 1936 in honor of M. French Gilman, Death Valley naturalist, "whose intelligent and persistent search for this seemingly lost plant has resulted in its rediscovery"; L., yellowish). **Fl.**: yellow. One of the rarest of desert plants. Collected in Furnace Creek by members of the Death Valley Expedition in 1891 and seen only a few times since. In March 1939 vigorous plants were actually abundant in some of the canyons to the east of Furnace Creek Wash. It is a yellowish-green annual with juicy green leaves having a slightly acid taste and lying flat to the ground. When in flower, the plant superficially resembles *Lepidium flavum*. When the fruiting stage is reached, the leaves turn clear golden yellow and the stems turn up, a position probably aiding in the dispersal of the seeds. "Apparently," wrote Dr. Coville, "*Gilmania* is a plant in process of extinction through the extreme dryness of Death Valley. Its seeds, like those of many desert annuals, evidently are able to lie dormant in the ground for several years. Some of them germinate after a good rain if the temperature conditions are suitable, but these germinated seeds apparently do not produce fruiting plants unless the seedlings are boosted to suitable size and vigor by a second rain, adequate in amount. In most years any little *Gilmania* plants that have been able to start will die before they produce seeds from lack of a second rain adequate in amount and properly timed. The continued existence of this species apparently depends on the dormancy of a sufficient number of seeds to carry it over unfavorable years to years of adequate and properly-timed double-rains. If Death Valley becomes drier and drier, and years with suitable double-rains become more and more infrequent, the vitality of the old *Gilmania* seeds in the soil will ultimately be insufficient to span these longer periods of years when no new seeds are produced, and extinction, which is now a menace, will become a fact."

80. WOOLLY-HEADS. *Nemacaulis denudata* (Gr., thread stem; L., denuded). **Fl.:** pinkish. Annual, with bare, reddish stems and woolly flower clusters. Sandy beaches, Los Angeles and southward; western edge of Colorado D.; to Baja Calif.

81. WILD RHUBARB. *Rumex hymenosepalus* (Old L. name used by Pliny; "having membranous sepals"). **Fl.:** pinkish. A coarse, herbaceous, somewhat reddish perennial, with acid sap. The leaves are wavy-margined; the compact panicle of fruit is delicately pinkish. A distinguishing feature of the fruit is the absence of the little, grain-like protuberance which in many of the rhubarbs shows up between the wings of the fruit. Common in dry washes and plains from the western and southern Mohave D. and the Colorado D.; to Colo. and N.Mex.

CHENOPODIACEAE. Pigweed Family

82. CATERPILLAR GREASEWOOD, BLACK GREASEWOOD, BIG GREASEWOOD, TRUE GREASEWOOD. *Sarcobatus vermiculatus* (Gr., flesh thicket; L., wormy). Caterpillar greasewood is one of the most conspicuous and widely distributed of desert bushes in the arid regions of Nevada and Utah, but in California it is largely confined to the Death Valley area. The bright green foliage makes a vivid contrast with the gray-hued shad-scale and the purplish-brown inkweeds with which it is often associated on clay flats or salt-encrusted playas surrounding dry lakes. Sarcobatus Flats in Nevada was so named because of the unusual abundance of this shrub on the peculiar clay dunes prevalent there. Though some cattlemen consider it valuable browse in the fall and early spring, caterpillar greasewood has proved to be highly poisonous to sheep; whole bands in spring may succumb to its toxic effects. To the geologist this plant is a trustworthy indicator of ground water. Bailey greasewood, *S. Baileyii* (after Vernon Bailey, eminent naturalist and authority on mammals, 1864–), with short, staminate spikes and leaves covered with branched hairs, is reported from gravelly soils of the Death Valley region. It differs from *S. vermiculatus* in its more intricate and compact growth, more spiny branchlets, and smaller leaves.

70. *Chorizanthe rigida*
71. *Chorizanthe Watsoni*
72. *Chorizanthe Thurberi*
73. *Chorizanthe corrugata*
74. *Chorizanthe spinosa*
75. *Oxytheca perfoliata*
76. *Oxytheca trilobata*
77.* *Oxytheca luteola*

78. *Pterostegia drymarioides*
79. *Gilmania luteola*
80. *Nemacaulis denudata*
81. *Rumex hymenosepalus*

83. INKWEED, TORREY SEA-BLITE, IODINE WEED. *Suaeda Torreyana ramosissima* (Arabic name; Dr. John Torrey [see **91**]; L., very much branched). Inkweed thrives only in soils containing both salt and alkali, and generally occurs around the edges of wet-type dry lakes in which moisture is near the surface. Inasmuch as it cannot stand much salt, it is limited to a belt outside the *Allenrolfea* thickets on the true salt flats such as occur in Death and Panamint valleys. On the Colorado Desert inkweed is associated with mesquite and quailbrush, especially in the Coachella Valley. The sooty-green and brown plants (2–3 ft. high) are very noticeable against the gray-green of the quailbrush. Although it is strong in tannic acid, which acts as an astringent, the plant produces dysentery in animals occasionally feeding upon it. The name "inkweed" was given because a poor sort of black ink can be made from the herbage; the Cahuillas extracted a black dye for use in art work.

84. ALKALI-WEED. *Nitrophila occidentalis* (Gr., alkali-loving; L., Western). **Fl.**: sepals pinkish or whitish, calyx lobes straw-colored. A low, perennial, fleshy-leaved, green herb, 4–15 in. high, arising from a taproot deep-seated in moist soils impregnated with black alkali. The perfect flowers and brown fruits generally occur in threes in axils of the leaf-like bracts. Deserts east of the Sierra Nevada; to Ore. and Nev.

85. RED MOLLY. *Kochia americana* (W. D. J. Koch, 1771–1849, German botanist; of America). **Fl.**: white-tomentose. Red molly is a perennial herb with many erect branches (5–11 in. high) arising from a woody crown. The grayish, narrowly linear leaf blades in this species are oval in cross section, as contrasted with the oblong ones of *K. californica* (of the western Mohave D.; to eastern Nev.), which are flat. Rare alkaline plant of the northern Mohave region; to Nev., Wyo., N.Mex., and northeastern Calif. *K. trichophylla*, a native of China, is the compact, lively-green Summer Cypress of formal gardens.

86. PICKLEBUSH, PICKLEWEED. *Allenrolfea occidentalis* (Robert Allen Rolfe, 1855–1921, botanist of Kew Gardens and long the leading authority on orchids, founder and editor of the *Orchid Review;* L., western). Pickleweed is unique because of its almost leafless, cylindrical stems, made up of joints which appear like elongate green beads. It possesses a large taproot which deeply penetrates the moist alkaline soils of dry-lake borders and brackish streams. In the large flat areas north of the Salton Sea, where the earth is usually damp owing to rising ground water, it is associated with inkweed. This small, scrubby species can endure more alkali and salt than any other desert plant. The older portions of the stem turn dark brownish-green in summer, making the plants very conspicuous against the dazzling white of the saline playas surrounding dry lakes.

87. PATATA. *Monolepis Nuttalliana* (Gr., "one-scale," in allusion to the single sepal; Thomas Nuttall, 1786–1859, distinguished naturalist and author of *The Genera of North American Plants.* He came to the United States from England in 1808 and explored the Arkansas and Missouri rivers. Nuttall was the naturalist, "Old Curious," of Dana's homeward voyage commemorated in *Two Years Before the Mast.* He later became Professor of Natural History at Harvard and curator of the botanical gardens). **Fl.**: reddish. Pale green annual, 5–10 in. high, with fleshy leaves. Moist, alkaline places, cismontane and desert valleys, and barren clay hills at low elevations; to Alberta, Mo., Tex.

88. BRACTSCALE. *Atriplex Serenana* (Gr., "not" + "to nourish" —because it robs the soil—also the Latin name for "orache," *Atriplex hortensis* of the botanists; [obscure]). Annual herb of erect habit, 8 in. to 2½ ft. tall, now invading the Mohave Desert area along its western margin. The plants remain succulent late in the season, making them favorable breeding sites for the sugar-beet leaf hopper from spring to autumn. Frequent in alkaline soil in cismontane southern Calif.; along roadsides in the Mohave D.; to Nev.

89. WHEELSCALE. *Atriplex elegans fasciculata* (L., slender; L., growing in bundles). The English name "wheelscale" alludes to the orbicular, flattish fruiting bracts. Erect annual herb, 4 in. to 2 ft. high, branched from the base. In Arizona it is a valuable forage plant. Among the Pima Indians its tender shoots were used as greens. Wheelscale is found in moderately saline and alkaline places, such as in the Salton Sink and the Colorado R. bottom. Both Calif. deserts.

90. MOHAVE SALTBUSH. *Atriplex spinifera* (L., thorn-bearing). A very woody, stout-branched shrub, taller than broad. During the late summer the leaves drop and the small stems which bore them become modified into rigid spines. It is partial to strongly alkaline soils. Middle and western Mohave D. to the San Joaquin Valley.

91. TORREY SALTBUSH. *Atriplex Torreyi* (John Torrey). Principally a Nevada species extending southwest to the Mohave D. of Calif. and east to southwestern Utah. The gray-scurfy branches

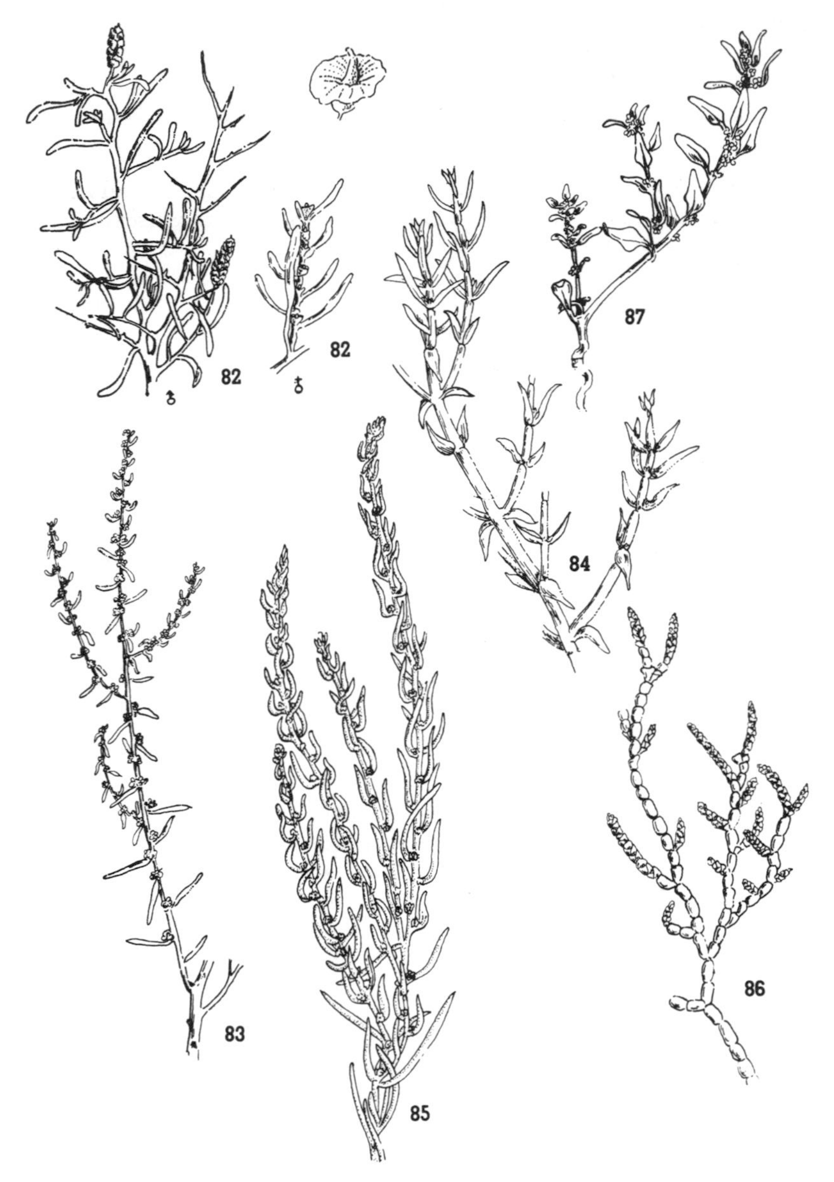

82. *Sarcobatus vermiculatus*
83. *Suaeda Torreyana ramosissima*
84. *Nitrophila occidentalis*
85. *Kochia americana*
86. *Allenrolfea occidentalis*
87. *Monolepis Nuttalliana*

and sharply angled twigs make up large, almost impenetrable bushes, 3–6 ft. tall. As the numerous leaves and bracts drop off, the small branchlets become tough spines. Torrey saltbush is ranked as an intense halophyte and is generally found in places having more than ordinary supplies of moisture. Dr. John Torrey, New York botanist, specialist in mosses, in whose honor Sereno Watson named this plant, was the describer of many of the species collected by Frémont in the Western deserts. He visited California in 1865, collecting in the region about Santa Barbara and in the Sierra Nevada.

92. CATTLE SPINACH, ALLSCALE. *Atriplex polycarpa* (Gr., many-fruited). Perennial shrub, common on both deserts, most often in strongly alkaline soils. Its squat form, fascicled leaves, and light tan-colored branches make it easy to identify. Its usual plant associates are creosote bush, mesquite, palo verde, and ocotillo. The common name, "cattle spinach," was given because of its high value as a browse plant. Flats and low hills of both deserts; to Baja Calif., central Calif., and Utah.

93. ARROW SCALE. *Atriplex Phyllostegia* (Gr., leaf-roofed). Erect annual herb, 4–16 in. high, with somewhat succulent leaves and arrow- or spear-shaped fruiting bracts. It often occurs in "clan-like" groups in strongly alkaline soils along with caterpillar greasewood, rabbit brush, and salt grass. Western and middle Mohave D. to San Joaquin Valley; to Ore., Utah.

94. WEDGE SCALE. *Atriplex truncata* (L., "cut off," because of the broadly cuneate fruiting bracts, which are truncate, i.e., cut off squarely at the summit). An erect, annual herb, widely distributed in the Great Basin states and known in Calif. from the western Mohave D. It often grows in depressions white with alkaline salts.

95. REDSCALE, RED ORACHE. *Atriplex rosea* (L., rosy). An introduced annual weed, native to the Old World but now abundant in western United States. It occurs as a rounded bush along roadsides and in waste places. When dried it is one of our common tumbleweeds. Redscale is one of the important host plants of the leaf hopper (*Eutettix tenella*), which carries the curly-top disease of the sugar-beet. The young plants are satisfactory for pigs but somewhat toxic to sheep. In Greece potash is made from it.

96. QUAILBRUSH, LENSCALE.* *Atriplex lentiformis* (L., lens-shaped). This large, dense saltbush is very common on the flats of

* This and other unhyphenated names of the Atriplexes are those coined by H. M. Hall in his monumental treatise, *The Phylogenetic Method in Taxonomy.*

88. *Atriplex Serenana*
89. *Atriplex elegans fasciculata*
90. *Atriplex spinifera*
91. *Atriplex Torreyi*
92. *Atriplex polycarpa*
93. *Atriplex phyllostegia*
94. *Atriplex truncata*
95. *Atriplex rosea*

the west end of the Salton Sea, also along the lower Colorado River. The plants are sometimes 10 ft. high and 12–15 ft. across. The name "quailbrush" was given because of the acceptable cover and protection the broad, rounded bushes provide for quail. Rabbits and other animals also seek shelter from predators beneath its branches. In the Coachella Valley it is a favorite hide-out for the road runner. Its young shoots were boiled by the Pimas and eaten as greens. The Cahuillas use its seeds for food, first grinding them and then cooking the meal in salted water. Low alkaline flats, Colorado and Mohave deserts, to central Calif.; Utah and Mex.

97. PARRY SALTBUSH. *Atriplex Parryi* (C. C. Parry; see **24**). A plant of restricted range but abundant on alkaline flats of the Mohave D.; to western Nev. It is readily distinguished from other saltbushes of the area by its closely set, nearly heart-shaped leaves. The rounded bushes are 8–16 in. high, with slender, rigid thorns about the size of small darning needles. The flower clusters appear early in March.

98. WINGSCALE, HOARY SALTBUSH. *Atriplex canescens* (L., gray). Erect, woody shrub, with gray-scurfy branches. The fruiting bracts have conspicuous wings arising from the middle of the exposed face of the seed and from this the English name "wingscale" is derived. It is one of the most widely distributed of American saltbushes and is exceptionally valuable for grazing. Its seeds are ground into meal by the desert Indians. The Zuñi Indians ground the roots and blossoms, moistened them with saliva, and used the mixture to cure ant bites; when this powder was not at hand they applied bruised fresh blossoms. Alkaline and sandy soils from western Tex. to the Colorado and southern Mohave deserts.

99. NARROW-LEAVED WINGSCALE. *Atriplex canescens linearis* (L., "linear," pertaining to the leaves). Leaves ½–1½ in. long and usually not more than ⅛ in. wide. The very narrow fruiting bracts have 4 deeply toothed wings. Colorado D.; to Mex.

100. SHADSCALE. *Atriplex confertifolia* (L., crowded-leaved). From Idaho southward to eastern California and northern Arizona, shadscale predominates over all other species of saltbushes but is absent from the Colorado Desert. It is a very woody, spiny shrub of rounded outline, 8–30 in. high. Once the leaves drop, the twigs soon change to spines. It tolerates heavy-textured soils containing amounts of salt harmful to most plants.

101. DESERT HOLLY. *Atriplex hymenelytra* (Gr., "membrane-covered," with reference to the broad, membranous fruiting bracts). Widely distributed but nowhere found over large areas. A striking plant especially in spring when in prime growth, or again later when the leaves take on a pinkish tinge. The blossoming period is earlier than that of any other saltbush, the flowers appearing as early as February. By late March the female plants hang heavy with large, light green, disk-shaped, somewhat succulent fruiting bracts. The silvery leaves make it a desirable Christmas decoration and much of it is sold for that purpose. First described from specimens collected by Arthur Schott of the Mexican Boundary Survey, in whose honor *Parosela Schottii* (see **226**) was named. Both Calif. deserts; to Utah and Mex.

102. SPINY HOP-SAGE. *Grayia spinosa.* (Asa Gray, 1810–1888, who laid the foundations of systematic botany in this country and founded the Gray Herbarium of Harvard University. During his twenty-eighth year, good-natured Gray wrote home from Europe: "Hooker has a curious new genus of Chenopodiaceae from the Rocky Mountains which he wishes to call Grayia! I am quite content with a Pigweed, and this is a very queer one." Dr. Gray visited California in 1885 and in company with S. B. Parish botanized in the San Bernardino Mountains. Also, L., thorny.) **Fl.**: inconspicuous. Because of the peculiar green of its somewhat fleshy, gray-tipped leaves and their tendency to assume a pinkish tint, hop-sage is readily distinguishable from other brush with which it is associated. It is a small shrub, 1–3 ft. high, with somewhat mealy herbage, often brilliant rose-purple fruits, and twigs which are spine-like at the tips. "It is eaten," says Dayton, "by all classes of livestock and is considered good both for cattle and sheep, which crop the buds and leafy twigs with avidity and fatten notably on the copious harvest of flat, winged fruits. The female plants thus are the most desirable for forage." Common on mesas and flats about the Mohave D. at 2,500–7,500 ft.; to eastern Wash., Wyo., and Colo. Whitewater Canyon on the Colorado D.

103. WINTER FAT. *Eurotia lanata* (Gr., "mold," in reference to the hairy covering; L., woolly). Erroneously called white sage. The most valuable winter grazing plant in the Great Basin. It is common from the Mohave Desert of California to the sagebrush plains of Nevada and Idaho. In some parts of Nevada, there are plains of thousands of acres covered by no other plant. "Such *Eurotia* areas," says Dr. Schantz, "can be detected at great distances because of the uniform white or light gray color which at a distance appears much like snow." Winter fat has a deep taproot which enables it to withstand drought to a remarkable degree. From the woody base numerous erect, herbaceous stems arise 1–2½ ft. high. The white to rusty, long-haired, fruiting involucres always make the plants conspicuous among green shrubs and grasses with which they grow. The fresh root is chewed and used as a remedy for burns by the Zuñis.

104. RUSSIAN THISTLE, PRICKLY SALTWORT. *Salsola Kali tenuifolia* (L., dim. of *salsus*, "salted," in allusion to the salty soil in which it grows; *Kali* is the Arabic name for a plant now of the genus *Salicornia;* "thin-leaved"). This obnoxious Asiatic weed, erroneously given the misleading name of "thistle," has spread over most of the Mohave Desert and often occurs in such abundance as to make wide areas a refreshing green in midsummer when all other plants have turned brown. In Europe the ashes of this plant were once much used in the production of an impure carbonate of soda known as "Barilla." This is the common tumbleweed in cismontane southern Calif.; to Canada and central U.S.

AMARANTHACEAE. Amaranth Family

105. HONEY-SWEET. *Tidestromia oblongifolia* (Ivar Tidestrom, author of *A Flora of Utah and Nevada,* now connected with Department of Botany, Catholic University of America; L., oblong-leaved). **Fl.**: yellowish. White-woolly perennial, with widely branching stems, forming broad, rounded, or mat-like plants, 9–18 in. high, and up to 1 yd. across. Sandy washes and alkali flats of both deserts; southern Nev., Ariz. Very common in Death Valley after summer rains, blooming in mid-September and rejoicing in the heated blasts of air which then sweep over the basin floor. Its name, "honey-sweet," alludes to the sweet-scented, yellow flowers, which often occur in such profusion as to make the plant pronouncedly decorative. The stems and leaf-like involucral bracts often turn reddish in age.

106. FRINGED AMARANTHUS. *Amaranthus fimbriatus* (Gr., "unfading flower," in reference to the length of time some of the species retain their bright color; L., "fringed," in reference to the fringed sepals of the pistillate flowers). **Fl.**: rose or lavender. Handsome annual, several-stemmed from the base. 1–2 ft. high, often blooming after summer rains in the Joshua Tree Nat. Mon. and at Jacumba. Sand and gravels of both Calif. deserts; Utah and Mex.

96. *Atriplex lentiformis*
97. *Atriplex Parryi*
98. *Atriplex canescens*
99. *Atriplex canescens linearis*
100. *Atriplex confertifolia*
101. *Atriplex hymenelytra*
102. *Grayia spinosa*
103. *Eurotia lanata*

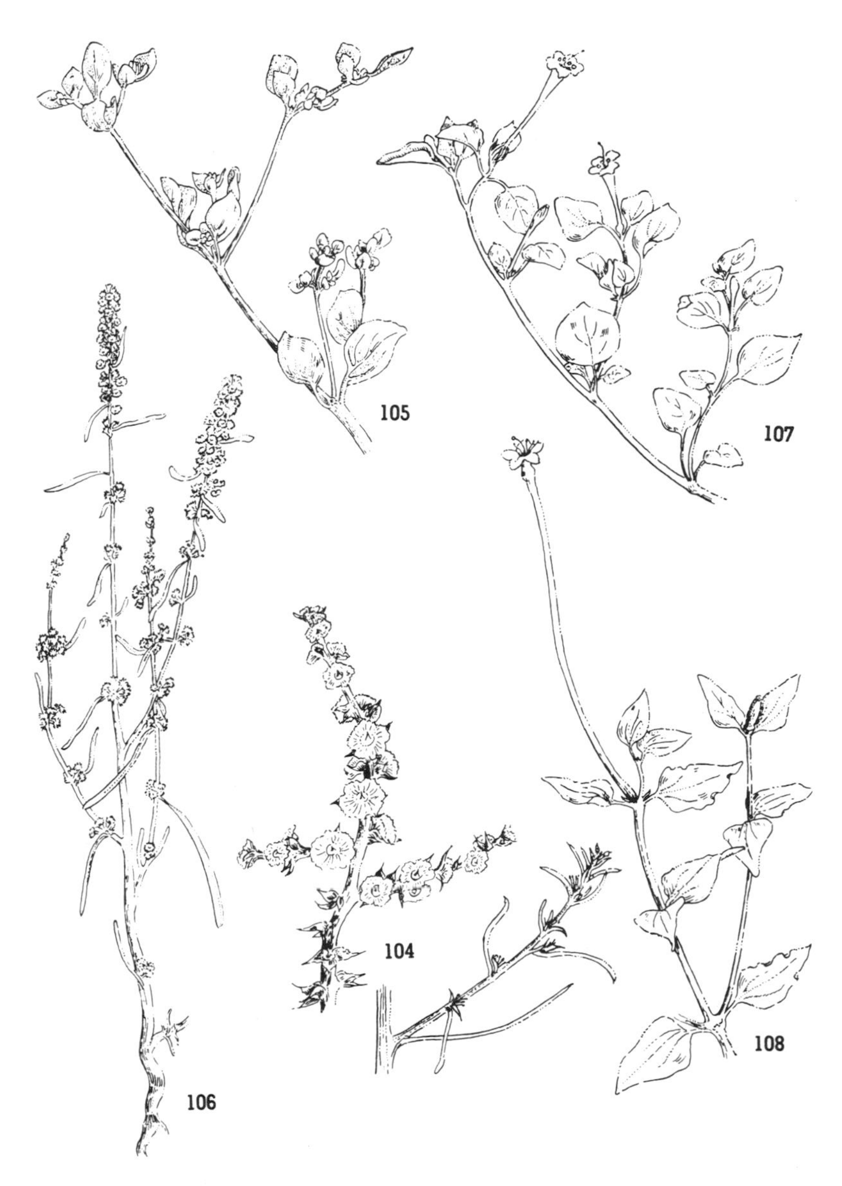

104. *Salsola Kali tenuifolia*
105. *Tidestromia oblongifolia*
106. *Amaranthus fimbriatus*
107.* *Selinocarpus diffusus*
108.* *Acleisanthes longiflora*

NYCTAGINACEAE. Four-o'Clock Family

107. DESERT WING-FRUIT. *Selinocarpus diffusus* (Gr., parsley fruit; L., diffuse). **Fl.**: greenish-white. Spreading perennial, reaching the eastern borders of California near Ivanpah. The leaves are leathery, not fleshy as in some species, and the fruits are conspicuously winged. Dry slopes of the creosote-bush belt in eastern Calif.; to Nev., southern Utah, and Tex.

108. YERBA DE LA RABIA. *Acleisanthes longiflora* (Gr., without something which closes, i.e., without an involucre; L., long-flowered). **Fl.**: white. Low, slender-stemmed perennial, immediately attention-compelling because of the long-tubed perianth. The leaves are thick and fleshy. Stems 4–12 in. long. In Calif. known from the Maria Mts.; to Tex. and Mex.

109. STICKY-RING, WET-LEAF. *Boerhaavia annulata* (Hermann Boerhaave, 1669–1738, Professor of Botany and Medicine at the University of Leyden, and founder of modern clinical methods of medical instruction; L., ringed). Named "sticky-ring" because on each internode there is a very odd and noticeable reddish-brown, broad ring exuding a mucilaginous fluid. It is a perennial, with bright green leaves and erect, stout stems, 1–3 ft. tall. Writing of its discovery Dr. Coville of the Death Valley Expedition said: "The plant was first seen Jan. 20, 1891 as we descended Long Valley, on our entrance into Death Valley. The few specimens growing there had put out a few large radical leaves, but had not yet flowered. In this condition the plant strongly suggests a begonia." Confined to dry sand washes entering the Death Valley Sink.

110. WEST INDIAN BOERHAAVIA. *Boerhaavia caribaea* (Caribbean). **Fl.**: calyx red. Perennial, with branching stems, 4 in. to 2 ft. long, generally lying flat to the ground. Along the edge of washes, Colorado D., San Jacinto Valley; to Tex., Fla., W.I., and Mex.

111. CREEPING STICKY-STEM. *Boerhaavia spicata Torreyana* (L., spiked; J. Torrey—see **91**). Similar to *B. intermedia;* but the stems are very glandular and the flowers are generally borne singly at the ends of the thread-like branchlets.

112. FIVE-WINGED RING-STEM. *Boerhaavia intermedia* (L., "intermediate," i.e., in form between two other species). Low annual, spreading or ascending, with thick, fleshy, gray-green leaves and small 5-angled fruits, which are truncate at the apex. Stems 9–15 in. long. Rare, southern Colorado D.; to Tex. and Mex.

113. LARGE-BRACTED BOERHAAVIA (not illustrated). *Boerhaavia Wrightii* (C. Wright—see **209**). An annual, 4–20 in. high, with stems erect and glandular and the leaves wavy-margined. The fruits are 4-ribbed, the bracts large and persistent. Black Point, Palo Verde Valley, Orocopia and Chuckawalla mountains; to Texas.

114. WISHBONE BUSH. *Mirabilis Bigelovii retrorsa* (L., wonderful; Dr. J. M. Bigelow [see **338**]; L., "turned-back," because of the rough retrorse hairs of the upper stems). **Fl.**: white. This low, rounded perennial has dark green herbage and has been given the appropriate name of "wishbone bush" in allusion to its form of branching. It is widely distributed on gravelly flats and rocky surfaces of both deserts.

115. GIANT FOUR-O'CLOCK. *Mirabilis Froebelii.* (Julius Froebel, 1805–1893, nephew of Friedrich Froebel, founder of the kindergarten. Compelled to flee Germany because of participation in the Revolution of 1848, he found refuge in this country, settling in or near San Francisco, where he became a member of the California Academy of Sciences. He traveled widely in Central America, northern Mexico, and Far Western United States. He died in Zurich.) **Fl.**: pale or bright purple. A perennial, often forming large round mats "as big as a wagon-wheel," and attractive because of its show of handsome, sweet-scented, night-blooming flowers. What appears to be the corolla is only a bright-colored funnel-form calyx. The large flowers are set in clusters in the green, cup-like involucres. The herbage is very glutinous. Stony mesas and desert washes of mountains of Inyo Co. south to the western edge of the Colorado D.; also eastern Mohave D.; to Nev.

116. PALE-STEMMED FOUR-O'CLOCK. *Mirabilis aspera* (L., "rough," harsh to the touch, with reference to the grayish herbage). **Fl.**: white to pinkish. Erect or ascending, much-branched perennial, found about rocks and among shrubs. The flowers open in late afternoon. Dry hills of the western edge of the Colorado D. and the Mohave D. Providence Mts.

117. SMALL-FLOWERED ABRONIA. *Abronia micrantha* (Gr., from *abros*, "delicate," referring to its involucre; Gr., "small-flowered"). **Fl.**: greenish-white or reddish. Annual, with stem forming a low, leafy mat. The thick leaves are leaden-green in color, and the orbicular, pinkish fruits have three membranous, strongly net-veined wings. Locally known from the sand dunes near Kelso, Mohave D.; Ark. and southwestern Colo.

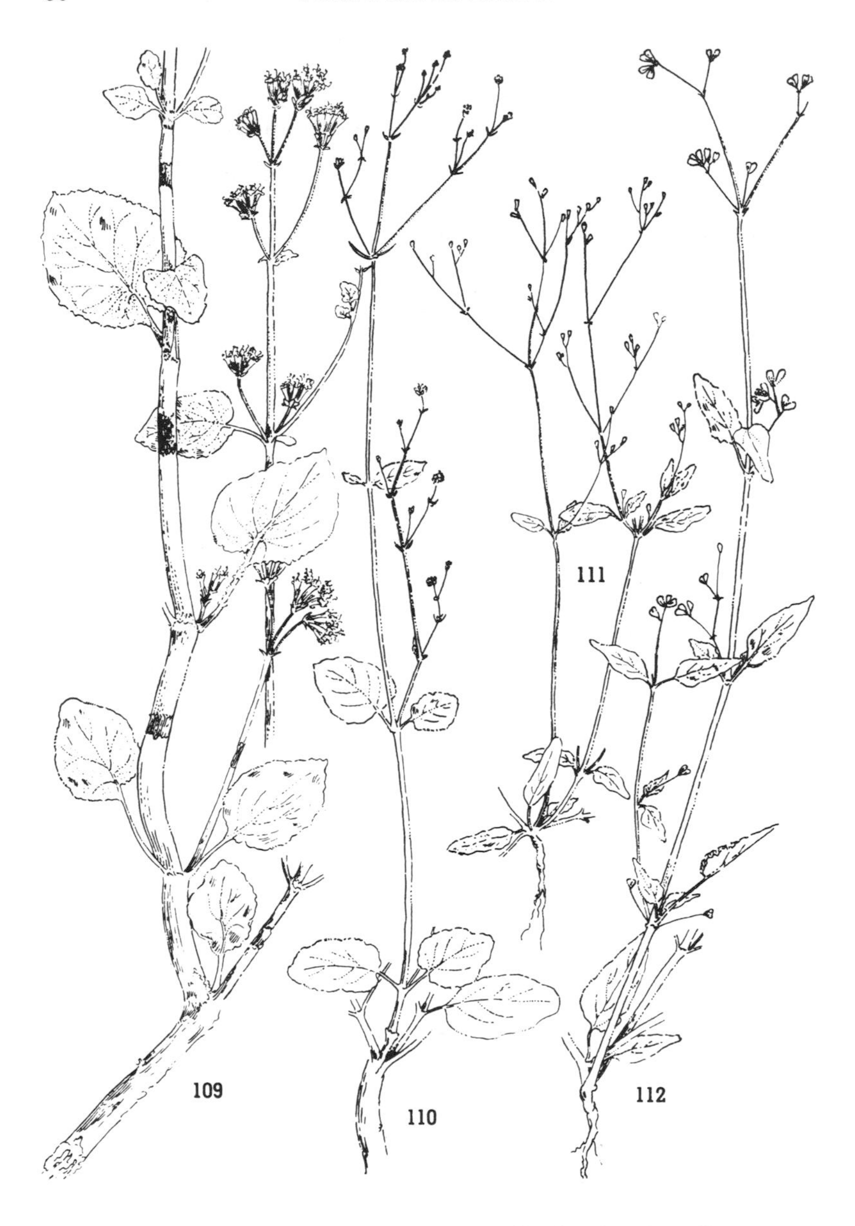

109. *Boerhaavia annulata*
110. *Boerhaavia caribaea*
111. *Boerhaavia spicata Torreyana*
112. *Boerhaavia intermedia*

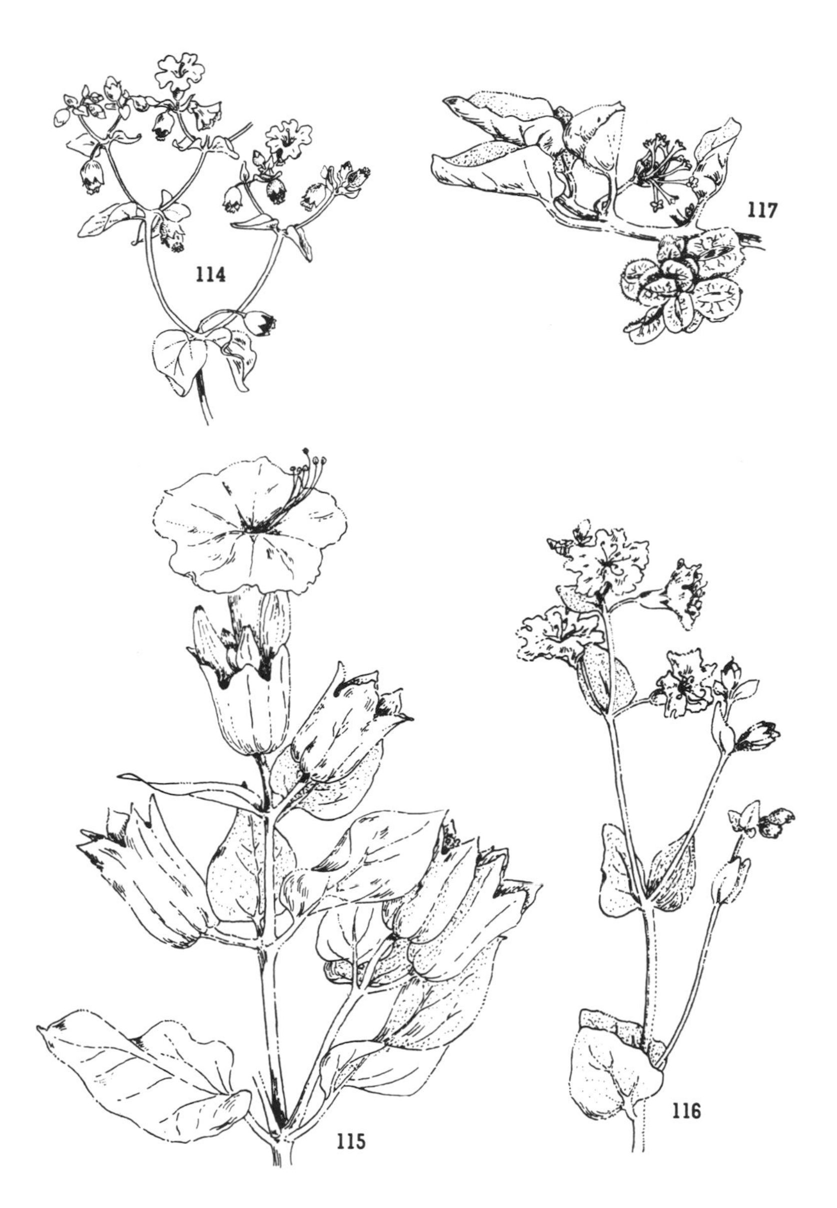

114. *Mirabilis Bigelovii retrorsa*
115. *Mirabilis Froebelii*
116. *Mirabilis aspera*
117. *Abronia micrantha*

118. HAIRY SAND-VERBENA. *Abronia villosa* (L., hairy). **Fl.**: calyx, rose-purple. Trailing annual partial to dunes and sand flats, where it makes a fine show of color, beginning in early spring. From the flowers comes a most delicate and agreeable fragrance, particularly noticeable at night. Open sandy places, Mohave and Colo. deserts; to Ariz. and Utah. The var. *aurita* has unpitted seeds, with five wings prolonged above the body of the seed. Palm Springs region to San Jacinto, Temecula, and Anaheim.

119. MOHAVE SAND-VERBENA. *Abronia pogonantha* (Gr., bearded-flowered). **Fl.**: calyx, rose-pink or light lavender. Trailing annual, with stems only 8–12 in. long. The reticulate seed wings are generally but two in number. Sandy places, Mohave D., above 2,500 ft.

120. WHITE FOUR-O'CLOCK. *Mirabilis tenuiloba* (L., "thin-lobed," because the involucres are cleft into lanceolate or linear lobes). **Fl.**: white or pink. Perennial herb, with several-to-many branches, 1–1½ ft. high. The fruit is smooth and brown. A favorite habitat is beside or between rocks, and also about shrubs. Occasional, western edge of Colorado D.; to Baja Calif.

121. PINK THREE-FLOWER. *Allionia pumila* (Charles Allioni, 1725–1804, Professor of Botany at Turin, author of *Flora of Piedmont,* and exponent of the natural classification of plants; L., little). **Fl.**: pale pink to bright purple. The fruit of this *Allionia* is club-shaped and wingless, whereas in *A. incarnata* the fruit is winged. The stems are erect, 6–12 in. high, from a stout, woody root. The small flowers wilt quickly, remaining open but a brief period. Slopes and canyons of the transition zone between the creosote and juniper belt. Western Mohave D., to southeastern Nev. and Utah.

122. GREEN SAUCERS. *Allionia comata* (L., clothed with hair). **Fl.**: light green, tinged with pink. The several stems spring annually from a perennial root. A Great Basin species known in California from collections recently made by the author in the juniper belt of the Ivanpah Mts. Called "green saucers" because of the greenish, wide-flaring flowers.

123. WINDMILLS. *Allionia incarnata nudata* (L., embodied in flesh., i.e., "flesh-colored"; and "stripped"). **Fl.**: white to rose. A spreading annual, with slender stems 1–2½ ft. long. The prostrate stems are strongly viscid (sticky), hence often covered with bits of sand and small flakes of mica. What on first sight seems to be the flower is really a group of three flowers. Occasional, dry sandy benches, Colorado and Mohave deserts; to Nev.

AIZOACEAE. Carpetweed Family

124. LOWLAND PURSLANE. *Trianthema Portulacastrum* (Gr., "three-flowered," alluding to a frequent arrangement of the flowers in some species; L., *Portulaca,* + star in the sense of a flower, i.e., with flower like that of *Portulaca*). **Fl.**: calyx, purple within. A diffusely branched, somewhat succulent annual, often plentiful, particularly after summer rains, in sands and clays along the borders of dry lakes such as Silver and Cronese dry lakes of the Mohave D.; Ariz. to Fla., West Indies; Old World.

PORTULACACEAE. Purslane Family

125. BITTERROOT. *Lewisia rediviva* (Capt. Meriwether Lewis, leader of Lewis and Clark Expedition; L., "reviving," alluding to the plant's ability to revive after long periods of drought). **Fl.**: petals bright rose to white. Our drawing was made from specimens from the crest of the Inyo Mountains in the juniper-piñon forests near Westgard Pass (after A. L. Westgard, who as representative of the American Automobile Association and the U.S. Bureau of Public Roads led a transcontinental tour crossing the White Mountains in 1913), where members of the Death Valley party collected it nearly fifty years ago. *Lewisia* is the state flower of Montana. The club-shaped leaves dry up as soon as the flowers appear at the ends of the brittle-jointed stems. The farinaceous roots, in spite of their bitterness, were once a prized food of the Indians. They generally boiled them with other edible plants into a soup. Mountains of northern Mohave D.; to Mont., Colo., and B.C.

126. DESERT POT-HERB. *Calandrinia ambigua* (J. L. Calandrini, botanist and mathematician of Geneva, who made the "sensible suggestion" to Charles Bonnet, writer on the function of leaves, "that the underside of the leaf absorbed dew that rose [!] from the soil"; L., uncertain, with respect to relationship). **Fl.**: petals, white. A very succulent, brilliant green, tufted plant, which the pioneers and Indians used for greens. Its spongy, juicy leaves have a cool, pleasant, and slightly salty flavor. Occasional in somewhat alkaline washes and on barren, alkaline, clay hills of the Colorado D. and Death Valley region.

127. SAND-CRESS. *Calyptridium monandrum* (Gr., a "small covering," in reference to the petals, which in drying close in and form a little "cup" carried atop the growing capsule; Gr., one-sta-

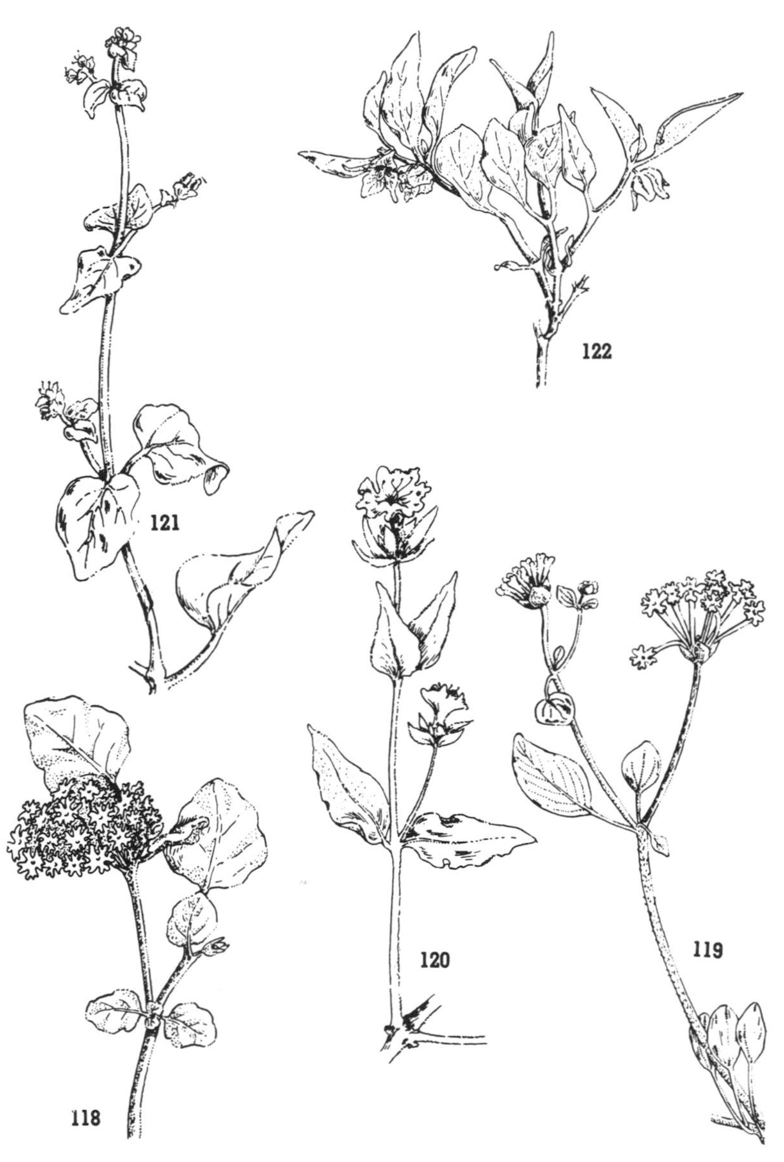

118. *Abronia villosa*
119. *Abronia pogonantha*
120.* *Mirabilis tenuiloba*
121. *Allionia pumila*
122. *Allionia comata*

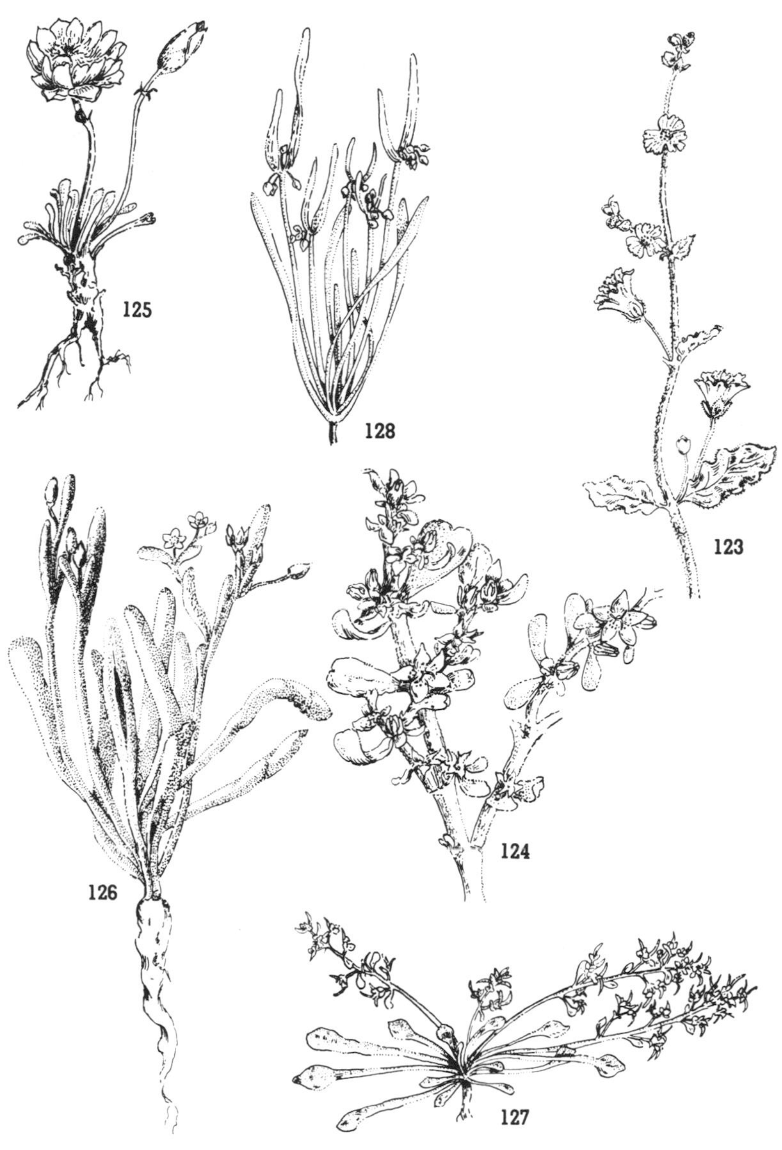

123. *Allionia incarnata nudata*
124. *Trianthema Portulacastrum*
125. *Lewisia rediviva*
126. *Calandrinia ambigua*
127. *Calyptridium monandrum*
128. *Montia spathulata tenuifolia*

mened). **Fl.:** white. A sprawling, fleshy-leaved plant of the desert sands. The stems and leaves in age turn distinctly reddish. Usually below 5,000 ft.; cismontane southern Calif. and deserts, to central Calif.; Ariz. and Baja Calif.

128. NARROW-LEAVED MINER'S LETTUCE. *Montia spathulata tenuifolia* (G. Monti, 1682–1760, Italian botanist; L., spatulate; L., narrow-flowered). **Fl.:** white. Pale-herbaged, succulent annual. A good salad herb. Mountains of the western and middle Mohave D.; to Wash. Sometimes with it is the coastal miner's lettuce, *M. perfoliata.*

CARYOPHYLLACEAE. Pink Family

129. RIXFORD ROCKWORT. *Scopulophila Rixfordii* (Gr., lover of rocky peaks; G. P. Rixford, father of Dr. Emmet Rixford, surgeon and conchologist. He worked for many years for the U.S. Department of Agriculture and many plants were introduced into cultivation in California by him). **Fl.:** calyx thin, translucent, with central, elongate, green spot; petals none. The many erect, somewhat fleshy-leaved stems, 3–8 in. long, spring from a chunky root-crown, the upper parts of which are often white with peculiar, dense tufts of wool. Crevices of vertical limestone cliffs are its favorite habitat. Inyo Mts.; eastward to Nev.

130. MOUNTAIN CATCH-FLY. *Silene montana* (Gr., "saliva," referring to the viscid stems of some species which entrap small flies, or perhaps from Silenus, Asiatic woodland deity, who taught men to play reed instruments; L., "mountain"). **Fl.:** greenish-white. Dry slopes, 4,000–7,000 ft., of the juniper belt of the mountains of the northern and eastern Mohave D.; to Nev. and northern Calif.

131. DESERT SANDWORT. *Arenaria macradenia Parishorum* (L., sand; Gr., "large-glanded," because of the well-developed basal glands of the stamens; W. F. and S. B. Parish—see **603**). **Fl.:** light greenish-yellow. Found in clefts of rocks in the low, dry hills and occasionally in the higher mountains of the Mohave D.; also the desert slopes of the San Gabriel, San Jacinto, and Santa Rosa mountains. *A. macradenia,* with larger white flowers, is found at higher elevations. A single flower is shown on the right in our illustration.

132. FROST-MAT, CHAFF-NAIL. *Achyronychia Cooperi* (Gr., "chaff claw" or "fingernail," in reference to the chaffy calyx; Dr. J. G. Cooper, geologist of the Geological Survey of California, who collected plants in the Mohave Desert between Cajon Pass and Camp Cady in 1861). **Fl.**: bright silvery-white, green at base. A flat, spreading plant, with a very long taproot. The mats, which appear as if covered with frost crystals, are sometimes 1 ft. across. Sandy areas of the Colorado and eastern Mohave deserts; to Ariz. and Baja Calif.

RANUNCULACEAE. Crowfoot Family

133. PARISH LARKSPUR. *Delphinium Parishii* (Gr., "dolphin," referring to the nectary, which bears a resemblance to imaginary figures of the dolphin; S. B. Parish—see **603**). **Fl.**: "light but lively sky-blue." The only desert larkspur, but very widely distributed and quite plentiful. It occurs on both deserts, but is not known from the most southern portion of the Colorado D. The type specimens were taken at Agua Caliente, now Palm Springs.

134. DESERT WINDFLOWER. *Anemone tuberosa* (Gr., "wind," "because its petals flaunt about with the wind, in which it delights in spite of the delicacy of its petals"; L., "tuberous," referring to the root). **Fl.**: sepals rose-color. A perennial herb, known on the California deserts from upper, dry slopes of the Providence and Panamint mountains; to Utah and N.Mex.

135. MOUSE-TAIL. *Myosurus cupulatus* (Gr., "mouse's tail," with reference to the appearance of the pistillate spike; L., cupped). **Fl.**: whitish. A queer little annual, occasionally found in spring in shady places, 3,500–5,000 ft. altitude, Joshua Tree Nat. Mon.; eastward to N.Mex.

BERBERIDACEAE. Barberry Family

136. DESERT BARBERRY, FRÉMONT HOLLYGRAPE, YELLOW-WOOD. *Berberis Fremontii* (the Arabic name of the fruit, signifying a shell; J. C. Frémont—see **315**). **Fl.**: yellow. A Mohave D. shrub, particularly abundant in the New York Mts. The berries, at first blue, turn dull brown at maturity. Strip the bark, and you will notice the characteristic deep-yellow wood from which the Navahos extracted a yellow dye. The crushed berries were sometimes used by the Indians to paint their skin and their ceremonial objects a deep purple.

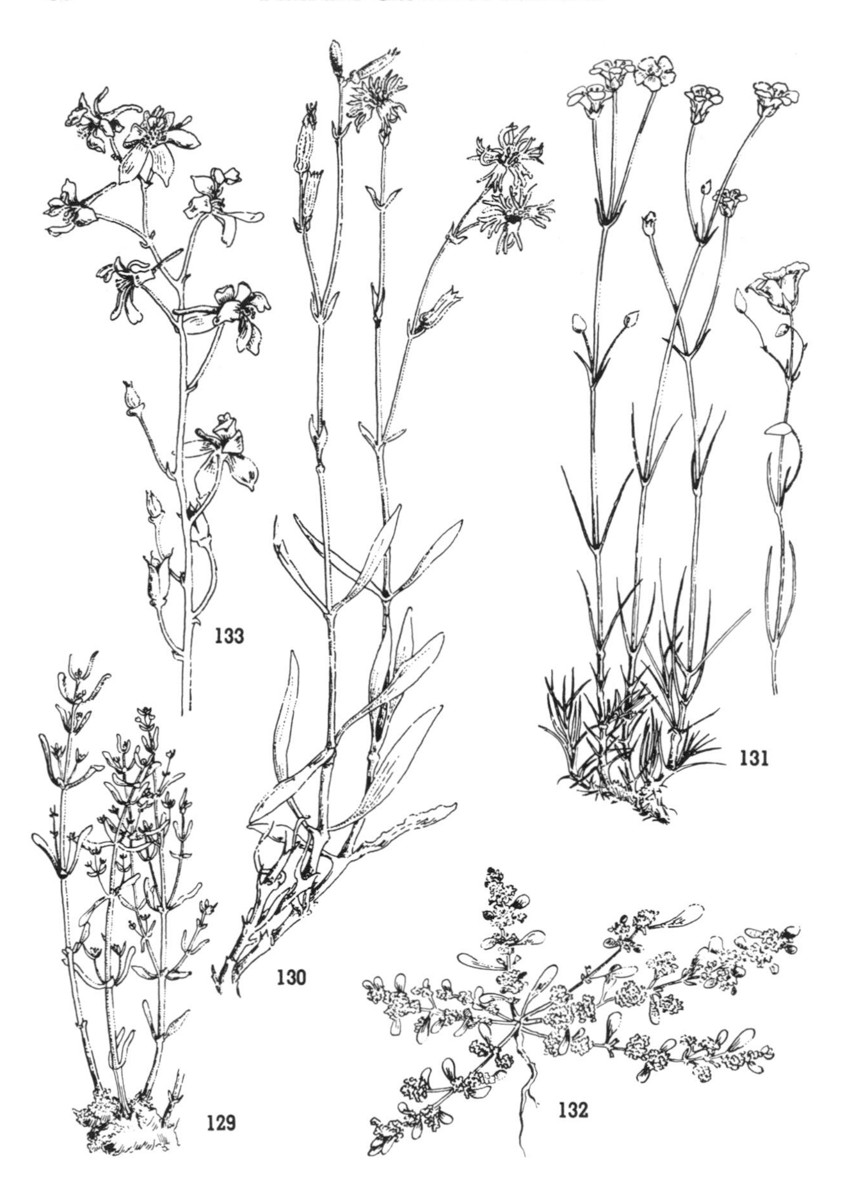

129. *Scopulophila Rixfordii*
130. *Silene montana*
131. *Arenaria macradenia Parishorum*
132. *Achyronychia Cooperi*
133. *Delphinium Parishii*

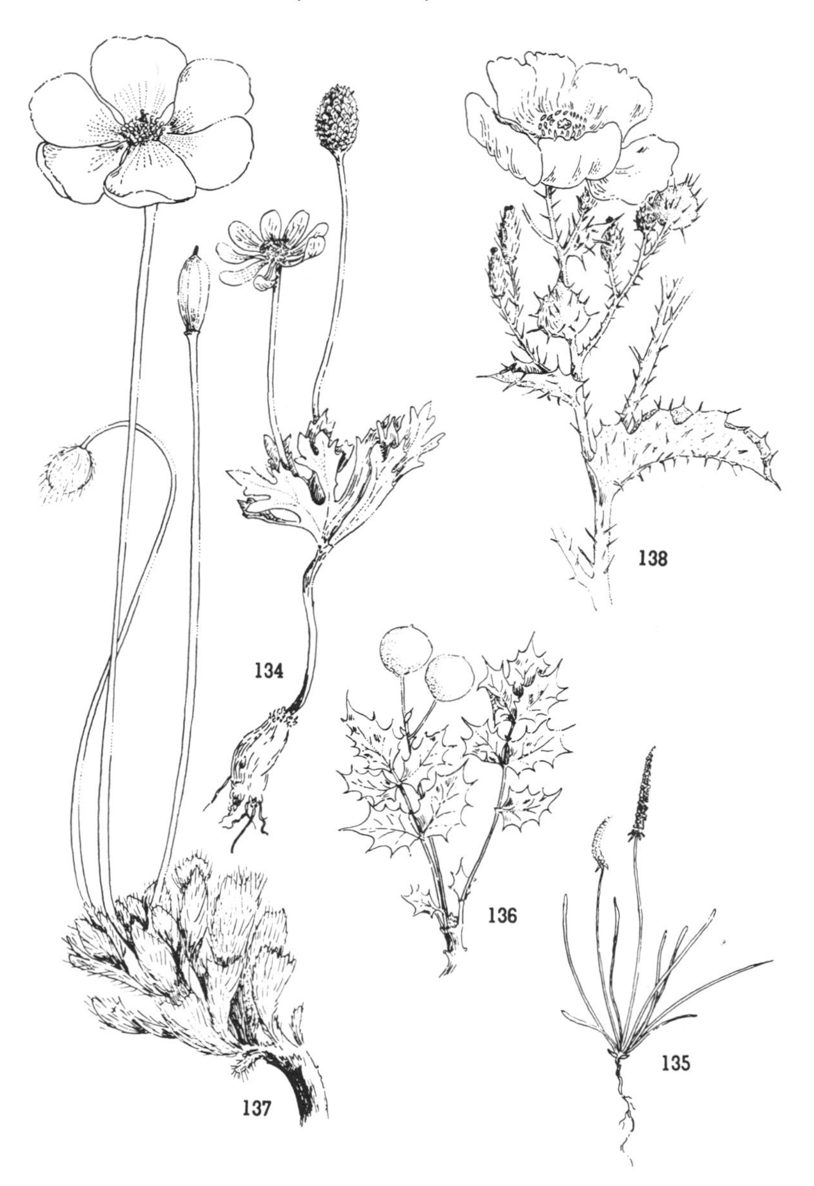

134.* *Anemone tuberosa*
135.* *Myosurus cupulatus*
136. *Berberis Fremontii*
137. *Arctomecon Merriami*
138. *Argemone intermedia corymbosa*

PAPAVERACEAE. Poppy Family

137. BEAR POPPY, DESERT POPPY. *Arctomecon Merriami* (Gr., "bear poppy," referring to hairiness; Dr. C. Hart Merriam, Chief of U.S. Biological Survey from 1885 to 1910, authority on birds and mammals of North America, geographic zoölogy, botany, and the linguistics of Pacific Coast Indian tribes). **Fl.**: white. A rare species of the Death Valley area blooming in April.

138. PRICKLY POPPY. *Argemone intermedia corymbosa* (Gr., *argema,* a disease of the eye, for which the juice of the plant was thought to be specific; L., "intermediate," with relation to two other species; L., "flower-clustered"). **Fl.**: white. This rare, small-flowered prickly poppy was first collected more than fifty years ago (1884) on the mid-Mohave Desert by Mrs. M. K. Curran (later Dr. Katherine Brandegee—see **248** and **602**). "Its obovate, rarely more than sinuate-margined leaves," says Dr. Carl Wolf, "make it easy to segregate" from the more widespread, stout-stemmed, and larger-flowered California prickly poppy (*A. platyceras*); great fields of the latter often cover hundreds of acres of alkaline gravels north of Harper Dry Lake on the Mohave Desert. The flowers of both species are fragrant; when plucked, an acrid, orange-colored juice exudes from the injured stems. The seeds of *Argemone* are said to be more powerfully narcotic than opium.

139. CREAM-CUPS. *Platystemon californicus crinitus* (Gr., flat stamen; of California; L., long-haired). **Fl.**: cream. Many-stemmed, exceedingly hairy annual. Western edge of both deserts and adjacent mountains.

140. DESERT GOLD-POPPY. *Eschscholtzia glyptosperma* (Dr. J. F. Eschscholtz, 1753–1831, Russian traveler and naturalist and surgeon on the exploring ship "Rurik," which under the command of Otto von Kotzebue undertook in 1815 a world voyage of discovery. Another naturalist, Adebeck von Chamisso, and Dr. Eschscholtz botanized in San Francisco Bay in October of that year. Gr., "carved-seeded," the ashen-gray globose seeds being coarsely pitted). **Fl.**: yellow. The erect flower stems are often very numerous; there may be as many as thirty-five on a single plant. Petals over 1 cm. long. Found over most of the Mohave area and northeastward into Utah.

141. LITTLE GOLD-POPPY. *Eschscholtzia minutiflora* (L., having very small flowers). **Fl.**: yellow. A common species of the southern portion of the Great Basin. The flower size is somewhat variable, with early flowers of robust specimens the largest.

142. PYGMY POPPY. *Canbya candida.* (Wm. M. Canby, 1831–1904, botanist and founder of the Society of Natural History of Delaware, which now owns his herbarium of about 30,000 specimens. He traveled extensively over all the United States, including Alaska. Although he carried on a large correspondence, he did not publish extensively. In appreciation of Mr. Canby's work in furthering the park system of Wilmington, Del., a massive stone bench, suitably inscribed, has been erected there on a prominent site overlooking the Brandywine. It was he who financed Pringle's early work in Mexico—see **705.** Mr. Canby was a man of great refinement—a fact which led to Asa Gray's appropriate pun: "Always just as nice as can be." L., glistening white.) **Fl.**: white. Described by Dr. C. C. Parry from specimens collected by Dr. Edward Palmer "near the head of the Mohave R.," 1876—see **474.** A charming little plant only 1 in. high, a mere pygmy among papavers. Most frequently collected in sands on the mesas north of Barstow. Confined to the Mohave D. of Calif.

CAPPARIDACEAE. Caper Family

143. BLUNT-LEAF STINKWEED. *Cleomella obtusifolia* (Gr., diminutive of *Cleome;* L., blunt-leaved). **Fl.**: yellow. Our most common stinkweed, found in alkaline soils of flats and valleys of both Calif. deserts to Inyo Co.; and Ariz.

144. BLADDER-POD. *Isomeris arborea* (Gr., "equal parts," probably in allusion to regular petals and the equal length of the stamens and pistil; L., tree-like). **Fl.**: yellow. Very common, much-branched, rounded shrub, up to 4 ft. high. Bladder-pod blooms throughout the year whenever moisture is available and lends a cheerful bit of color wherever found. Because of its foul-smelling leaves it is sometimes referred to as a "skunk among plants." It is one of those plants which are as much at home on the dry gullies and washes about the western edge of our California deserts as on the sea bluffs. The var. *globosa,* with subglobose fruits, is known from the western Mohave D.; the var. *angustata,* with narrow fruits, is a Colo. and eastern Mohave D. form.

145. YELLOW SPIDERWORT. *Cleome lutea* (Anc. Gr. name of a mustard plant, from *kleio,* "to shut," with reference to parts of the flower; L., yellow). **Fl.**: yellow. By midsummer many of our northern desert roadsides and flats are masses of yellow, owing to this 3-leaved annual, which often grows up to 2 ft. in height. The genus to which it belongs is a large one of about seventy species, mostly annual herbs. From Inyo Co.; to Wash. and Colo.

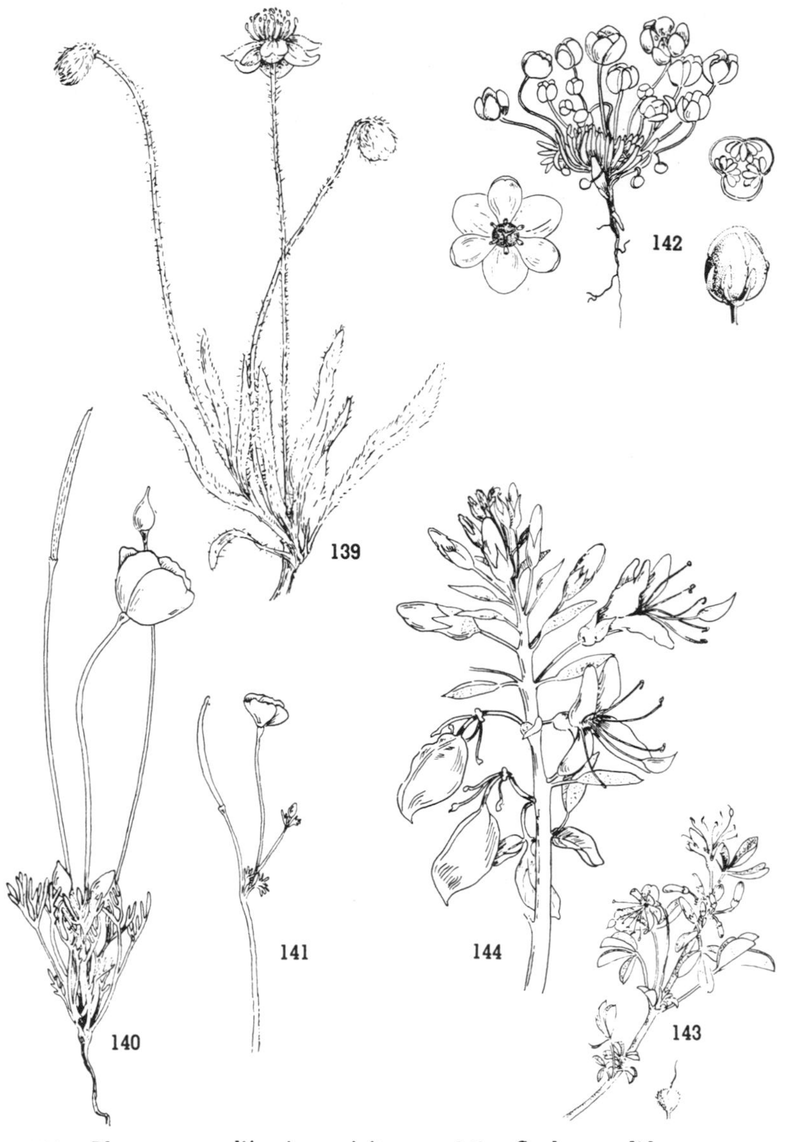

139. *Platystemon californicus crinitus*
140. *Eschscholtzia glyptosperma*
141. *Eschscholtzia minutiflora*
142. *Canbya candida*
143. *Cleomella obtusifolia*
144. *Isomeris arborea*

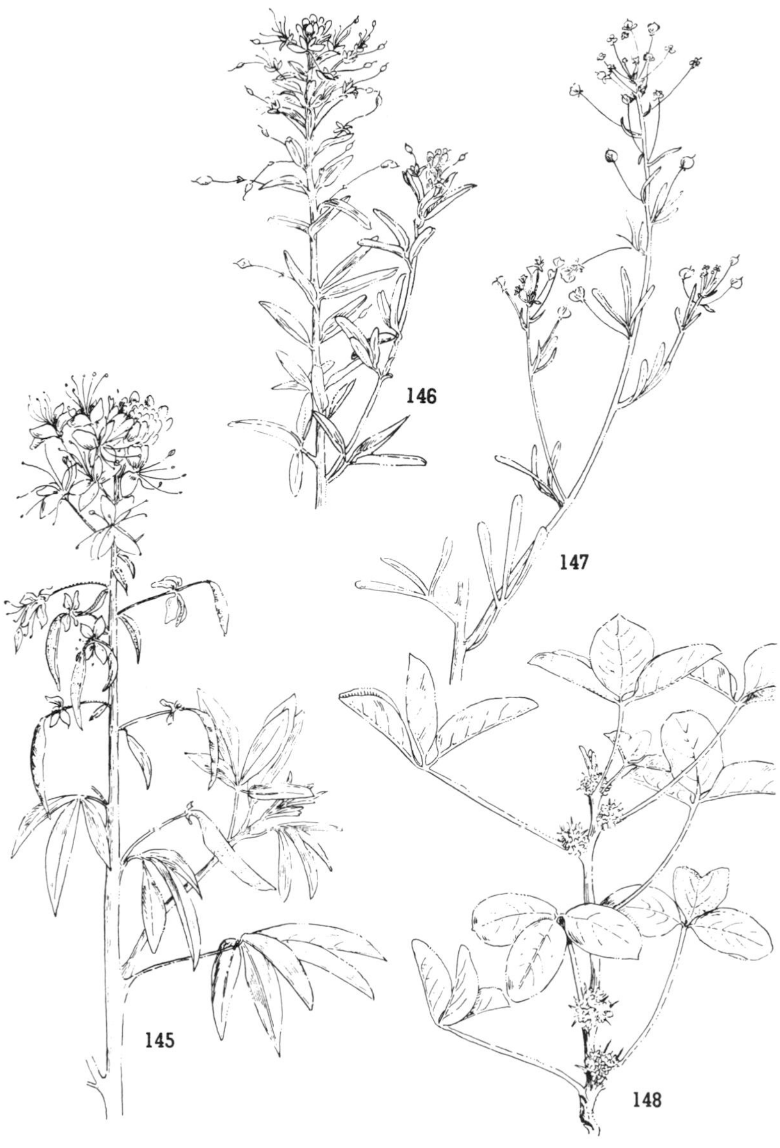

145. *Cleome lutea*
146. *Cleomella oocarpa*
147. *Cleomella parviflora*
148. *Oxystylis lutea*

146. MOHAVE STINKWEED. *Cleomella oocarpa* (diminutive of *Cleome;* Gr., egg-fruited). **Fl.:** yellow. Erect, 8 in. to 2 ft. high; common about alkaline seeps near Cushenbury Springs on the Mohave D.; north to Owens Valley; Ore. and Nev.

147. SMALL-FLOWERED STINKWEED. *Cleomella parviflora* (L., small-flowered). **Fl.:** yellow. Simple or branched, low-statured (4–8 in.) annual, found on alkaline soils of the Mohave D. and north to Owens Valley; and Nev.

148. FALSE CLOVER. *Oxystylis lutea* (Gr., sharp style; L., yellowish). **Fl.:** yellow. Another of the many odd plants of the Death Valley region collected by Frémont on the Amargosa River, where it is still abundant. It is an annual with bright green leaves and stout, unbranched stem, often reaching a height of 3 ft. The dried plants, bearing numerous rounded clusters of capsules with spiny styles, are most singular and remain standing a year or more. It was first erroneously described as a Crucifer.

CRUCIFERAE. Mustard Family

149. DESERT PLUME. *Stanleya pinnata* (Edward Stanley, Earl of Derby, 1773–1849, ornithologist, and once president of the Linnean Society; L., feathered). **Fl.:** yellow. Why Dr. C. Hart Merriam referred to this as a "miserable crucifer" is hard to conjecture, for it is one of the handsomest of plants. On first sight it much resembles a caper and was so first described under the genus name *Cleome.* Desert plume is a flexuous-stemmed perennial, 2–5 ft. high. The several stems come from a woody root and are seldom strictly erect. It is principally found in sandy washes, though not confined to them. Mohave D.; east to Utah and S.D. Closely related to this and often growing with it is the short-lived *Stanleya elata,* from ranges of the northern Mohave Desert. It is a less handsome but often taller, more erect plant, with entire or few-lobed, thick, leathery leaves and light-yellow flowers. The pods are 3–4 in. long and almost thread-like. In Death Valley both species of *Stanleya* are called "Paiute cabbage"; but that English name properly belongs to *Caulanthus inflatus,* which is not found in Death Valley. Mr. French Gilman informs me that the Indians use the young leaves of *Stanleya* for greens. They bring the leaves to a boil and then drain off the water and finish cooking them in fresh water. The Indians say they make them sick if the first water is not thrown out. This is probably due to the fact that the plants carry a percentage of the poisonous mineral selenium.

150. HEART-LEAVED TWIST-FLOWER. *Streptanthus cordatus* (Gr., "twisted-flowered," referring to the petals; L., heart-shaped, in reference to the leaf). **Fl.**: basal portion of calyx brilliant lemon-yellow, upper parts maroon, petals deep maroon. A stout-stemmed crucifer, 1–3 ft. high, common in the arid piñon forests of the Inyo Mts. and mountains of the eastern Mohave D.; to Colo.

151. ALKALI CRUCIFER. *Thelypodium affine* (Gr., "woman foot," referring to the stipitate ovary; L., "bordering upon"). **Fl.**: white. Stout-stemmed, much-branched biennial, often 3–5 ft. high. Found about alkaline seeps and springs of the south and west Mohave D.

152. INFLATED-STEM. *Caulanthus crassicaulis* (Gr., stalk flower; L., thick-stemmed). **Fl.**: petals purplish. A short-lived perennial, 1–2½ ft. high, consorting with plants of the dry open piñon belt, and conspicuous because of its gray-green, velvety stems and white-woolly calyces. Eastern Mohave D.; to Utah and Ida.

153. HAIRY-STEMMED CAULANTHUS. *Caulanthus pilosus* (L., hairy). **Fl.**: sepals deep purple or green, petals whitish. A biennial, from the arid mountains of Inyo Co.; to Ore. and Ida. It is seen at its best late in May on the Darwin Mesa, where the plants advertise their presence by their tall stems, which are often up to 3½ ft. high.

154. HALL CAULANTHUS. *Caulanthus Hallii* (H. M. Hall—see **275**). **Fl.**: sepals yellowish-green, petals yellowish-white. Slender-stemmed annual, of the Colorado D. from the Joshua Tree Nat. Mon. to San Felipe.

155. BIG-LEAVED CAULANTHUS. *Caulanthus glaucus* (L., covered with a white bloom). **Fl.**: greenish-yellow. A most handsome species, 1–1½ ft. high, of mountains of the Death Valley Nat. Mon. and western Nev.

156. COOPER CAULANTHUS. *Caulanthus Cooperi* (J. G. Cooper—see **132**). **Fl.**: lavender, aging white. An annual species, with auriculate leaves, often growing in the protection of shrubs. Common on the Mohave D., infrequent on the Colorado D.; to Ariz. and Nev.

157. LONG-BEAKED TWIST-FLOWER. *Streptanthella longirostris* (diminutive of *Streptanthus;* L., long-beaked). **Fl.**: sepals green-

164. The tall, hollow-stemmed squaw cabbage is one of the most remarkable plants of the Mohave plains.
Photo by Avery Edwin Field

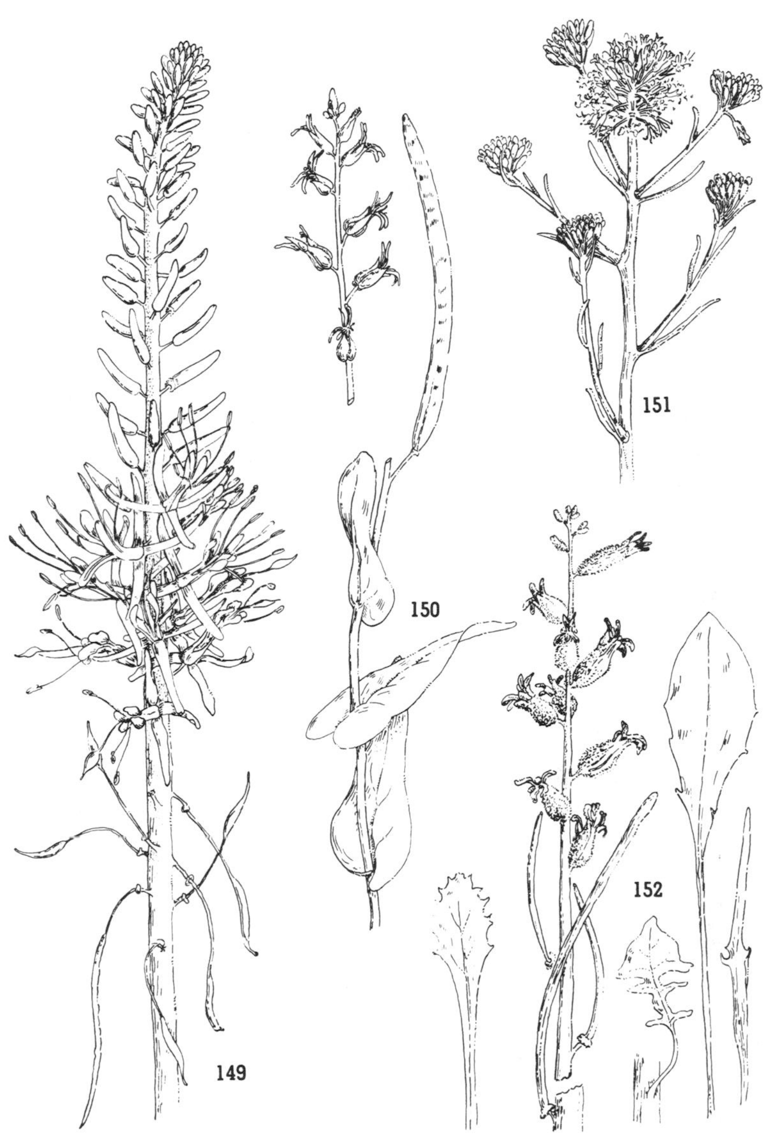

149. *Stanleya pinnata*
150. *Streptanthus cordatus*
151. *Thelypodium affine*
152. *Caulanthus crassicaulis*

ish, or with some purple; petals yellowish. An annual, with blue-green herbage, which often retreats for protection to the shelter of woody plants. We generally find it growing up through the shrubs or immediately beside them. 8–20 in. tall. Common on both Calif. deserts; to Wyo., N.Mex. The two illustrations show variations in leaf and fruits.

158. ROCK MUSTARD. *Sisymbrium diffusum Jaegeri* (Gr., name of some member of the mustard family; L., diffuse; Edmund C. Jaeger). **Fl.**: white. A perennial, generally found in the shelter of rocks in canyons of the piñon-covered mountains of the eastern and northern Mohave D.

159. YELLOW TANSY MUSTARD. *Descurania pinnata* (F. Descurain, 1658–1740, apothecary; friend of Antoine and Bernard Jussieu, the great French botanists; L., "feathered," with reference to the leaves). **Fl.**: yellow. A fine, hairy annual, growing in abundance about the base of cat's-claw and creosote bushes along the western edges of the deserts. More widespread is the somewhat similar, white-flowered *D. brachycarpon.*

160. HAIRY-LEAVED CAULANTHUS. *Caulanthus lasiophyllus* (Gr., hairy-leaved). **Fl.**: light yellow. Early spring annual, 6–18 in. high. The leaves are all petioled and the flowers quite small. Very common under and about shrubs on desert flats; widespread over the Great Basin.

161. COULTER LYRE-FRUIT. *Lyrocarpa Coulteri* (Gr., lyre fruit; Thos. Coulter—see **755**). **Fl.**: tawny yellow. A perennial herb, 1½–2 ft. high, seen only in the wild picturesque canyons along the southwestern border of the Colorado D.; to Sonora and Baja Calif.

162. SLENDER KEEL-FRUIT. *Tropidocarpum gracile* (Gr., "keel fruit," referring to the keeled capsular valves; L., slender). **Fl.**: yellow. A weak-stemmed annual. Common in grassy and open places below 3,500 ft. in the western portions of both deserts; on the western Mohave D. it is partly replaced by the var. *dubium*, with one-celled pods.

163. SPECTACLE-POD. *Dithyrea californica* (Gr., "two shields," in reference to the double, spectacle-like fruit; of California). **Fl.**: white. This sweet-flowered crucifer is very abundant in sandy places in the low deserts. It was once called "biscutella" (lit., "double little shield") because of its resemblance to a certain European species of that name. Described and illustrated in Hooker's *London Journal of Botany* as long ago as 1845.

164. SQUAW CABBAGE (see p. 76). *Streptanthus inflatus* (L., inflated). **Fl.**: at first deep purple, later white. In mid-spring hundreds of acres of its green and lemon-yellow, inflated stems studded with purple blossoms make a sight worth going many miles to see. Later the fields of barren plants are almost as startling as we view the dead stalks standing like whitish stakes driven in the Mohave sands. The pioneer women and Indians made a tasty stew by boiling the young stems with meat; the sweet stems were also eaten green. Common, generally in aggregations, on open flats, above 2,000 ft., of the middle and western Mohave D.

165. DESERT ALYSSUM. *Lepidium Fremontii* (Gr., "small scale," probably from its once-supposed virtue in curing leprosy or diseases forming scales on the skin; Capt. J. C. Frémont—see **315**). **Fl.**: white. A rounded, bushy, green-stemmed perennial, with many branches, 8–20 in. long, densely covered in spring with numerous fragrant flowers. Common both in sand washes and on rocky surfaces of the creosote-bush belt. Rare on the Colorado D., but occurring abundantly over a large part of the Mohave D.; to Nev., southern Utah, and Ariz. Not a true *Alyssum* but so called by many because of its superficial resemblance to plants of that genus.

166. YELLOW PEPPER-GRASS. *Lepidium flavum* (L., yellow). **Fl.**: yellow. Low, yellowish-green annual; about the first flower to bloom in spring on the Mohave Desert. Commonly found in the open or snuggling close under the protective covering of shrubs in the broad washes and flats of both deserts; to Baja Calif. and Nev. A good salad plant, much resembling water cress in flavor but a bit more spicy.

167. PRINCE'S ROCK-CRESS. *Arabis pulchra* (the Arabic name, meaning "from Arabia," one district of which, Petraea, where a plant of this genus is found, being a stony desert; L., beautiful). **Fl.**: purple-rose. A well-marked, handsome species, having pods covered with small, stellate hairs. Frequents rocky hillsides of the Mohave D. and western Colorado D.; to Ore., Utah.

The var. *gracilis*, with smaller rose-colored flowers and hairless pods, is entirely Mohavean.

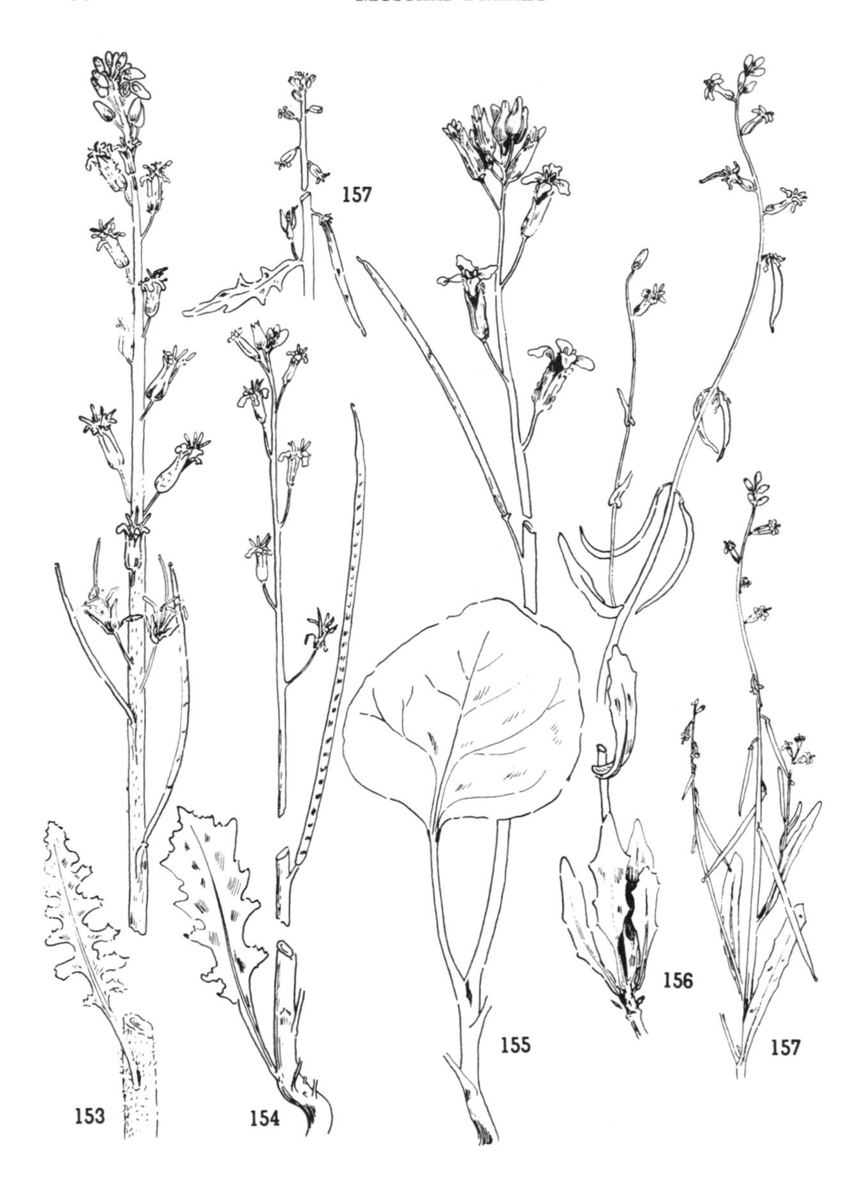

153. *Caulanthus pilosus*
154.* *Caulanthus Hallii*
155. *Caulanthus glaucus*
156. *Caulanthus Cooperi*
157. *Streptanthella longirostris*

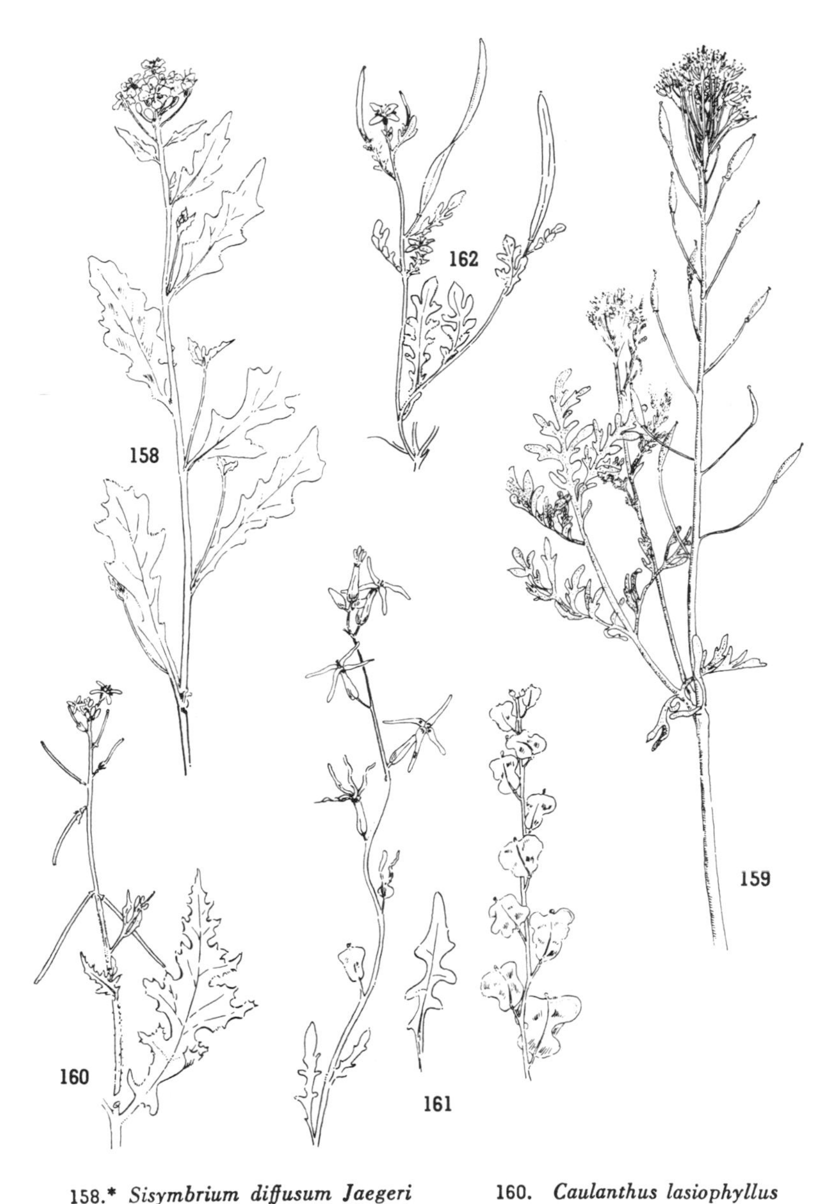

158.* *Sisymbrium diffusum Jaegeri*
159. *Descurania pinnata*, ×1
160. *Caulanthus lasiophyllus*
161. *Lyrocarpa Coulteri*
162. *Tropidocarpum gracile*

168. PALMER BEAD-POD. *Lesquerella Palmeri.* (Charles Leo Lesquereux, 1806–1889, American botanist, educated in Switzerland and brought to America by Louis Agassiz in 1848. When twenty-five years old he became totally deaf. Author of several works on American mosses, also a large number of monographs and reports of the fossil botany of North America, most of which were published in connection with state and national surveys; Edward Palmer—see **474.**) **Fl.**: yellow. An annual of the lower, sandy portions of the California deserts, such as Pinto Basin in the Joshua Tree Nat. Mon.; to Utah and Ariz. The common name alludes to the bead-like fruits.

169. KING BEAD-POD. *Lesquerella Kingii* (Clarence C. King—see **544**). **Fl.**: yellow. A perennial species of the piñon-juniper belt of the higher mountains of the eastern Mohave D. The herbage is densely covered with silvery, stellate hairs.

170. GLAUCOUS-VALVED ROCK-CRESS. *Arabis glaucovalvula* (L., blue-gray valved). **Fl.**: purplish. A perennial, with several erect stems, sometimes 10 in. tall. The herbage is distinctly gray-green, tinged with blue. It is sometimes found in the open but more often growing up through bushes. Plains and dry slopes of mountains of the Mohave D., up to 5,000 ft.

171. WEDGE-LEAVED DRABA. *Draba cuneifolia integrifolia* (name used by Dioscorides for a cress, from Gr., "sharp," because of the burning taste of its leaves; L., wedge-leaved; L., entire-leaved). **Fl.**: white. This fine little draba is abundant in early spring about rocks in the mountains near Palm Springs. Its leaves are dotted with unique stellate hairs. It also is found in shaded places on the Mohave D.; to Ariz., Nev., and Mex.

172. FRINGE-POD. *Thysanocarpus curvipes eradiatus* (Gr., fringe fruit; L., curved-footed; L., without rays). **Fl.**: white or purplish. This broad-leaved fringe-pod, with large, rayless capsules, is rare but is widely distributed on both California deserts, principally in the mountains. The linear-leaved, small-capsuled species, *T. lacinitus,* having many varieties, is another common desert fringe-pod.

173. DESERT WALLFLOWER. *Erysimum asperum* (Erysimon—an old Gr. name for a crucifer used in medicine to draw blisters—from eryo, "to draw"; L., "rough," with reference to the herbage).

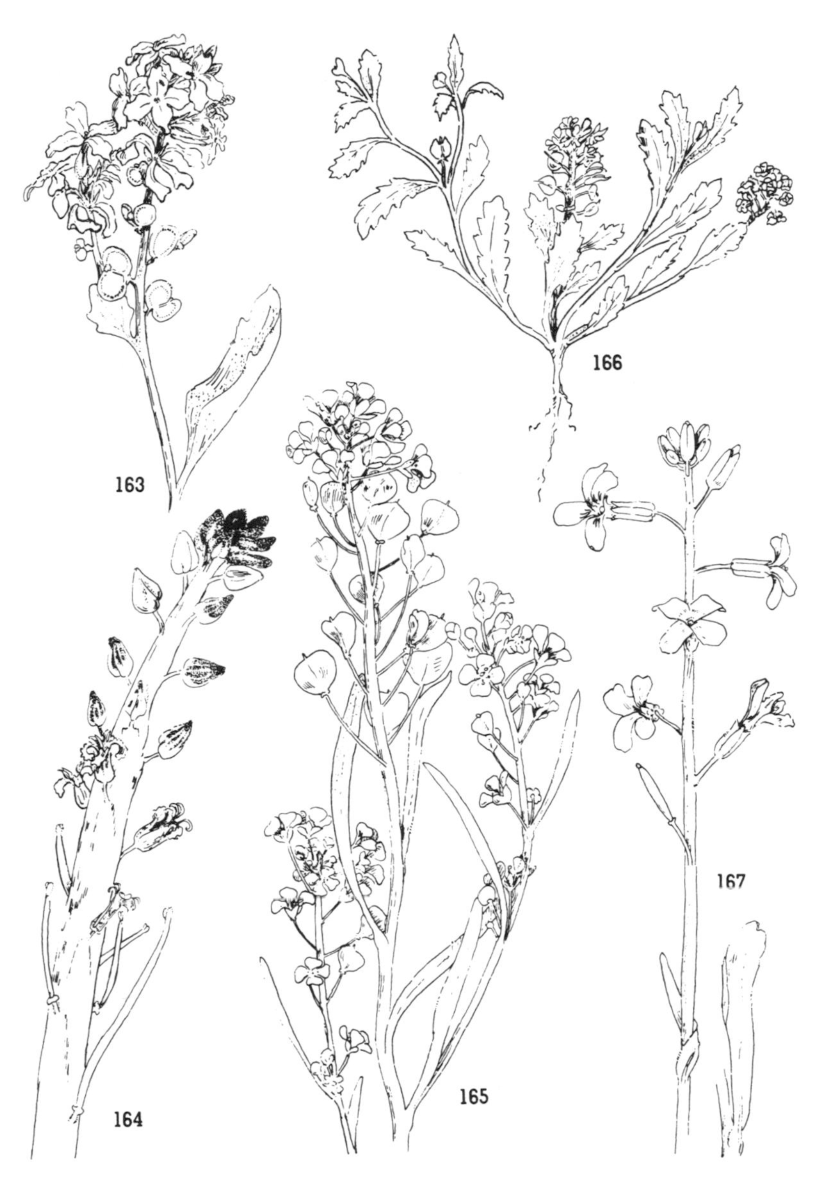

163. *Dithyrea californica*
164. *Streptanthus inflatus*
165. *Lepidium Fremontii*, ×1
166. *Lepidium flavum*
167. *Arabis pulchra*

Fl.: lemon-yellow. One of the early-blooming species. The flowers are very showy and have a delicious, spicy fragrance. The plants, 1–2 ft. tall, frequently grow up through shrubs. Quite common in washes and on benches of the western and mid-Mohave D.; Panamint Mts.; to Rocky Mts. and Wash.

RESEDACEAE. Mignonette Family

174. LINEAR-LEAVED CAMBESS. *Oligomeris linifolia* (Gr., little or few parts; L., linear-leaved). **Fl.**: white. Erect annual, 6 in. to 2 ft. high, with somewhat juicy, dark green leaves and small, mignonette-like seed vessels, which reveal its close relation to the common garden mignonette. The seeds are very small and jet black. Plentiful on barren clay and alkaline soils of the Colorado D.; to Baja Calif. and Tex.

CRASSULACEAE. Stonecrop Family

Seen in Anza Borrego, Plum Canyon

175. ROCK ECHEVERIA. *Echeveria saxosa* (Atanasio Echeverria, eminent Mexican botanical artist. DeCandolle, who described the genus, thought the artist spelled his name with but one *r*. L., of rocky places). **Fl.**: yellow, turning reddish. This is the prevalent succulent of the rocky hills and mountains of the Mohave Desert. It is very common in the Joshua Tree Nat. Mon.; north to the Panamint Mts. On the dry desert slopes of the Laguna and Cuyamaca mountains occurs *E. lagunensis,* with rich brick-red flowers.

SAXIFRAGACEAE. Saxifrage Family

176. OAK-BELT GOOSEBERRY. *Ribes quercetorum* (Anc. Arabic name for a plant properly belonging to a species of *Rheum;* L., "of an oak wood or oak forest," where first found). **Fl.**: yellow. Dry slopes of mountains on the western edge of the Colorado D.; north to Tuolumne Co. and south to Baja Calif.

177. WHIPPLEA. *Whipplea utahensis* (Lieut. A. W. Whipple, commander of the Pacific Railroad Expedition from the Mississippi River to Los Angeles, 1853–54; of Utah). **Fl.**: white. Low, twiggy shrub, from high desert ranges near the Nevada line and eastward to Utah. The fruit is a 3-valved capsule with one seed in each cavity.

CROSSOSOMATACEAE. Crossosoma Family

178. BIGELOW RAGGED ROCK-FLOWER. *Crossosoma Bigelovii* (Gr., "fringe-body," in reference to fimbriate seed appendage; Dr. J. M. Bigelow—see **338**). **Fl.**: white. This small shrub when in flower ranks as one of the handsomest plants of the desert area. It is seen at its best in the narrow, high-walled canyons of the western border of the Colorado D.; to Baja Calif. and Ariz. It comes into flower late in March. Recently collected at Ashford Canyon in the Black Mts. of Death Valley Nat. Mon. With another species found on Catalina and adjacent islands it represents a unique family of plants probably of great antiquity. The discontinuous and spotted distribution suggests that the family was once much more widespread.

ROSACEAE. Rose Family

179. COVILLE SERVICEBERRY. *Amelanchier alnifolia Covillei* (the Savoyan name for the medlar, a plant of a related genus; L., alder-leaved; Dr. Frederick V. Coville, 1867–1937, leader of the Death Valley Expedition and long connected with the U.S. National Herbarium). **Fl.**: white. Shrub, 3–10 ft. high, with dark gray twigs and green leaves; generally found scattered among other shrubs in dry, rocky stream beds or on slopes. It is ranked among the most valuable browse plants for deer. This variety of the common serviceberry is distinguished by its obtuse leaves, pointed at the apex, and small flowers. Desert slopes of the San Bernardino, Clark, Panamint, and Granite mountains.

180. ROCK FIVE-FINGER. *Potentilla saxosa* (L., diminutive of *potens*, "powerful," alluding to the reputed medicinal value, though most of the species are merely mild astringents; L., growing among rocks). **Fl.**: light yellow. In shady, almost vertical rock crevices, up to 6,000 ft. in the mountains of Inyo Co., Joshua Tree Nat. Mon.; south to Baja Calif.

181. TURF SPIRAEA. *Spiraea caespitosa* (an old Gr. name used by Theophrastus, probably from Gr., *speira*, "a coil," alluding to the fitness of some of the plants for making garlands; L., turfy). **Fl.**: white. A most singular, gray-green plant, forming large mats which spread irregularly like turf over the almost vertical surfaces

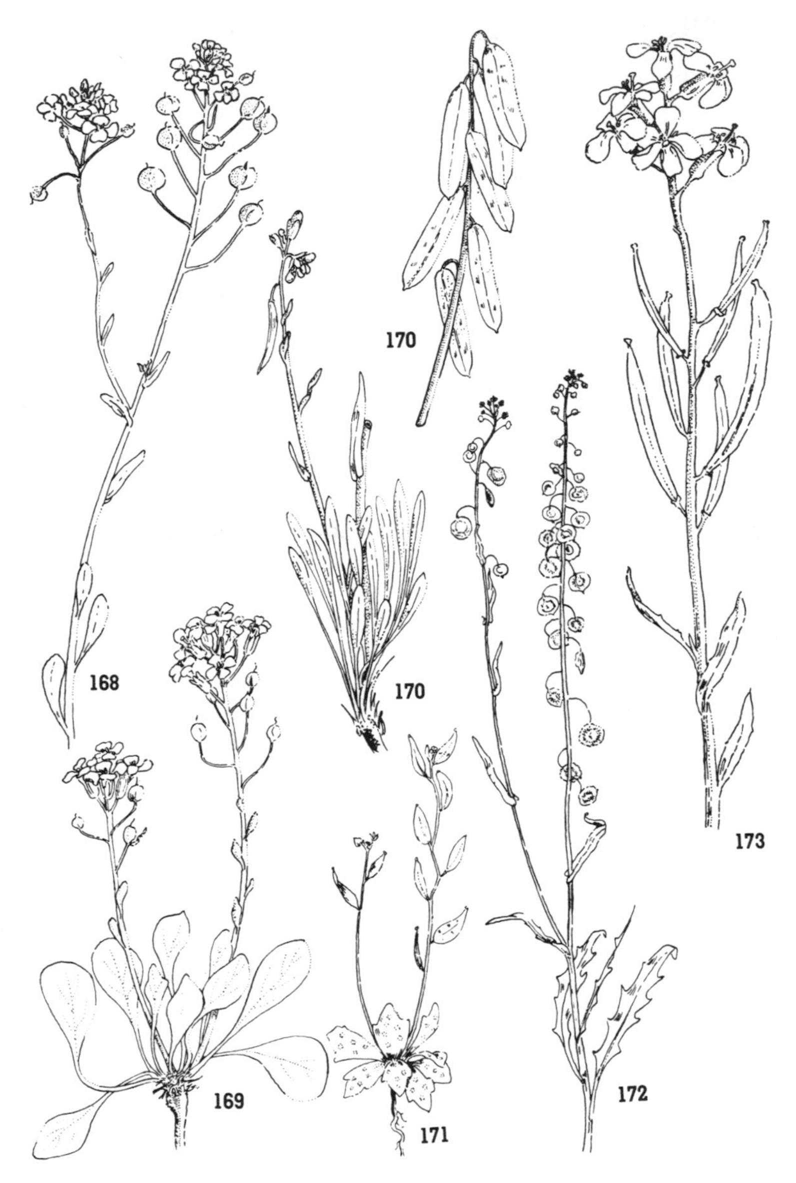

168. *Lesquerella Palmeri*
169. *Lesquerella Kingii*
170. *Arabis glaucovalvula*
171. *Draba cuneifolia integrifolia*
172. *Thysanocarpus curvipes eradiatus*
173. *Erysimum asperum*

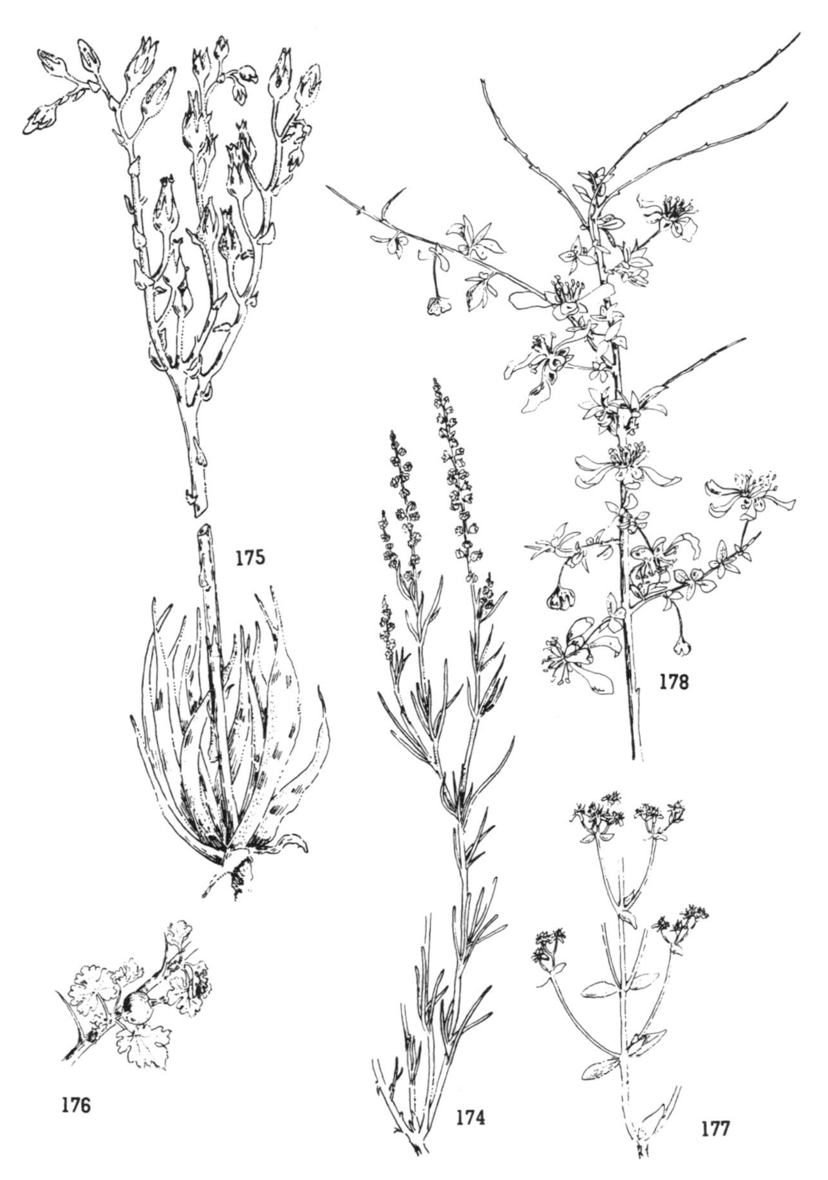

174. *Oligomeris linifolia*
175. *Echeveria saxosa*
176. *Ribes quercetorum*
177.* *Whipplea utahensis*
178. *Crossosoma Bigelovii*

of rocks. The woody roots strike deep into crevices for their moisture. Limestone cliffs of the ranges of the northern and eastern Mohave D.; to Ore. and N.Mex.

182. MOHAVE ROSE. *Rosa mohavensis* (Gr., *rhodon*, rose; L., of the Mohave D., where the type was taken). **Fl.**: pink. This is the only rose known from the California deserts. It is confined to moist places on the desert slopes of the San Bernardino and San Gabriel mountains. First collected at Cushenbury Springs, the place which yielded so many new species to S. B. Parish, early explorer of California desert plants—see **603.** It differs from *Rosa californica* "in the smaller size of all its parts, and in the absence of infrastipular spines or of any glandular or hirsute adornment."

183. SHRUBBY CREAM-BUSH. *Holodiscus dumosus* (Gr., "entire disk," to distinguish it from related genera with lobed disks; L., bushy). **Fl.**: pinkish. Low, compact shrub, with arching twigs. Dry rocky slopes and ridges of the upper piñon-juniper belt, eastern Mohave D. to Rockies.

184. CLIFF-ROSE. *Cowania Stansburiana* (J. Cowan, British merchant and amateur botanist; Captain Howard Stansbury, leader of a government exploration and survey of the Valley of the Great Salt Lake of Utah in 1850). **Fl.**: cream-colored. A small bush, 3–8 ft. high, bearing handsome flowers on rather short peduncles. It is common on many of the desert ranges, where in May it blooms in such profusion that the flowers almost completely hide the deep green of the foliage. Though the leaves are very bitter, it is a valuable evergreen browse plant for both deer and cattle. Some of the Great Basin Indians made cloth, cordage, mats, and even sandals from the silky inner bark. Dry hillsides and canyons from eastern Mohave D.; to Utah, Colo., and Mex.

185. ANTELOPE-BRUSH, BITTER-BRUSH. *Purshia glandulosa* (Frederick T. Pursh, 1774–1820, botanical explorer and author of *Flora Americae Septentrionalis*, 1817; L., provided with glands). **Fl.**: yellow or cream-colored. One of the most widely distributed of Western shrubs but confined with us to the piñon-juniper areas of ranges bordering the Mohave Desert. It is considered to be the most important of Western browse species. Both cattle and deer forage on it extensively throughout the year. In dry summers the leaves turn yellowish-brown, making it very conspicuous among other dark-foliaged plants.

186. BLACKBRUSH. *Coleogyne ramosissima* (Gr., sheathed ovary; L., very much branched). **Fl.**: yellow, aging brownish. Blackbrush is a shrub ordinarily about 2 ft. high, with short, rigid, twiggy branches. It is an important zone plant, growing with *Grayia* in a belt just above the upper limit of the creosote bush, thus marking the upper division of the Lower Sonoran Life Zone. It often occurs in such pure stands as to give a peculiar blue-gray or pale purple appearance to wide areas on the benches and slopes of the desert ranges. It should be noted that the flowers are without petals. Mountains of the Mohave D., up to 5,000 ft., also western Colorado D.; to Colo. and Ariz.

187. APACHE-PLUME, PONIL. *Fallugia paradoxa.* (Fallugius, Florentine abbot and botanist, who flourished near the end of the seventeenth century; Gr., contrary to opinion, i.e., to the describer's first opinion. The plant's habit was so widely different from any species recorded up to that time that David Don thought it might prove to be a distinct genus; but upon careful examination of the sexual parts he could discover nothing to warrant the creation of a new genus.) **Fl.**: white. In contrast to *Cowania, Fallugia* bears its white flowers on very long peduncles. It is a much smaller bush than *Cowania*, seldom being more than a meter high. Inasmuch as it is a palatable evergreen, it ranks well as a winter range plant. It is valued also as an erosion-control species. Because of some fancied resemblance of the reddish, feathery-tailed seed clusters to the plumed war bonnets of the Apache Indians, the common name "Apache-plume" was applied. The Tewa Indians used it in making brooms and arrow-shafts. In the higher arid ranges of eastern California it grows among piñons and junipers; to Ariz., Colo., and western Tex.

188. BIRCH-LEAVED MOUNTAIN MAHOGANY. *Cercocarpus betuloides* (Gr., "shuttle fruit," referring to the form of the achene and its tail; akin to *Betula*, the birch). This tall, slender shrub enters the high mountainous portions of the Joshua Tree Nat. Mon. from the west. Very variable. Cismontane chaparral belt of Calif.; Baja Calif.

189. LITTLE-LEAVED MAHOGANY. *Cercocarpus intricatus* (L., "involved," "complicated," in reference to the branching). **Fl.**:

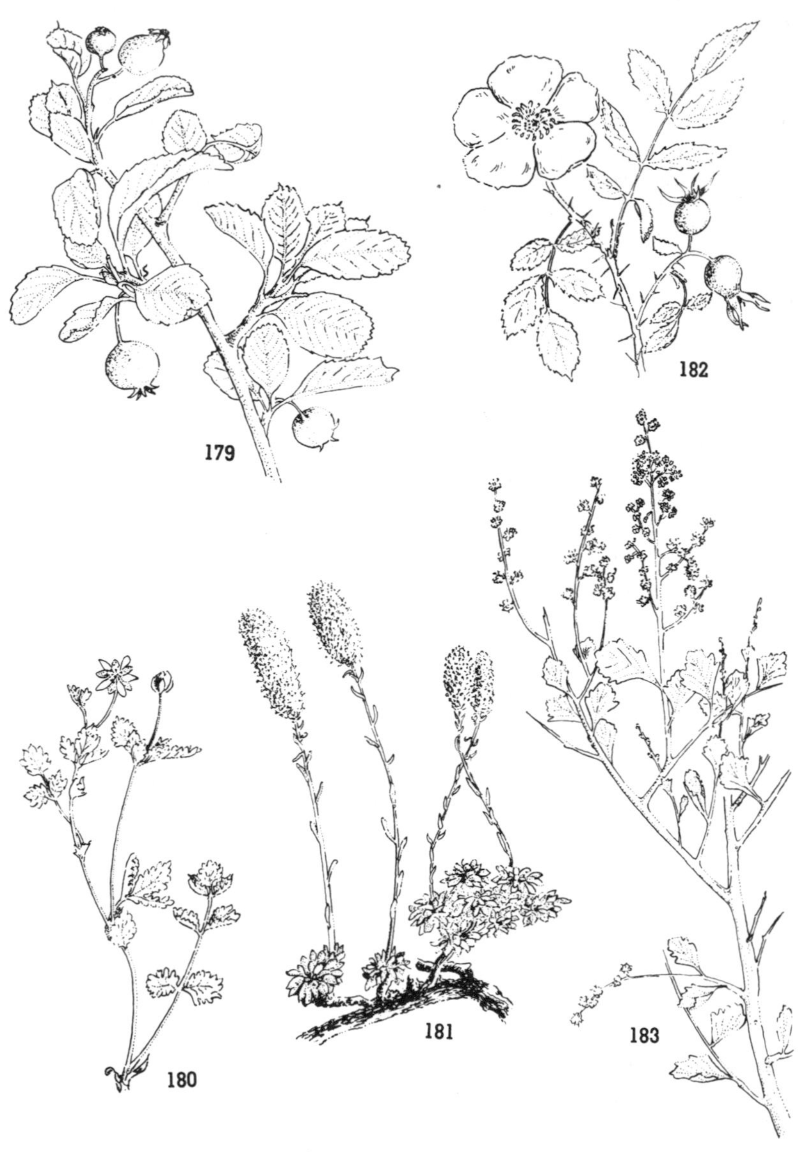

179. *Amelanchier alnifolia Covillei*

180. *Potentilla saxosa*

181. *Spiraea caespitosa*

182. *Rosa mohavensis*

183. *Holodiscus dumosus*

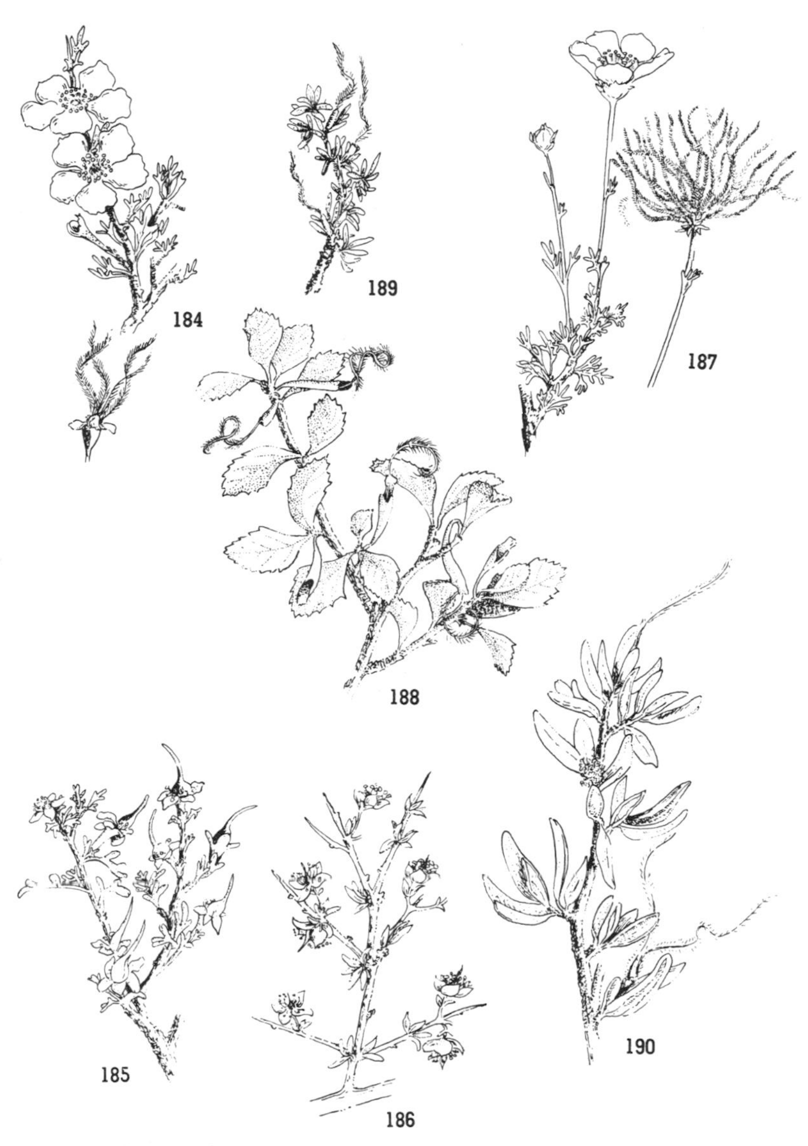

184. *Cowania Stansburiana*
185. *Purshia glandulosa*
186. *Coleogyne ramosissima*
187. *Fallugia paradoxa*
188. *Cercocarpus betuloides*
189. *Cercocarpus intricatus*
190. *Cercocarpus ledifolius*

greenish. A low, intricately branched, long-lived, dark green shrub, often finding lodgment in crevices of walls of vertical limestone cliffs and steep mountain slopes of the piñon belt. Eastern Calif.; Utah, Nev., and Ariz.

190. CURL-LEAF. *Cercocarpus ledifolius* (L., *Ledum*-leaved, i.e., with leaf like *Ledum*, the Labrador tea). **Fl.**: greenish. Typically a desert-range shrub or tree (6–18 ft. high), often in association with single-leaf piñon. On hot days a peculiar but most agreeable sweet scent comes from the leaves. The hard wood makes unusually good fuel, the flame being hot and practically smokeless. Go into the curl-leaf thickets when you will during summer, and you will always find them much frequented by birds, some in search of the seeds, others hunting bark insects or finding shelter from predators. Dry, rocky slopes, 4,000–9,000 ft., along the southwestern edge of the Mohave D., east slope of the Sierra Nevada to northern Calif.; Wash., Nev., and Colo.

191. WILD PEACH, NEVADA WILD ALMOND. *Prunus Andersonii* (L., plum; Dr. C. L. Anderson, 1827–1910, physician and naturalist, long a resident of western Nevada and central California). **Fl.**: pink. In early spring large areas on the northern desert mountains are made pink by the abundant plum-like blossoms of this thorny, spreading shrub (1–6 ft. high). It is later, when devoid of flowers, easily detected as a *Prunus* by the abundance of webs of the tent caterpillar, *Malacosoma.* Flowers appear from April to June. Piñon belt, Coso Mts. and mountains bordering Death Valley; north to Ore.

192. DESERT APRICOT. *Prunus eriogyna* (Gr., actually meaning "woolly woman," but of course here referring to the hairy pistil). **Fl.**: white to rose. This spiny, twiggy plum is remarkably showy in spring when covered with its dainty, perfumed flowers. The small, apricot-like fruits are thin-meated but were eaten by the Cahuilla Indians. Coyotes sometimes use them in default of better food. Mountain slopes of the western edge of the Colorado D., below 3,000 ft.

193. DESERT RANGE ALMOND, WILD ALMOND. *Prunus fasciculata* (L., "growing in bundles," because of the clustering of the leaves). **Fl.**: white. This tough, branchy shrub is found growing in clumps on rocky slopes and in canyons of the upper creosote-bush and juniper belt. It is quite common and widely distributed over the desert ranges of all the southern Great Basin area. Like other species of *Prunus,* it is almost invariably infested with tent caterpillars, which build large, gray tents of webbing in the forks of the branches. The small, almond-like fruits often hang in unbelievable numbers. Very variable in leaf form, as shown by the drawings of two different plants.

194. DESERT SWEET. *Chamaebatiaria millefolium* (similar to Chamaebatia; L., thousand-leaved). **Fl.:** white. Stout, erect, branching shrub, 2½ to 6 ft. high, with fragrant herbage. Eastern slopes of the Panamint Mts.; north to Ore., east to Wyo.

LEGUMINOSAE. Pea Family

195. BORDER PALO VERDE (see p. 94). *Cercidium floridum* (Gr., a weaver's instrument, which the pods resemble; L., flowered). **Fl.:** yellow. The name "palo verde," meaning "green stick," is of Spanish-Mexican origin and refers to the very noticeable green color of the trunk of this drought-resistant tree of the hot southern deserts. The broad-crowned trees, 12–20 ft. high, when in full leaf and flower are indeed a handsome spectacle. The pods of this species are somewhat flat. The dead wood makes only fair fuel, generally being spongy from the work of powder beetles. The seeds are made into a meal by the Indians. Common along dry stream-ways at low altitudes. Colorado D.; to southern Ariz., Sonora, and Baja Calif.

196. LITTLE-LEAVED HORSE BEAN. *Cercidium microphyllum* (Gr., small-leaved). **Fl.:** yellow. Rare tree, 15–20 ft. high, or often only a small shrub. The bark of the erect trunk and the spreading branches are bronzy green. It is distinguished from the border palo verde by its more numerous leaflets (4–8 pairs) and its cylindrical pods. Whipple Mts.; to Ariz., Sonora, and Baja Calif.

197. SCREW BEAN, TORNILLA. *Prosopis pubescens* (Anc. Gr. name for the butter-burr, *Petasites vulgaris,* one of the *Compositae;* L., "downy," in reference to the leaves). **Fl.:** yellow. A deciduous shrub or small, erect tree occasionally 20 ft. high, most often met with in dense thickets, especially on rich sandy or loamy bottom lands along canals and rivers. Flowering begins in early spring, and the peculiar pods, appearing in bunches of from two to ten, ripen during summer. The beans, green or ripe, are eaten by grazing animals. Indians and Mexicans eat the raw pods or grind the ripe ones into

Photo by Herbert C. Little

195. *Above*: BORDER PALO VERDE, with the usual litter of dead branches about its base. *Below*: LITTLE-LEAF PALO VERDE (*Cercidium microphylla*)

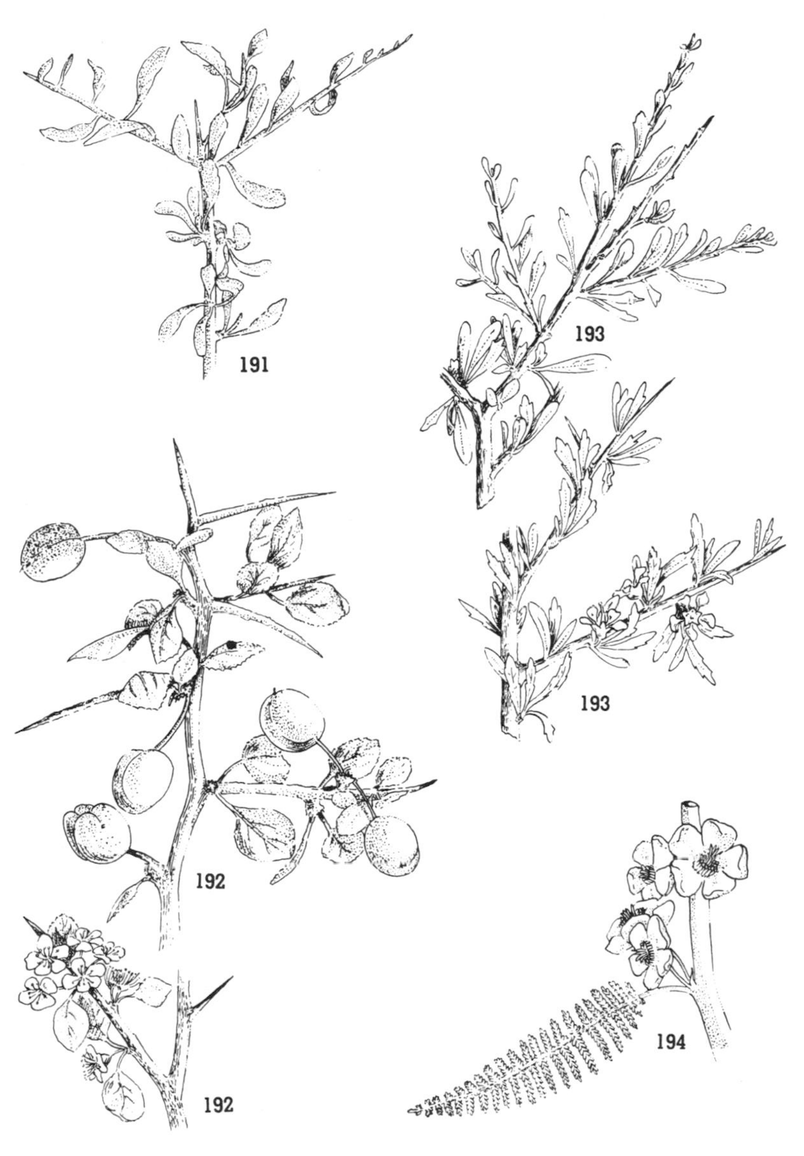

191. *Prunus Andersonii*
192. *Prunus eriogyna*
193. *Prunus fasciculata*
194.* *Chamaebatiaria millifolium*, ×1

meal. The hard wood they utilize for fuel and fence posts. The shreddy bark of the screw bean is often sought by the crissal thrasher as lining material for the coarse-twigged nest which it builds in quailbrush or other saltbushes. Where screw bean and mesquite grow together, the screw bean trees are readily distinguished by their gray-barked twigs. This is particularly true in winter when the trees are barren. The mesquite twigs are distinctly brownish-red. Colorado and Mohave deserts as far north as Death Valley; east to Tex. and Mex.

198. MESQUITE. *Prosopis juliflora glandulosa* (L., "catkin-flowered," in reference to the cylindrical spike; L., provided with glands). **Fl.**: yellow. Mesquite is a many-branched shrub or small tree, 15–20 ft. tall, sometimes so largely buried under the sand of dunes that only 2–3 ft. of the brownish branch tips protrude. It is a water-indicating plant of highest value, its long roots penetrating at times 50–60 ft. to moisture. The flowers appear from April to June and are much visited by bees. The sweet-meated pods ripen in September or October and are eaten by numerous mammals, including domestic livestock. Among many Indians and Mexicans they still provide a staple food. Dr. C. C. Parry (see **24**), writing in the *San Francisco Bulletin,* of the food of the early Cahuilla Indians said: "A due mixture of animal and vegetable diet is also secured in the mesquite bean, the pods of which are largely occupied with a species of weevil. The whole pod and its contents are pounded into a fine powder, only the woody husk of the seed being rejected. The process of baking is equally primitive. A squaw takes, generally from her head, a cone-shaped basket of close texture; the meal, slightly sprinkled with water is packed in close layers into this hat or pot as the case may be; when full it is carefully smoothed off and then buried in the sand exposed to a hot sun. The baking process goes on for several hours, till the mass acquires the consistency of a soft brick, when it is turned out, and the hat resumes its proper position on the head. The solid cake so made (if we could forget the process) is sufficiently palatable, containing a gummy sugar which dissolves in the mouth and is unquestionably nutritious." The Mexicans make a beverage from the strained cooked beans. Among certain of the Cahuilla Indians the eight seasons of the year were named in relation to the development of the bean. Large baskets woven of *Artemisia ludoviciana,* or of arrowweed, and mounted on layers of brush or on a rude scaffold, served as mesquite-bean granaries in almost every

Indian settlement. They were from 3 to 6 ft. across and half as high. From the mesquite bark, rubbed, pounded, and pulled until it became soft, were made diapers for the babies and skirts for the women. The larva of the mesquite girdler, *Oncideres pustulatus,* a small gray beetle, burrows beneath the bark, often killing the trees. A powder beetle, *Megacyllene antennatus,* works in the dry wood, soon reducing it to worthless "sponge and powder." The wood of the mesquite, especially that of the roots, is a valuable source of fuel. Considerable open forest of both mesquite and screw bean was covered by the inundation of waters forming the Salton Sea in 1906. The sand hummocks in which mesquite so often grows form retreats for many small rodents, and their numerous burrows run through the mass of roots and sands in all directions. Mesquite is a particularly handsome plant when first the new bright green leaves and myriads of blossoms appear. Colorado and Mohave deserts to Death Valley; to La., Baja Calif., and Sonora. Our mesquite, sometimes referred to under the name *P. chilensis,* is now known to be a tree quite distinct from the Chilean species.

199. CAT'S-CLAW. *Acacia Greggii.* (Gr., *akakie,* from *ake,* a point, in allusion to the thorns; Josiah Gregg, 1806–1850, frontier trader and author, and because he possessed some knowledge of the elements of medicine and surgery, "dubbed doctor." His books are considered frontier classics. He traveled widely in northern Mexico and reached California in 1849. While on an exploring expedition there during the following year, he died, worn out from hunger and exposure.) **Fl.**: yellow. Cat's-claw is a spreading deciduous shrub or small tree, of rocky desert hillsides and washes. It is extraordinarily spiny and most appropriately called "tear-blanket," "wait-a-minute," and "devil's-claw" by the pioneers. The flowers are very fragrant and are an important source of high-grade honey. In mid-August the light green pods begin to turn reddish and, if abundant, make a fine show of color. The seeds were formerly used as food by some of the Arizona and Mexican Indians. A gum much like gum arabic exudes from the bark and is used locally in Mexico. The highest thorny branches are a favorite nesting site for the verdin, while the lower ones are often a "haven" for pursued jack rabbits. Mistletoe-infested trees occasionally develop large spindle-shaped swellings like those common on ironwood (see **237**). At low altitudes, on rocky hillsides, and in ravines and canyons of the Colorado D. to the southern Mohave D.; east to Tex. and Mex.

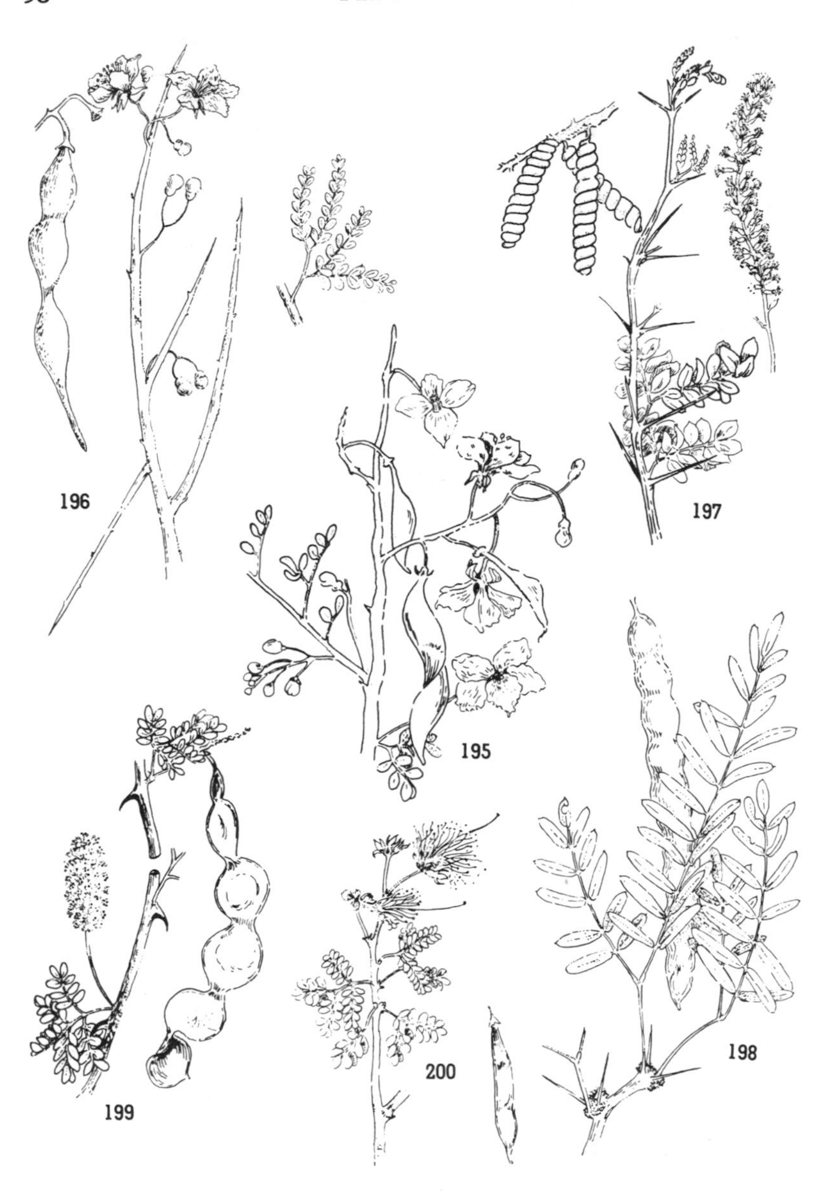

195. *Cercidium floridum*
196.* *Cercidium microphyllum*
197. *Prosopis pubescens*
198. *Prosopis juliflora glandulosa*
199. *Acacia Greggii*
200. *Calliandra eriophylla*

200. HAIRY-LEAVED CALLIANDRA. *Calliandra eriophylla* (Gr., beautiful stamens; Gr., "woolly-leaved," the leaf being fine-hairy). **Fl.**: scarlet. A low, rounded, thornless shrub, with dark green, acacia-like leaves. In California it is especially plentiful on the east side of the Chocolate Mountains, where it edges the banks of numerous narrow gullies trenching the long patina flats; known also from San Felipe Wash. The small corolla is crimson; the stamens are pure white at the base but tipped with scarlet. The brown, flat pods, about 2 in. long, are densely soft-hairy and thick-margined; when ripe they split and the valves bend wide apart. The leaves are very nutritious, and the prospector's donkeys eagerly seek out the plants. In drought the leaves enter a state of droop or long-continued wilt.

201. PIN-POINTED CLOVER. *Trifolium gracilentum* (L., three-leaved; L., slender). **Fl.**: purple. A handsome, dainty clover, found frequently in coastal southern Calif. and occasionally on the western Mohave D., as near Victorville; to Wash.

202. MULTIFLOWERED CLOVER. *Trifolium involucratum* (L., with an involucre). **Fl.**: purple to pink. A perennial species, with many-flowered heads. Common below 8,000 ft., southern Calif. to Del Norte Co.; blooming occasionally along the desert's edge in April.

203. HAIRY LOTUS. *Lotus tomentellus* (From Gr., *Lo*, to cover, a word originally applied to a fruit which was said to make those who tasted it forget their home; L., dim. of *tomentum*, "a stuffing for cushions" but here taken to mean "covered with little hairs"). **Fl.**: yellow, often reddish on back. An annual, growing over much of the desert region of Calif.; Ariz. and Baja Calif.

204. STIFF-HAIRED LOTUS. *Lotus strigosus* (L., thin). **Fl.**: yellow, with reddish back. A common cismontane species reaching the desert's western borders.

205. BROOM DEER-WEED. *Lotus scoparius* (L., broom-like). **Fl.**: yellow. A southern California species found occasionally along the edge of the desert to northern California. The var. *breviolatus*, known as the short-winged deer-weed because the wings of the flowers are shorter than the keel, is confined to the western edge of the Colorado D.

206. HILL LOTUS. *Lotus humistratus* (L., spread out on the ground). **Fl.**: yellowish. Annual, with somewhat ascending or flat-

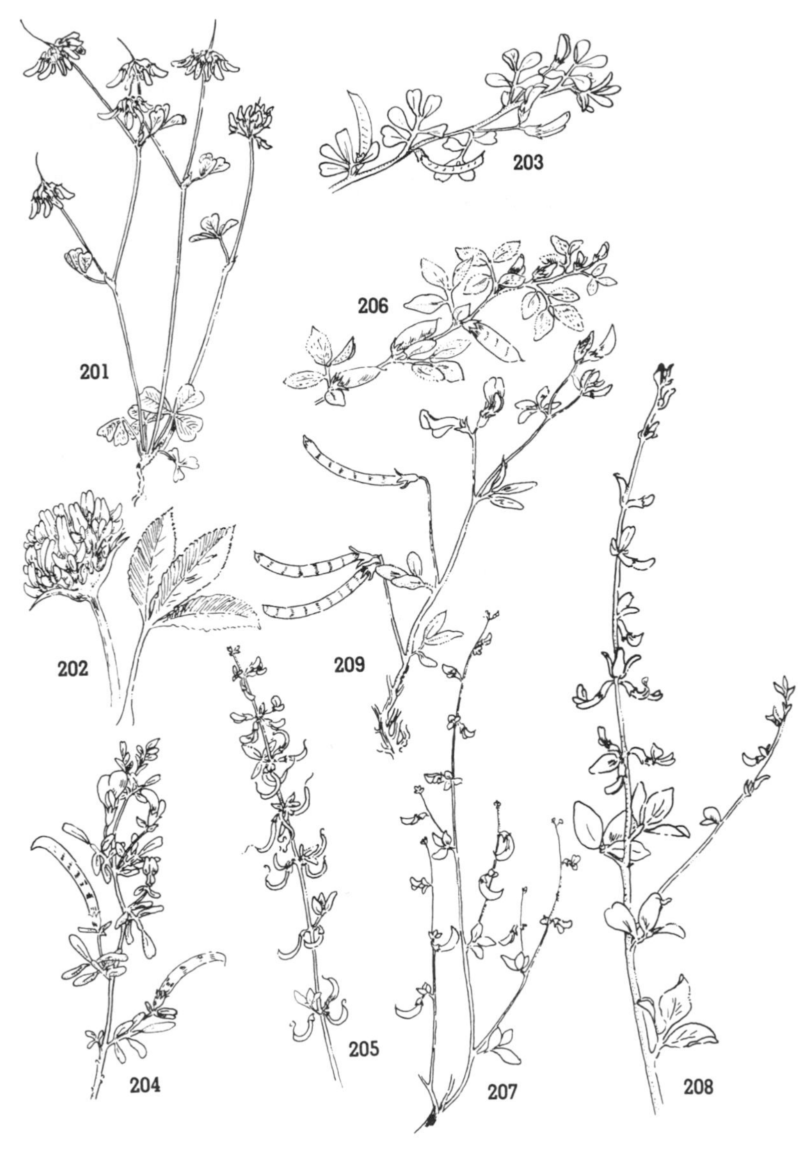

201. *Trifolium gracilentum*	204. *Lotus strigosus*	207.* *Lotus Haydonii*
202.* *Trifolium involucratum*	205. *Lotus scoparius*	208. *Lotus leucophyllus*
203. *Lotus tomentellus*	206. *Lotus humistratus*	209. *Lotus Wrightii*

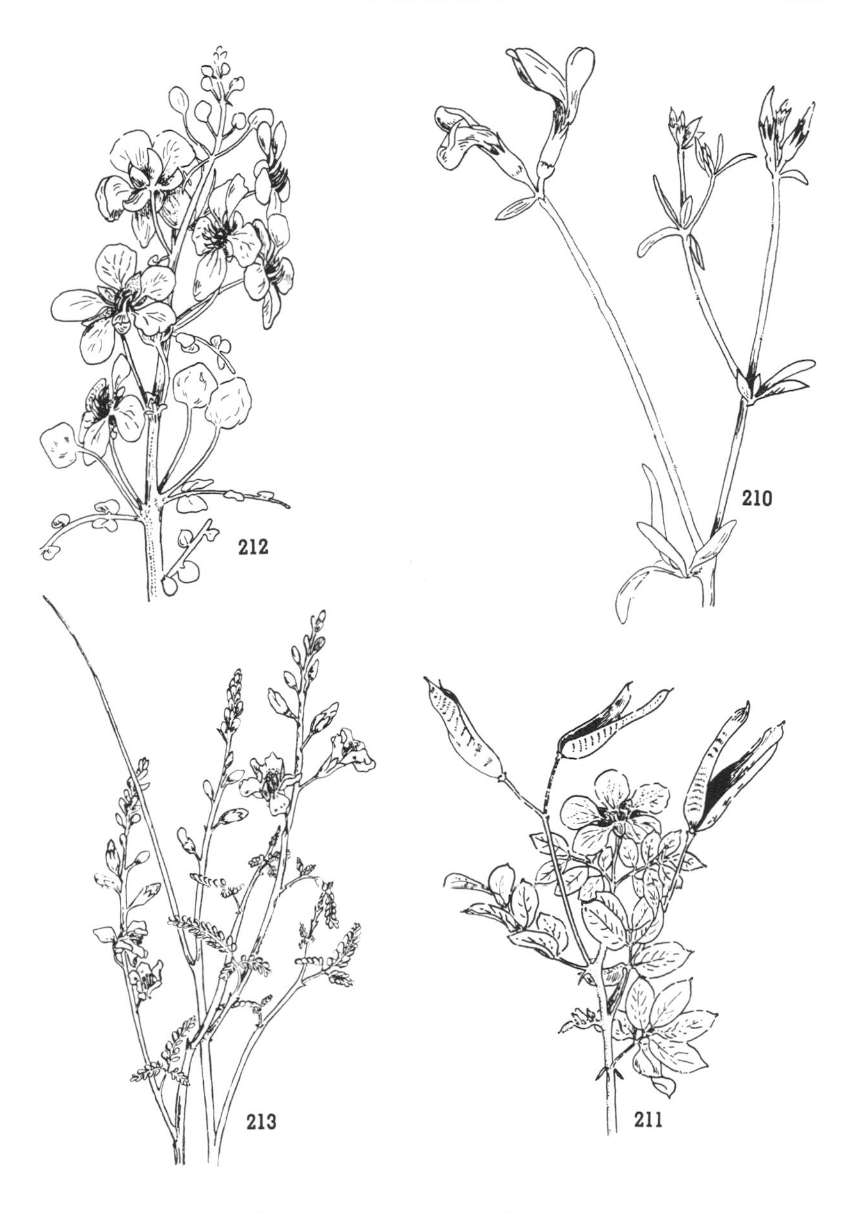

210. *Lotus rigidus*
211. *Cassia Covesii*
212. *Cassia armata*
213. *Hoffmannseggia microphylla*

lying stems, 4–12 in. long, and densely hairy leaves. Mohave and western Colorado deserts; to N.Mex.

207. PYGMY DEER-WEED. *Lotus Haydonii* (M. D. Haydon, friend of C. R. Orcutt, San Diego field botanist). **Fl.**: yellow. A low, erect, much-branched species, inhabiting the dry stony slopes of the western Colorado D.; to Baja Calif.

208. PALE-LEAVED DEER-WEED. *Lotus leucophyllus* (Gr., white-leaved). **Fl.**: pale yellow. Many-branched, light green perennial, with silky-haired herbage. Juniper-piñon belt along the western edge of the Mohave D. to the Joshua Tree Nat. Mon.

209. WRIGHT LOTUS. *Lotus Wrightii.* (Charles Wright, 1811–1885, who, as a young surveyor and teacher, collected many plants in Texas. These he sent to Asa Gray, thus opening "a correspondence destined to have important results for American botany." In 1849 he accompanied a battalion of United States troops from San Antonio to El Paso, gathering specimens all the way. Many of the plants proved to be new species. From 1853 to 1856 he was botanist of the North Pacific Exploring and Surveying Expedition. Wright collected in central California in 1855 and 1856.) **Fl.**: yellow. The stems are many, ascending to erect, from a single root crown. The herbage is covered with short hairs. The bractless umbels are but one- to two-flowered. Dry slopes of New York and Providence mountains; eastward to Nev., Colo., and N.Mex.

210. DESERT ROCK-PEA. *Lotus rigidus* (L., rigid, stiff). **Fl.**: yellow. A showy, somewhat woody species, with several erect stems, 1–2 ft. high. Rather common on the rocky hills and in canyons from Death Valley south to the Colorado D.; east to Utah and Nev.

211. COUES' CASSIA. *Cassia Covesii* (Gr., *Kasia,* of Dioscorides, from the Hebrew *Quelsi'oth;* Dr. Elliott Coues, 1842–1899, noted ornithologist who was stationed by the U.S. government at Fort Whipple in 1864, author of *Birds of the Colorado Valley*). **Fl.**: yellow. Quite different is this rare cassia from the common desert cassia, *Cassia armata.* The whole bush is densely white-hairy, and the racemes of flowers, instead of terminating the stems as in the common species, are borne in the axils of the leaves. Rare in dry, gravelly washes of the Colorado D.; to Ariz. and Mex.

212. DESERT CASSIA. *Cassia armata* (L., armed). **Fl.**: golden, sweet-scented. A low, many-branched shrub, of rounded form, appearing quite dead and leafless throughout most of the year but bursting into a riot of color over large areas in April and May if winter rains have been abundant. "This plant is particularly interesting," said the late Dr. Coville, "in the fact that while belonging to a characteristically tropical and subtropical genus, the species of which nearly all have a large leaf surface adapted to moist climates, this one departing from the typical form has developed thickened epidermal structures and small early deciduous leaflets, characters common in desert shrubs." Plants of the genus *Cassia,* alike as trees, shrubs, and herbaceous species, occur abundantly to the south in Mexico, some 200 species being known. Mohave and Colorado deserts; to Ariz.

213. SMALL-LEAVED HOFFMANNSEGGIA. *Hoffmannseggia microphylla* (J. Centurius, Count of Hoffmannsegg, 1776–1849, collaborator with J. F. Link on a flora of Portugal; Gr., small-leaved). **Fl.**: yellow to orange-red. This is a perennial shrub which occurs as a rounded bush, with several-to-many rush-like stems 2–3 ft. long. Common in barren, stony soils of the hills and canyon sides of the Colorado D.; to Sonora and Baja Calif.

214. ADONIS LUPINE. *Lupinus excubitus* (L., wolf, as these plants were once thought to rob the soil; L., vigilant). **Fl.:** blue or lilac. A most handsome perennial with greenish, silky-haired stems and leaves; found growing near sandy washes in high desert areas where junipers are prevalent. Common in the Joshua Tree Nat. Mon. The name "adonis" is suggested by its great beauty. Beside it to the right and numbered **214A** is *Lupinus sparsiflorus,* a wide-spread annual with blue flowers, the banner yellow-spotted.

215. SHOCKLEY LUPINE. *Lupinus Shockleyi* (Wm. H. Shockley, 1855–1925, mining engineer and plant collector in western Nevada and eastern California; he was the first to collect in the White Mountains of California). **Fl.**: deep blue or purplish, fading to whitish. Leaves basal; stems 4–8 in. high; the green parts of the plant, except for most of the upper surface of the leaves, covered with long, soft hairs. Only one flower of the raceme is open at a time. The chubby, 2–3-seeded fruits are covered with small, crystal-dewy pustules, giving a beaded appearance. The beads, on drying, appear as scales and are so described by those who make their descriptions only from dried plants. Sandy areas of Colorado and Mohave deserts; to Ariz. and Nev.

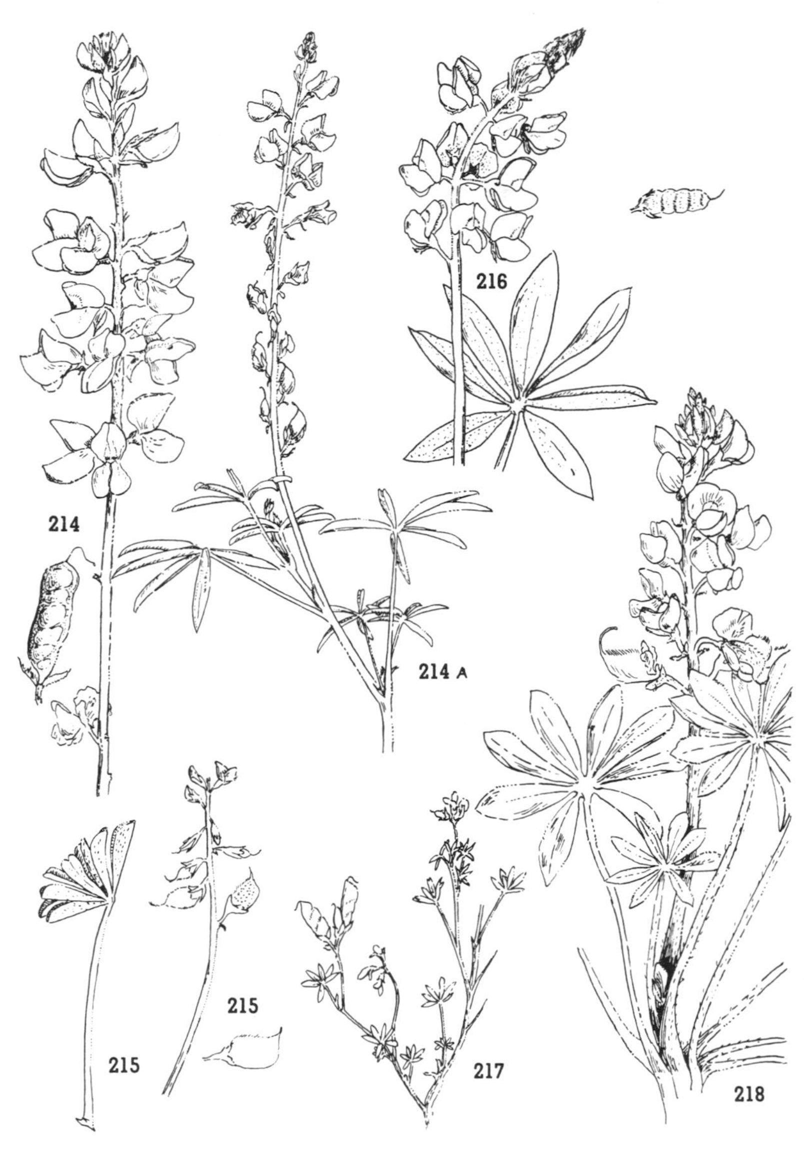

214. *Lupinus excubitus*
214A. *Lupinus sparsiflorus*
215. *Lupinus Shockleyi*
216. *Lupinus sparsiflorus arizonicus*
217. *Lupinus bicolor microphyllus*
218. *Lupinus odoratus*

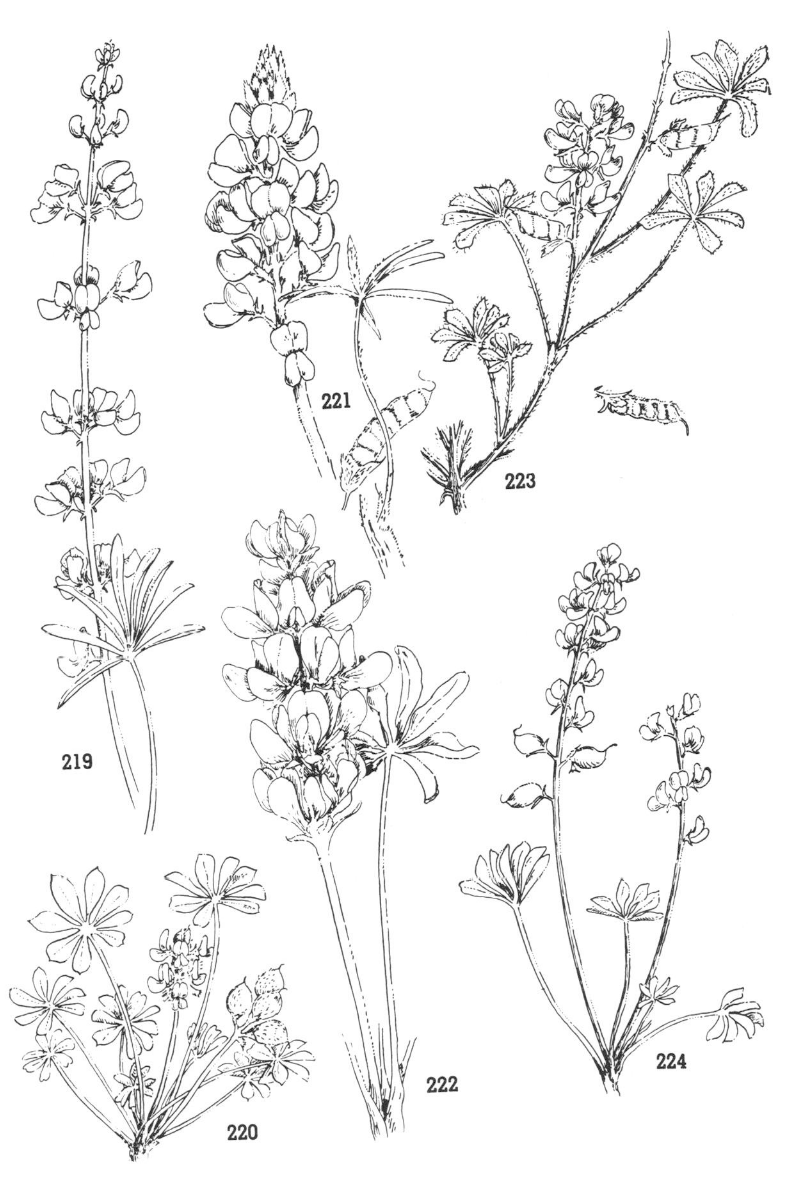

219. *Lupinus caudatus*
220. *Lupinus brevicaulis*
221.* *Lupinus Benthamii opimus*
222. *Lupinus horizontalis platypetalus*
223. *Lupinus concinnus*
224. *Lupinus rubens*

216. ARIZONA LUPINE. *Lupinus sparsiflorus arizonicus* (L., few-flowered; of Arizona). **Fl.:** reddish light blue or reddish-blue, with yellow on the banner. This annual lupine is one of the most common but is not a particularly handsome species except when seen in masses along roadsides as between Indio and Blythe. It is very variable as to both flower and leaf characters. Colorado D. to Ariz. and Sonora.

217. PYGMY-LEAVED LUPINE. *Lupinus bicolor microphyllus* (L., two-colored; Gr., small-leaved). **Fl.:** blue, with white on banner. A low, dainty-leaved, hairy annual common in southern California and occasional on the desert's edge, as in the Joshua Tree Nat. Mon.

218. ROYAL DESERT LUPINE. *Lupinus odoratus* (L., fragrant). **Fl.:** royal purple. The most showy species of the desert area, and most handsome, with its light, clear green leaves and array of brilliant, royal-purple flowers. The young flowers are especially striking when their banners have an ivory-white spot at the base. This spot fades to blue with age. The stems bear long, sparsely placed, fine hairs. Often occurring in dense colonies on the western and northern Mohave D.; to Nev. and Ariz. The var. *pilosellus,* having the stems and petioles with a heavy covering of short, spreading hairs, is also found on the western Mohave D.

219. TAIL-CUP LUPINE, SILVER LUPINE. *Lupinus caudatus* (L., tailed). **Fl.:** blue, violet, or white. A Great Basin species entering the arid ranges of the Death Valley Nat. Mon. It is sometimes seriously poisonous to cattle, sheep, and horses. Called "tail-cup" in allusion to the backward-pointing tail of the upper part of the calyx.

220. SHORT-STEMMED BLUE LUPINE. *Lupinus brevicaulis* (L., short-stemmed). **Fl.:** yellowish, sometimes with blue wings. Densely hairy annual of the northern and eastern Mohave D.; to Ore., N.Mex., and Mex.

221. BENTHAM LUPINE. *Lupinus Benthamii opimus.* (George Bentham, 1800–1884, logician and botanist; at Kew Gardens from 1855 until the end of his active life. His most notable contribution was the *Genera Plantorum,* begun in 1862 and finished in collaboration with Sir Joseph Hooker in 1883, a work which established him as perhaps the greatest systematic botanist of his century. L., abundant.) **Fl.:** deep blue, with yellow spot on banner. Stout-stemmed, hairy annual, 1–2 ft. high; the flowers with slender, curved keels. Northwestern arm of the Mohave D., to middle Calif.

222. WIDE-BANNERED LUPINE. *Lupinus horizontalis platypetalus* (L., horizontal; Gr., wide-petaled). **Fl.**: light blue to pinkish-fawn. Colonies of this short-statured annual are common in open, sandy places of the Mohave Desert. The flowers, which occur in whorls, stand almost erect when expanded and have, when old, a peculiar papery appearance.

223. ELEGANT LUPINE. *Lupinus concinnus* (L., shapely, neatly put together, elegant, tasteful). **Fl.**: lilac, edged with red-purple. A species of coastal and central California extending to the desert. Its herbage is densely covered with white to brownish hairs. Notice that the flowers are not in whorls but occur irregularly along the stem. The var. *desertorum* of the Mohave Desert has reddish stems and flowers with yellow banners, faded blue wings, and red-tipped keels. The whole plant has a general appearance of pallidness, owing to the appressed white hairs. The var. *Orcutti,* with very small flowers, is short-statured.

224. YELLOW-EYES. *Lupinus rubens* (L., reddening). **Fl.**: deep violet-blue, "hyacinth purple," with yellow spot on the banner; this changes to black as the flower fades. Small, hairy annual. In sandy places on the Mohave D.; to Utah. The var. *flavoculatus* (L., yellow-eyed), called short-banner lupine, has longer flower stems and peduncles more widespreading than the species. It occurs in the Death Valley area; eastward to Nev.

225. CALIFORNIA DALEA. *Parosela californica* (an anagram of the name, *Psoralea;* of California). **Fl.**: purple. A somewhat erect, crooked-stemmed shrub, 2–5 ft. high, found both at low altitudes and on mountainous areas in and bordering the Colorado D. Its gnarled form, as typically found in the Whitewater Wash, is particularly ornamental.

226. INDIGO-BUSH. *Parosela Schottii* (A. Schott—see **734**). **Fl.**: deep indigo blue. Bush, 3–8 ft. high, with many very slender, light tan stems and contrasting, bright green leaves. The single-seeded pods are ornamented with numerous red, blister-like glands. A most handsome plant in flower. Common about Palm Springs and southward; to Ariz. and Baja Calif.

225. *Parosela californica*
226. *Parosela Schottii*
227. *Parosela Parryi*, ×1
228. *Krameria Grayi*, ×1
229. *Krameria parvifolia imparata*
230. *Parosela arborescens*

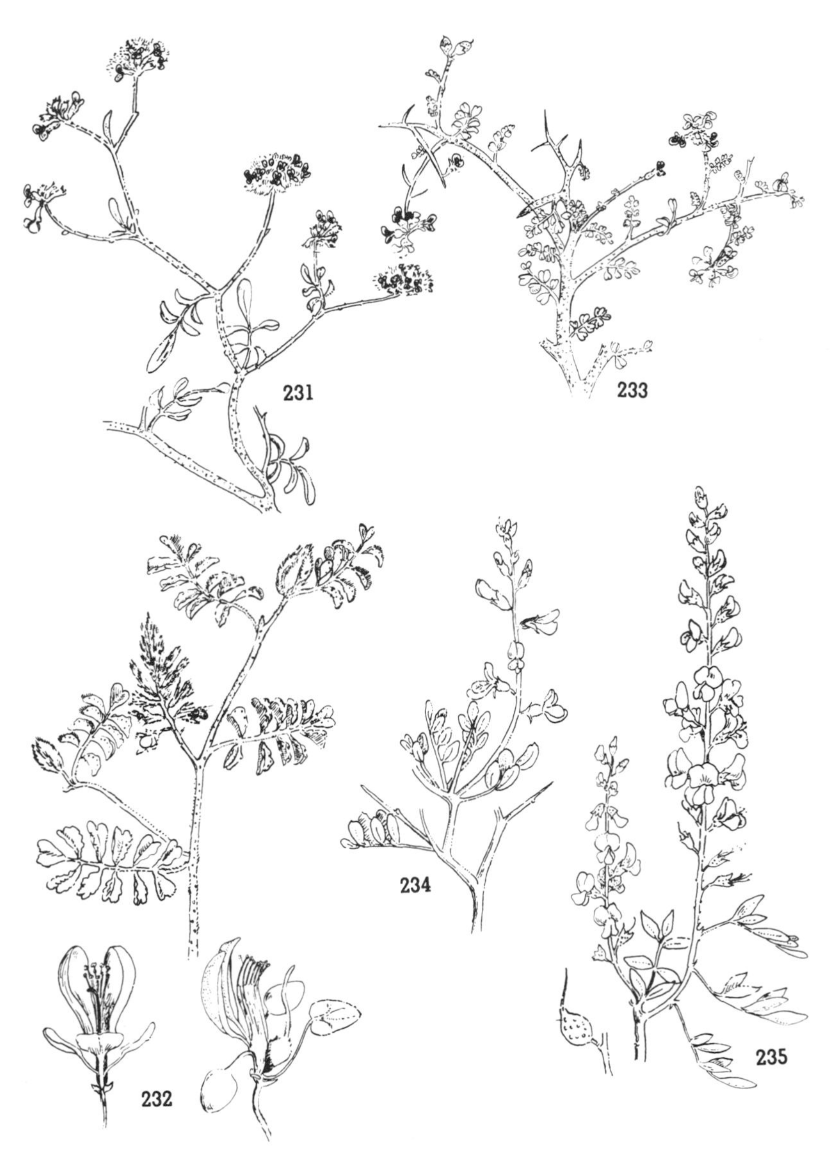

231. *Parosela Emoryi*
232. *Parosela mollis*
233. *Parosela polyadenia*
234. *Parosela Fremontii*
235. *Parosela Fremontii Saundersii*

227. PARRY DALEA. *Parosela Parryi* (Dr. C. C. Parry—see **24**). **Fl.**: purple. A low, slender, purple-stemmed, gland-dotted, somewhat woody herb, occasional in the eastern Mohave and northern Colorado deserts; to Sonora.

228. WHITE RATANY. *Krameria Grayi* (John George Henry and William Henry Kramer, father and son, Austrian botanists; Asa Gray—see **102**). **Fl.**: red-purple. The kramerias are low-branching shrubs, 1–2 ft. high, parasitic on the roots of other woody plants with which they are associated. In May the rigid branches are almost buried in masses of fragrant flowers. The Pima Indians of Arizona employed the powdered roots in the treatment of sores. A number of the Mexican kramerias yield reddish dyes which are used for coloring wool and skins, and one of the Peruvian species is used in coloring port wine. Examination of the flower with a lens will reveal many beauties which are invisible to the unaided eye. The petals are smaller than the sepals and the two lower ones are reduced to short, rounded, fleshy scales. The reddish, heart-shaped fruits are covered with delicate spines bearing a peculiar whorl of barbs at the tip. Colorado and southern Mohave deserts; to Tex. and Mex.

229. LITTLE-LEAVED RATANY. *Krameria parvifolia imparata* (L., small-leaved; L., unprovided). **Fl.**: red-purple. A sprawling shrub seldom over a foot high. In general it much resembles the white ratany, but its leaves are narrower and its fruit spines are barbed all along their upper third. It is most common on limestone and volcanic soils of the eastern Mohave D.

230. MOHAVE DALEA. *Parosela arborescens* (L., tree-like). **Fl.**: deep royal blue. From a poor specimen mislabeled as to locality, this dalea was erroneously first described as a tree. It is, in fact, only a spinescent shrub not more than 5 (mostly 2–3) ft. high. The young herbage of spring is covered with a white tomentum, making a remarkably fine background for the dark blue flowers which profusely adorn the plant. North and east of Barstow, in sandy washes and on low, stony hills. The English name, "dalea," is also an old generic name for plants of this genus. Samuel Dale (1659–1739), in whose honor it was given, was an English botanist and physician who wrote several botanical treatises and a work on drug plants.

231. EMORY DALEA. *Parosela Emoryi* (Maj. W. H. Emory—see **522**). **Fl.**: purplish. Densely branched, rounded shrub, 1½–5 ft. high, having herbage with felt-like covering and numerous orange-colored glands. It gives off, when crushed, a distinctive, agreeable odor. The wounded flower heads yield a saffron-yellow dye which is used by Indians in art work. Colorado D.; to Ariz. and Baja Calif.

"Many of the bushes east of Yuma," says L. N. Goodding, "are parasitised by a flowering plant which is the smallest in existence. About all there is to a single plant is the blossom or fruit and this is the size of a pin head. The stems are, however, often heavily laden with these parasites. Curiously enough this parasite, a Pilostyles, is related to a plant in Java and Sumatra which has the largest flower of any plant in the world, it being 2–3 feet across. This is the Rafflesia. In this case also, practically the entire plant consists of the blossom."

232. SILK DALEA. *Parosela mollis* (L., pliant, soft). **Fl.**: white, tinged with pink. A handsome perennial species growing in rounded mats low to the ground. Both leaves and calyx are densely covered with soft, silky hairs. A strong but pleasant odor is emitted from the herbage when crushed. Colorado D.; to Sonora and Baja Calif. The var. *mollissima* (L., very soft), with calyx tube so long that it usually includes the corolla, is recognized by some botanists as a distinct species.

233. DOTTED DALEA, MULTI-GLAND DALEA. *Parosela polyadenia* (Gr., many-glanded). **Fl.**: violet-purple. A low-spreading, densely glandular shrub. Plains and hillsides of the creosote-bush belt from western Nev. to the northern Mohave D. (near Leach Springs).

234. FRÉMONT DALEA. *Parosela Fremontii* (John C. Frémont—see **315**). **Fl.**: royal purple. Low shrub, 1–3 ft. tall, with zigzag branchlets. The nearly glabrous leaves have leaflets which are mostly distinct from the rachis. Mountains of the northern Mohave D.; to southern Nev.

235. SAUNDERS DALEA. *Parosela Fremontii Saundersii* (Charles Francis Saunders, of Pasadena, California, intelligent and pleasing writer on Western plants). Low shrub, with bright green, gland-dotted leaves. The brown calyx is hairy within. Middle and northern Mohave D. The nearly related variety *Johnsonii* (J. E. Johnson—see **357**) is more leafy, and its leaves are narrowly linear instead of ovate-oblong. Both are confined to the lower, hotter portions of the Mohave D.

236. SMOKE TREE (*Parosela spinosa*), an inhabitant of dry sandy washes of the hot southern deserts

Photo by Gregory D. Hitchcock

237 and 342. IRONWOOD TREE growing with MUNZ CHOLLA in one of the washes of the Chocolate Mountains

Photo by Herbert C. Little

236. SMOKE TREE (see p. 112). *Parosela spinosa* (L., thorny). **Fl.**: bluish-violet. This spiny, almost leafless, gray-green shrub, when in full flower in June, is one of the handsomest of desert plants. It is wholly confined to sandy washes and is quite dependent upon summer cloudbursts as well as winter rains for supplies of moisture. The seeds readily sprout following showers and hot days. Their further development is, however, very dependent upon well-spaced, timely summer downpours. Most of the small plants succumb because of drought, and many of those remaining are suffocated under sands washed over them during subsequent summer sheet floods. They are always in the bottoms of loose sandy washes in the very path of the rush of waters. The smoke tree is generally short-lived, and its only use is for fuel, shade, or ornament. Low, practically frostless areas of the southern Mohave and Colorado deserts; to Sonora and Baja Calif. As in the case of crucifixion thorn and several other leafless, green-stemmed trees, seedling plants have broad, well-formed leaves and bear little resemblance to adult plants.

237. DESERT IRONWOOD (see p. 113). *Olneya tesota.* (Stephen Thayer Olney, 1812–1878, Rhode Island woolen manufacturer and botanist. His special study was the genus *Carex.* He did the Carices in Watson's *Botany of the King's Expedition* [1871] and Rothrock's *Report of Wheeler's Expedition* [1874]. Corruption of Spanish *tieso,* meaning "stiff," "firm.") **Fl.**: violet-purple. A wide-crowned tree of the hot, sandy canyons and washes. Normally the flowers appear along with the new green leaves in early June. The seeds, which mature in late summer, are roasted and eaten by the Indians, who prize them for their peanut-like flavor. The remarkably heavy wood makes excellent fuel but only when thoroughly dried. It is shunned by many because of the peculiar pungent, musty stench it gives off when burning. Many of the trees are infested with mistletoe (see **35**), which not only has a stunting effect upon the host branches beyond the site of the infection but also produces great fusiform swellings. Some of these grotesque, tumor-like malformations are 2–3 ft. in diameter and weigh from 400 to 800 pounds! For some unknown reason very few of the desert birds nest in ironwood. Colorado D.; to Ariz., Sonora, and Baja Calif., at low elevations.

238. BEAVER-DAM BREADROOT. *Psoralea castorea* (Gr., "scurfy," because the plants bear glandular dots and wart-like points; L., "of a beaver," since first collected near Beaver City, Utah, by Dr. Edward Palmer, in 1877). **Fl.**: white. The tuberous, starchy taproots furnished food for the Indians and early settlers. The pods are single-seeded. In sandy places near Barstow, Daggett, and Calico Mts.; Utah and Ariz.

239. COLORADO RIVER HEMP. *Sesbania macrocarpa* (Sesban, Arabic name for *Sesbania aegyptiaca* of Egypt and India; Gr., large-fruited). **Fl.**: corolla yellow, with brownish-purple spots. This tall annual (3–10 ft. high), found so commonly along canals in the Imperial Valley, is a migrant from Arizona. Its strong fibers were formerly used by the Yuma Indians in the making of cord, nets, and fishlines. The leaves are sometimes a foot long, and the pods may be up to 9 in. in length. Colorado River hemp is gaining favor among orchardists as a cover crop. During winter many of the straw-colored, dead plant skeletons are apparent along roadsides and on canal banks.

240. PRAIRIE CLOVER. *Petalostemon Searlsiae.* (Gr., "petal stamen," with reference to the peculiar union of these floral organs. Miss Searls, who, with her father, visited the Pahranagat Mountains in Nevada about 1870 and made a collection in the vicinity of the Pahranagat Mines. "A lady," said Asa Gray, "who braves the hardships of a journey to such a remote and inhospitable district and has the sense and spirit to make a collection of plants in a place far out of the track of any botanist deserves to have her name perpetuated in the annals of botany.") **Fl.**: rose. Perennial herb, with gland-dotted leaves and flowering stems 12–20 in. high. Rarely seen in Calif. (Pahrump Valley, Providence Mts.); but well known in southern Nev., Utah, and Ariz.

241. MOHAVE LOCOWEED. *Astragalus mohavensis* (Gr., one of the curved bones of the foot, which the simple dry fruit was thought to resemble; L., of the Mohave). **Fl.**: rose-purple. Dry slopes and valleys of mountains of the northern and eastern Mohave D.; to Nev.

242. GRAY LOCOWEED. *Astragalus calycosus* (L., prominent-calyxed). **Fl.**: purple. The taproot of this small, tufted perennial is large and strong; the leaves are grayish. Flowers appears in May and June. Piñon-juniper belt to higher altitudes of the Inyo and Panamint mountains; north and east to Nev. and Ariz.

243. DEATH VALLEY LOCOWEED. *Astragalus funereus* (pertaining to a funeral, i.e., of the Funeral Mountains). **Fl.**: lower parts white, upper parts tipped with bluish-violet. A low plant with a strong, tuberous root. Most of the parts are white or ashen, except the calyces, which are clothed with short, black hairs. The pods, most prominent late in the season, are shaggy with white hairs. Darwin, Death Valley area; to Tonopah and Rhyolite.

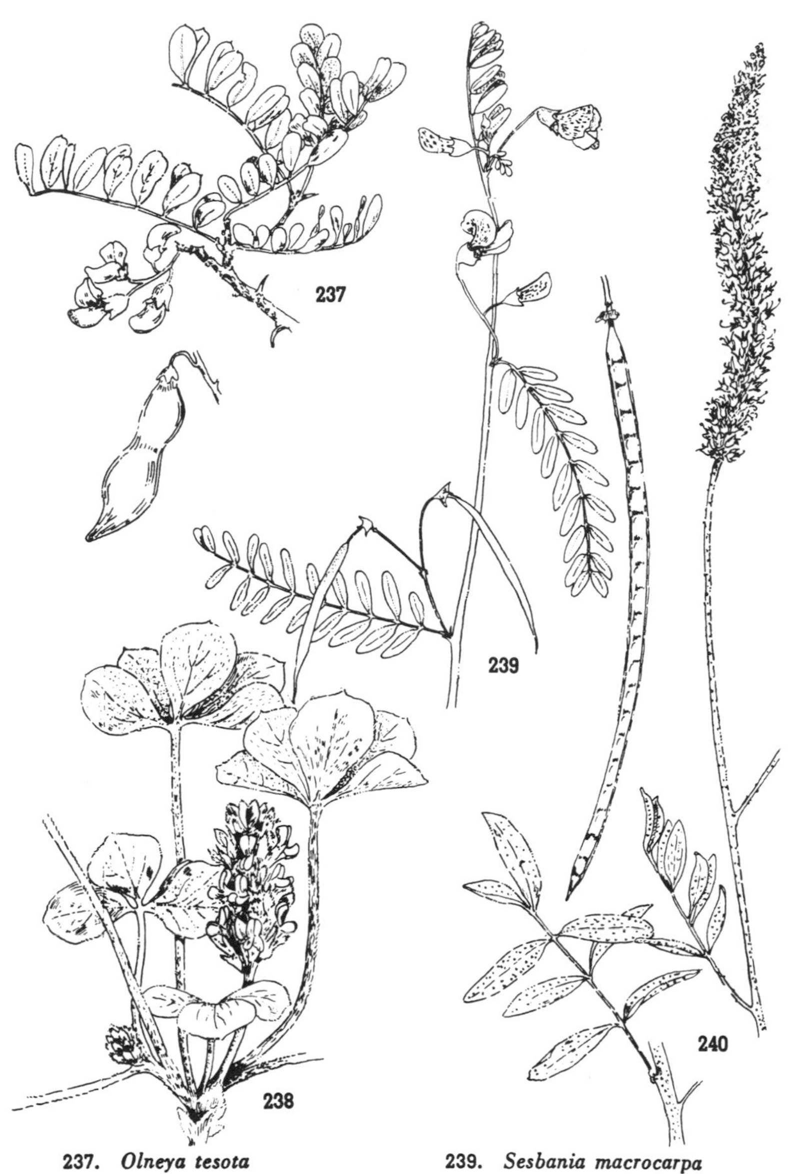

237. *Olneya tesota*
238.* *Psoralea castorea*
239. *Sesbania macrocarpa*
240.* *Petalostemon Searlsiae*

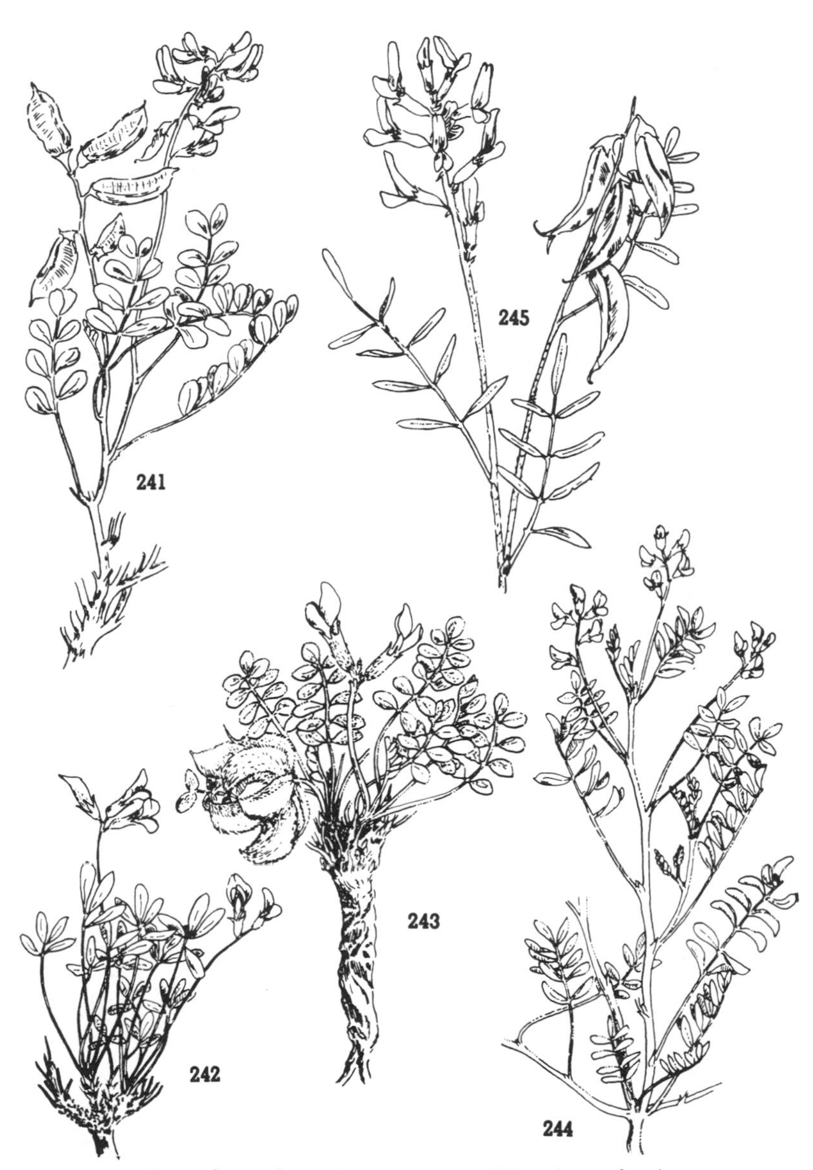

241. *Astragalus mohavensis*
242. *Astragalus calycosus*
243. *Astragalus funereus*
244.* *Astragalus Geyeri*
245. *Astragalus Casei*

244. GEYER LOCOWEED. *Astragalus Geyeri* (Charles A. Geyer, German botanist who traveled across the continent in 1843 with a party of missionaries. During 1844 he made explorations in Washington). **Fl.**: blue. Artemisia belt from Wyo. to Ore., eastern Calif., and Nev.

245. CASE LOCOWEED. *Astragalus Casei* (E. L. Case, who collected with J. G. Lemmon at Pyramid Lake, Nev.). **Fl.**: purplish. Distinguished by its strongly one-celled, obcompressed, strongly beaked pods. From Owens Valley east to western Nev.

246. VASEY LOCOWEED. *Astragalus Vaseyi* (George Vasey—see **514**). **Fl.**: purple. Western Colorado D. on dry, rocky slopes. The var. *Johnstonii,* also with purple flowers but with black-hairy calyx and papery pods, is found on dry slopes of the upper portions of the Little San Bernardino Mts.

247. BIG-PODDED LOCOWEED. *Astragalus oophorus* (Gr., "egg-bearing," because of the form of the fruits). **Fl.**: purplish. Owens Valley, Panamint Mts.; to Utah.

248. LAYNE LOCOWEED. *Astragalus Layneae* (Mary Katherine Layne, afterward Dr. M. K. [Curran] Brandegee, 1844–1920). **Fl.**: whitish, with purple tip. A widespreading perennial, with distinctive, strongly curved, hairy pods. Found in sand washes and on dry benches and slopes of the Mohave D. Flowers in April and May. Mrs. Brandegee, whose maiden name, Layne, was commemorated in this plant by Dr. Edward Greene, was best known as a botanical critic and indefatigable explorer. She was active in herbarium work both at the California Academy of Sciences and at the University of California. Her botanical trips, especially to the eastern Mohave Desert and to Nevada, were indeed many. Through the generosity of the railway companies she enjoyed the privileges of a general pass on the roads, which allowed her "to ride on anything from Pullman to engine." This was of inestimable value in that it allowed her, without much expense, to take greatest advantage of the short desert season. Mrs. Brandegee was the founder of California's first botanical club.

249. LAX-FLOWERED LOCOWEED. *Astragalus nutans* (L., nodding, drooping). **Fl.**: blue-lavendar, standard tipped with white, later drying blue. A perennial, diffusely branched species, 4–10 in. high; the calyx is black and hairy. The plants when mature are often a meter across. Stony places, Providence Mts. and vicinity.

250. NUTTALL LOCOWEED. *Astragalus Nuttallianus trichocarpus* (T. Nuttall—see **87**; Gr., hairy-fruited). **Fl.**: white, sometimes tipped with purple. An annual with several slender stems 4–12 in. long. The pods are covered with appressed, stiffish hairs. The fruits of the var. *canescens* are hairless. Open places, Colorado and eastern Mohave deserts; to Tex.

251. GOLDSTONE LOCOWEED. *Astragalus Jaegerianus* (Edmund C. Jaeger). **Fl.**: yellow, tipped with purple. A perennial, with long, many-branched stems, always found entangled in low bushes. Rarely collected. A species recently first collected by the author from the vicinity of Goldstone on the Mohave D. and recently described by Dr. Phillip Munz of Pomona College. The pods have a distinct longitudinal partition, making them 2-celled.

252. HOARY LOCOWEED. *Astragalus agninus* (L., pertaining to a lamb, i.e., fleecy). **Fl.**: white, purple-tipped. Sandy basins and valleys of the Colorado D. and central Mohave D.

253. CIMA LOCOWEED. *Astragalus cimae* (L., of Cima, a station on the Mohave Desert). **Fl.**: rich blue with white at center. A low, perennial species, with black hairs on the calyx. Rare, known only from mountains east of Cima in the eastern Mohave D.

254. DESERT RATTLE-POD (not illustrated). *Astragalus crotalariae* (Gr., rattling, in reference to the dry pods). **Fl.**: reddish-purple, with white stripes on banner. A coarse, light green, robust perennial, with several ascending stems, 8–30 in. high, and with leaves much resembling the Cima locoweed. The leaflets number 11 to 19. The most common locoweed of dry alkaline and sandy plains at low elevations on the Colorado D. The variety *Davidsonii*, with smaller flower and narrower pods, is known from western Mohave D.

255. TRIANGLE-PODDED LOCOWEED. *Astragalus bernardinus* (of Bernardino, i.e., of the San Bernardino Mountains). **Fl.**: purple. The pods, which are triangular in cross section, make this locoweed easy to identify. On dry slopes of the eastern San Bernardino, Little San Bernardino, and New York mountains; to the Charleston Mts. in Nev.

256. KEEL-BEAK. *Astragalus acutirostris* (L., pointed-beaked). **Fl.**: light violet. In granite gravels of the middle and western Mohave D., north to Inyo Co.; infrequent, western edge of Colorado D.

257. INYO LOCOWEED. *Astragalus malacus* (Gr., soft, with reference to the strongly hairy pods and herbage). **Fl.**: deep purple. Erect perennial, ½–1½ ft. high, with tufted basal leaves. On dry

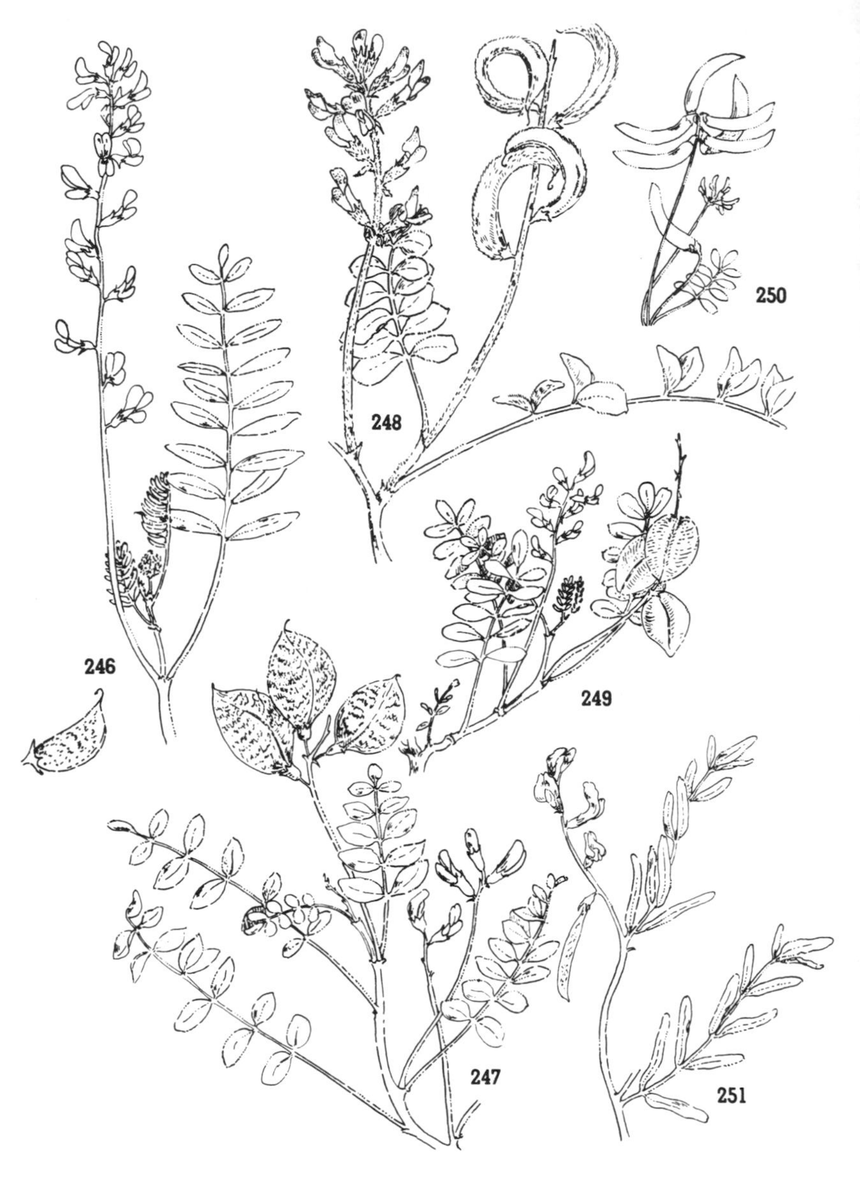

246.	*Astragalus Vaseyi*	249.	*Astragalus nutans*
247.*	*Astragalus oophorus*	250.	*Astragalus Nuttallianus trichocarpus*
248.	*Astragalus Layneae*	251.	*Astragalus Jaegeri*

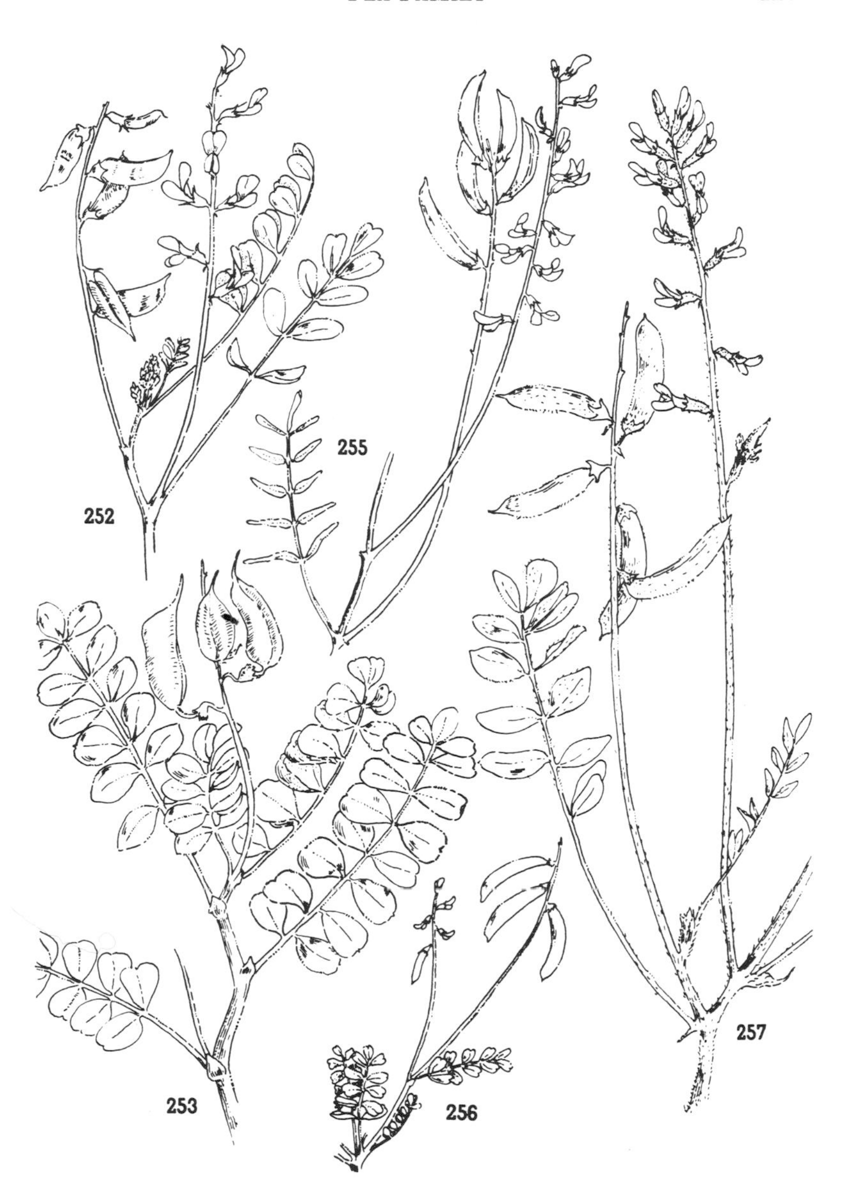

252. *Astragalus agninus*
253.* *Astragalus cimae*
255. *Astragalus bernardinus*
256. *Astragalus acutirostris*
257. *Astragalus malacus*

limestone hills, desert slopes, and mesas of the creosote-bush and artemisia belts, from the east side of Sierra Nevada; to Nev., Ida., and Ore.

258. PURSH LOCOWEED. *Astragalus Purshii longilobus* (Fredrick Pursh, 1774–1820—see **185**; L., long-lobed). **Fl.**: rose-lavender. A low, tufted perennial with densely hairy pods and hoary foliage. Western Mohave D. to Nev. The var. *tinctus,* with white flowers tipped with lavender on banner and wings, is known from the Providence Mts.; to Ore. and Nev.

259. NOTCH-LEAVED LOCOWEED. *Astragalus dispermus* (Gr., two-seeded). **Fl.**: lavender. The leaflets, blunt-ended and notched, are hairy both above and beneath. In this species the calyx is white with hairs, but in the nearly related *Astragalus didymocarpus* the calyx is black-hairy. Frequent on slopes and plains of both Calif. deserts; to Baja Calif. and Ariz.

260. DAPPLE-POD. *Astragalus lentiginosus Fremontii* (L., spotted; Capt. John C. Frémont—see **315**). **Fl.**: purple. The mottled, papery, bladder-like pod is probably the most characteristic feature of this plant. This species, and also the variety *nigricalycis* (with black-haired calyx and yellowish-white flowers) and the variety *albifolius* (with white flowers and silver leaves) are Mohave Desert plants.

261. SAND-FLAT LOCOWEED. *Astragalus insularis Harwoodi* (L., *insular*; Dr. R. D. Harwood, of San Diego State College). **Fl.**: purplish. A spreading annual, with purplish tinge—even the papery pods are purplish. Sandy flats of the Colorado D.

262. CLIFF LOCOWEED. *Astragalus atratus panamintensis* (L., clothed in black, in reference to the nigrescent calyx tube; of the Panamints). **Fl.**: yellowish-white, tipped with purple. Perennial, with minute, appressed hairs over the herbage. The narrow leaflets are whitish beneath. Cliffs of Panamint Mts. The var. *mensanus,* with ashen herbage, plain, whitish corolla, and flattened pods, is known from the Darwin Mesa. Superficially it resembles the Goldstone locoweed (**251**) and is often mistaken for it.

263. SCARLET LOCOWEED. *Astragalus coccineus* (L., scarlet). **Fl.**: scarlet. This is without doubt the handsomest *Astragalus* of the desert. It forms (Brandegee), "in favorable locations, hemispherical tufts a foot in diameter, the silvery leaves surpassed and nearly hidden by a profusion of bright red flowers." Found in the higher mountains of the northern Mohave D. south to the Joshua Tree Nat. Mon.; also in the Santa Rosa Mts. which border the Colorado D. on the west. A plant most worthy of cultivation in gardens.

GERANIACEAE. Geranium Family

264. DESERT HERON'S-BILL (see panel illustration heading the Preface, p. vii). *Erodium texanum* (Gr., "a heron," in reference to the long-beaked fruits; of Texas). **Fl.**: purple. Infrequent, spreading annual of the flat desert basins; to Tex. On desert flats, especially on the Mohave D., the common coastal heron's-bill, *Erodium cicutarium*, called filaree, often colors large areas rose-violet. The seeds of both species are stored in quantity by harvester ants of the genus *Messor*. The husks and spirally coiled beaks are discarded around the outskirts of their nest crater, forming a broad, circular cushion.

LINACEAE. Flax Family

265. BLUE FLAX. *Linum Lewisii* (L., flax; Capt. Meriwether Lewis—see **125**). **Fl.**: blue. A perennial widely distributed from the Missouri Valley to Calif. and N.Mex. Most frequently found on our deserts in eastern Calif. The related yellow flax, *L. puberulum*, with yellow flowers, is a pale green perennial, 4–8 in. high, with numerous leaves, especially crowded at the base. It occurs on the hills and limestone mountains of the eastern Mohave D.; to Colo. and Tex.

SIMARUBACEAE. Ailanthus Family

266. CRUCIFIXION THORN. *Holacantha Emoryi* (Gr., wholly a thorn; Maj. W. H. Emory—see **522**). **Fl.**: yellow. A southern Arizona plant which has spread westward to the low, hot deserts of eastern Calif.; locally abundant along the narrow, shallow washes east of the Chocolate Mts. The leaves are reduced to mere scales, and the flowers are of separate sexes, the anthers of the pistillate flowers being imperfect. The flowers appear late in May. Donkeys and goats are said to relish the clusters of small, dry, nut-like fruits but decline to eat the thorny branchlets. A large aggregation of this shrub once occupied much of the area now covered by waters of the Hayfields Reservoir.

POLYGALACEAE. Milkwort Family

267. SPINY MILKWORT. *Polygala subspinosa* (Gr., "much milk," because of the milky sap; L., "less spiny," apparently an epithet selected because some of the specimens before the describer were spiny while others were without spines). **Fl.**: pink-purple, with

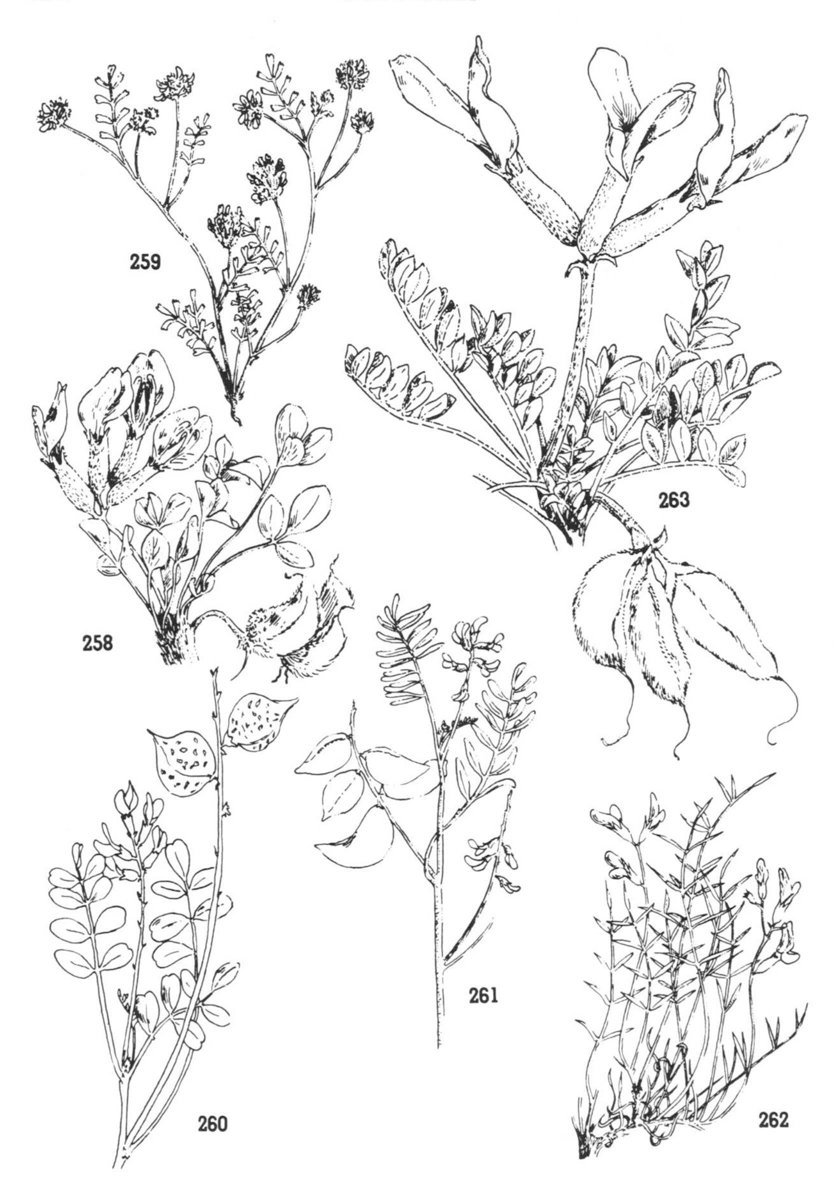

258. *Astragalus Purshii longilobus*
259. *Astragalus dispermus*
260. *Astragalus lentiginosus Fremontii*
261.* *Astragalus insularis Harwoodi*
262. *Astragalus atratus panamintensis*
263. *Astragalus coccineus*

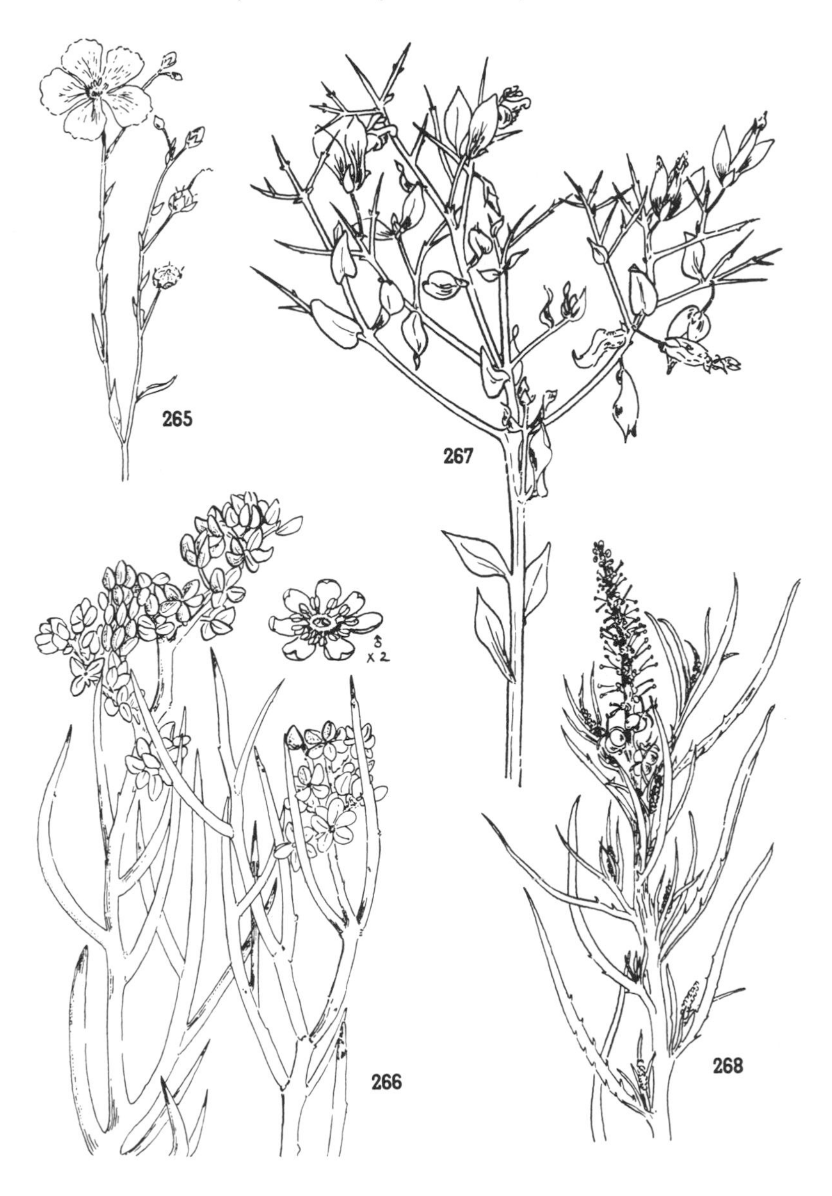

265. *Linum Lewisii*
266. *Holacantha Emoryi*
267. *Polygala subspinosa*
268. *Stillingia paucidentata*

yellowish beak. A spiny, woody-stemmed undershrub, 6–8 in. high. Rare, Chloride Cliff in Death Valley; to Nev. and N.Mex.

P. acanthoclada, known as thorn polygala, is an intricately branched spiny shrub up to 3 ft. high, with narrow linear sessile leaves. Rare on the California deserts. Otherwise known from southern Nevada northern Arizona, and Utah.

EUPHORBIACEAE. Spurge Family

268. TOOTH-LEAF. *Stillingia paucidentata* (Dr. Benjamin Stillingfleet, 1702–1771, botanist and author of "the first fundamental treatise on the principles of Linnaeus published in English"; L., few-toothed). **Fl.**: spikes wine-color to greenish. A rank-smelling herb growing in gravelly soils of western Mohave D., near the Ironwood Mts., and Mecca in the Colorado D. The many stems, from 8 in. to a foot high, grow anew each spring from a long, perennial, vertical root. When in full flower it is an appealing plant indeed.

269. TRAGIA. *Tragia ramosa.* (Hieronymus Tragus, Gr. name for Jerome Bock, 1498–1554, physician, scholar, clergyman, and one of the three "fathers of German botany," revered because he went directly to nature to describe plants for his celebrated *Kreuterbuch.* Among his curious beliefs were the ones that fungi represent the superfluous moisture of earth, trees, and rotten wood and that orchids have no seeds but arise from the excreta of birds. L., branching.) **Fl.**: greenish. A perennial herb, covered with stiff, stinging hairs. The male and female flowers are without petals and occur on the same plant. Occasional on the higher slopes and mountains of the eastern Mohave D.; to Mo. and Tex.

270. SAW-TOOTHED DITAXIS. *Ditaxis serrata* (Gr., "double-ranked," referring to the stamens; L., "serrate," with reference to the leaf margins and petals). **Fl.**: white. A low, drought-resisting herb, about 4 in. high, with dark green leaves; generally growing alone in sandy washes or on rocky benches. The saw-toothed edging of the coarse-hairy leaves, the white-margined sepals, and the densely hairy capsules offer a combination of characters which make recognition of this species easy. Eastern Mohave and Colorado deserts; to Sonora and Baja Calif. *Ditaxis californica* is much like it but does not have appressed hairs and the white margin of its sepals is more evident.

271. LANCE-LEAVED DITAXIS. *Ditaxis lanceolata* (L., "lance-shaped," in reference to the leaves). **Fl.**: white. A silvery-green perennial, found mostly among rocks. The several stems, 8–15 in.

long, spring from a coarse, woody taproot. Colorado D.; to Ariz. and Baja Calif.

272. CUT-LOBED SPURGE. *Euphorbia schizoloba* (Euphorbus, physician to King Juba; Gr., "cleft-lobed," with reference to the irregularly toothed glands of the inflorescence). A succulent, gray-green perennial, with several erect, often reddish stems, 6–16 in. high. Known from the ranges of the eastern and western Mohave D.; to Ariz. and Nev. It comes to perfection in the fine soils derived from lava beds such as occur at the south end of the New York and Providence mountains.

273. PARRY EUPHORBIA. *Euphorbia Parryi* (Dr. C. C. Parry —see **24**). **Fl.**: yellowish-green. Low annual, with yellowish, somewhat ascending to horizontal branches and finely granulate, obscurely angled seeds. Locally known from Kelso Sand Dunes of the mid-Mohave D.; to Nev. and Ariz.

274. DESERT POINSETTIA. *Euphorbia eriantha* (Gr., woolly-flowered). **Fl.**: greenish. The leaves are commonly bronze-green; this is true even of the involucral leaves, which on the true poinsettia are scarlet. The freely branching plants, 6–15 in. high, grow best in the shelter of small shrubs. Colorado D.; Sonora and Baja Calif.

275. PURPLE-BUSH. *Halliophytum Hallii.* (Hall's plant; of H. M. Hall. Curiously it carries in its scientific name a double reference to the scholarly botanist who with L. A. Greata first gathered it at Chuckawalla Bench while on his memorable buckboard journey to the lonely desert hinterland in 1905 in search of Compositae. The shrub was first described as a *Tetracoccus,* then as a *Securingia,* and at last, we hope, allowed permanent repose in the genus *Halliophytum.* Here is the account of its discovery, taken from Dr. Hall's own notebook: "Apr. 20 I took Mr. Louis A. Greata aboard at Coachella. He was living here with his family for his health. At Mecca we visited the Date Farm. . . . . At Canyon springs we drank alkali water. Here Greata wanted to turn back saying it was too far between water-holes and the water bad if you found it! But Chuckawalla Bench next day cheered him up. It was glorious! Ocotillas stretching their fingers skyward; creosotes perfuming the air; red-flowered cereus blushing in the barrancas; barrel cacti all over the bench; the new Tetracoccus coming out to meet us in at least two places—one in flower and the other in fruit; and annuals of colors of the rainbow everywhere!") This shrub once thought to be rare is now known not only from the Cottonwood, Chukawalla, and Choco-

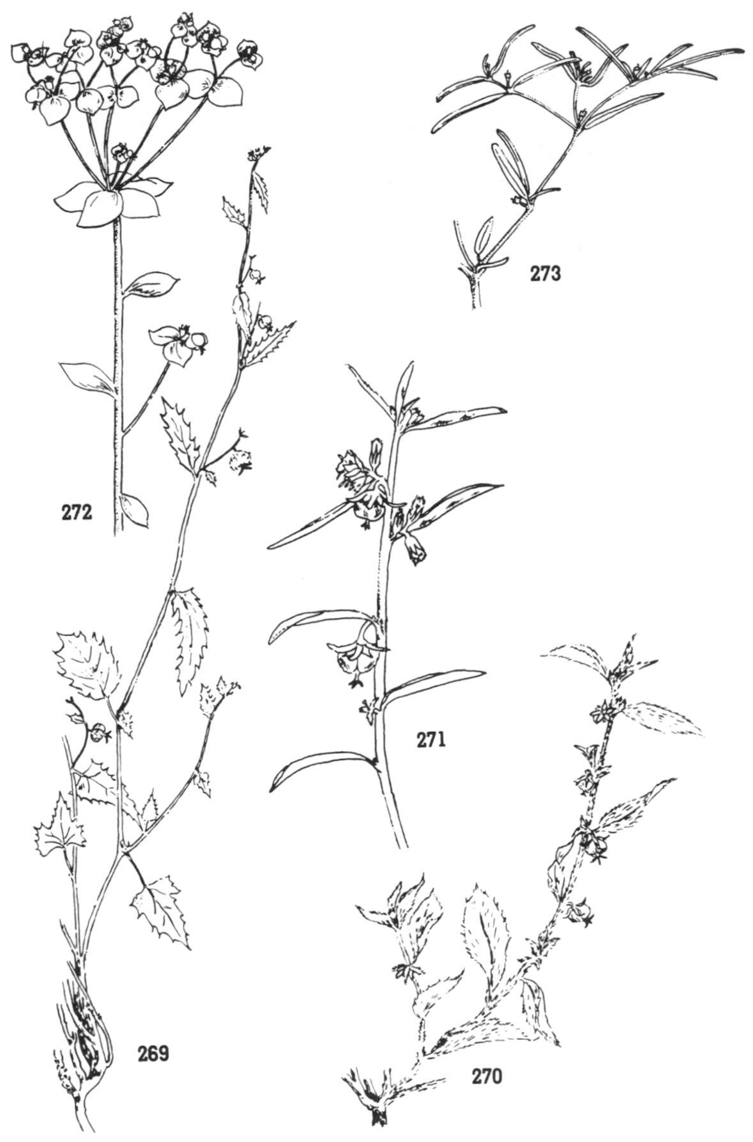

269.* *Tragia ramosa*
270. *Ditaxis serrata*
271. *Ditaxis lanceolata*, ×1
272. *Euphorbia schizoloba*
273. *Euphorbia Parryi*

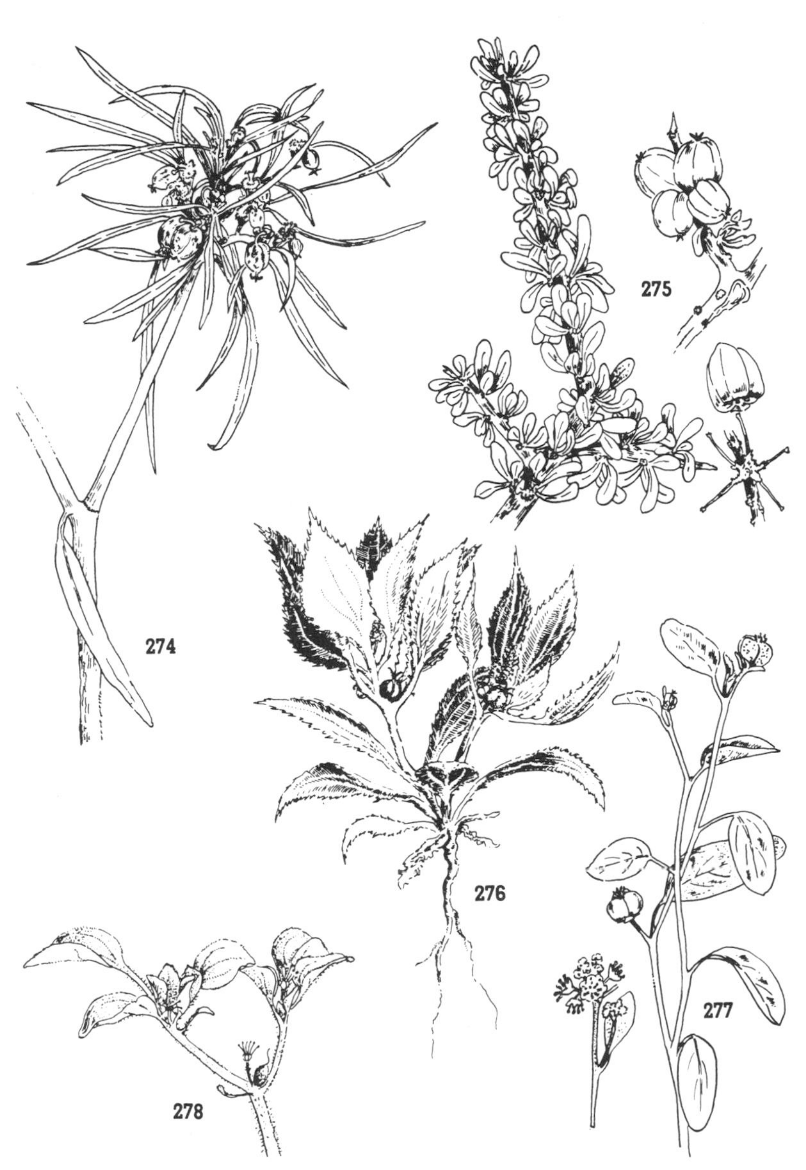

274. *Euphorbia eriantha*, ×1
275. *Halliophytum Hallii*
276. *Stillingia spinulosa*
277. *Croton californicus mohavensis*
278. *Eremocarpus setigerus*

late mountains but also along the Colorado River in California and in the marvelously picturesque Kofa range of Arizona. It forms a compact, rigid, divaricately branched bush, 1–4 ft. high. During much of the year the leaves are a dark purple, markedly contrasting with the gray stems. The blossoms are quite inconspicuous, of two sexes on separate plants, or, as the botanists say, dioecious. The green or reddish fruits which often hang on in great numbers are very rigidly attached to the woody stems. Both leaves and fruit are pleasantly strong-odored when crushed.

276. BROAD-LEAVED STILLINGIA. *Stillingia spinulosa* (L., minutely spiny). **Fl.**: greenish. Compact, rounded, green annual, 4–6 in. tall. The flower spikes bear curious trumpet-shaped glands. Mohave and Colorado deserts; into Nev. and Ariz.

277. DESERT CROTON. *Croton californicus mohavensis* (Gr., "tick"—also the Greek name for the castor bean, which has seeds of tick-like appearance; L., of California and of Mohave [Desert]). **Fl.**: greenish. A very common, pale, olive-green, perennial herb common to sandy areas. The stems when drying in late summer have a peculiar way of bending up and inward to form curious, pointed, broom-like tufts. The green plant, when crushed, has a strong but pleasant aroma. Both deserts; to Ariz. *Croton californicus* is called by the Spanish "El Barbasco," meaning a poisonous herb which fishermen use to narcotize fish. By native Californians it was applied as a hot poultice for rheumatic pains. A salve made by mixing the crushed or powdered dry leaves with tallow was also in repute as a pain reliever.

278. TURKEY MULLEIN, DOVE WEED. *Eremocarpus setigerus* (Gr., "solitary fruit," because the ovary contains but a single seed; L., "bristle-bearing," in reference to the herbage). **Fl.**: greenish, with prominent yellow stamens. Shallow-rooted, silver-green annual, common in fields and along roads in interior California. On the desert it assumes a low, compact form. The stems and leaves contain a poison used by the California Indians to stupefy fish; the herbage was bruised and then thrown into the water. Pacific Coast weed reaching the western arm of the Mohave D. near Randsburg and at Victorville.

279. SMALL-SEEDED SAND-MAT. *Euphorbia polycarpa* (Gr., many-seeded). Colorado and eastern Mohave deserts; to Ariz. and Baja Calif. The hairy-stemmed sand-mat, var. *hirtella,* occurs with the species. In both the stipules are always hairy and the seeds oblong and 4-angled. Closely related is *E. pediculifera* (**279A**), with cylindrical, grooved seeds and "flowers" with conspicuous purple-brown glands. The whole plant is hoary, owing to the numerous bent hairs covering its leaves and stems. This species is confined to the southern Colorado D.

280. RIB-SEEDED SAND-MAT. *Euphorbia glyptosperma* (Gr., "carved-seeded," in reference to the seeds, which are "sharply cross-ribbed and notched at the angles"). Imperial Valley; to Ariz. and eastward.

281. DEATH VALLEY SAND-MAT. *Euphorbia vallis-mortae* (L., of Death Valley). Like so many plants of the Death Valley area, this flat-lying spurge has taken on a very distinctive ashen-gray appearance. The seeds are quadrangular. Western Mohave D. from Coolgardie and Red Rock Canyon to Owens Valley.

282. FENDLER SPURGE. *Euphorbia Fendleri* (August Fendler —see **538**). This is a short, brown-stemmed perennial species, with several upright stems. Common in the juniper-piñon belt of the Little San Bernardino Mts. and the mountains of the eastern Mohave D.; to Kan. and Sonora, Mex.

283. NEW MEXICO SAND-MAT. *Euphorbia neomexicana* (of New Mexico). A flat, spreading species, with maroon leaves. Southeastern Colorado D.; to Ariz. and N.Mex.

284. SONORAN SAND-MAT. *Euphorbia micromera* (Gr., finely divided). Colorado D. to N.Mex. and Mexico. This and other milky-juiced sand-mats were considered valuable as snake-bite remedies by the Indians. The plants were crushed and applied as a poultice, or a tea to use internally was made from the stems.

285. BRISTLE-LOBED SAND-MAT. *Euphorbia setiloba* (Gr., "bristle-lobed," in allusion to the fimbriate appendages of the involucral glands). A hairy, spreading annual, with stems and leaves quite reddish in color. Mohave and Colorado deserts; to Ariz. and Mex.

286. SHRUB EUPHORBIA. *Euphorbia misera* (L., "poor," "wretched," i.e., with respect to appearance). **Fl.**: greenish. A small colony of this seashore spurge was located many years ago by the author on the Whitewater Bench in the Colorado D.; otherwise unknown from the desert. The plants have probably maintained their stand there since ancient times when marine waters occupied the trough between the Peninsula range and the Little San Bernardino Mts. At present it is otherwise confined to sea bluffs of southern and Baja Calif.

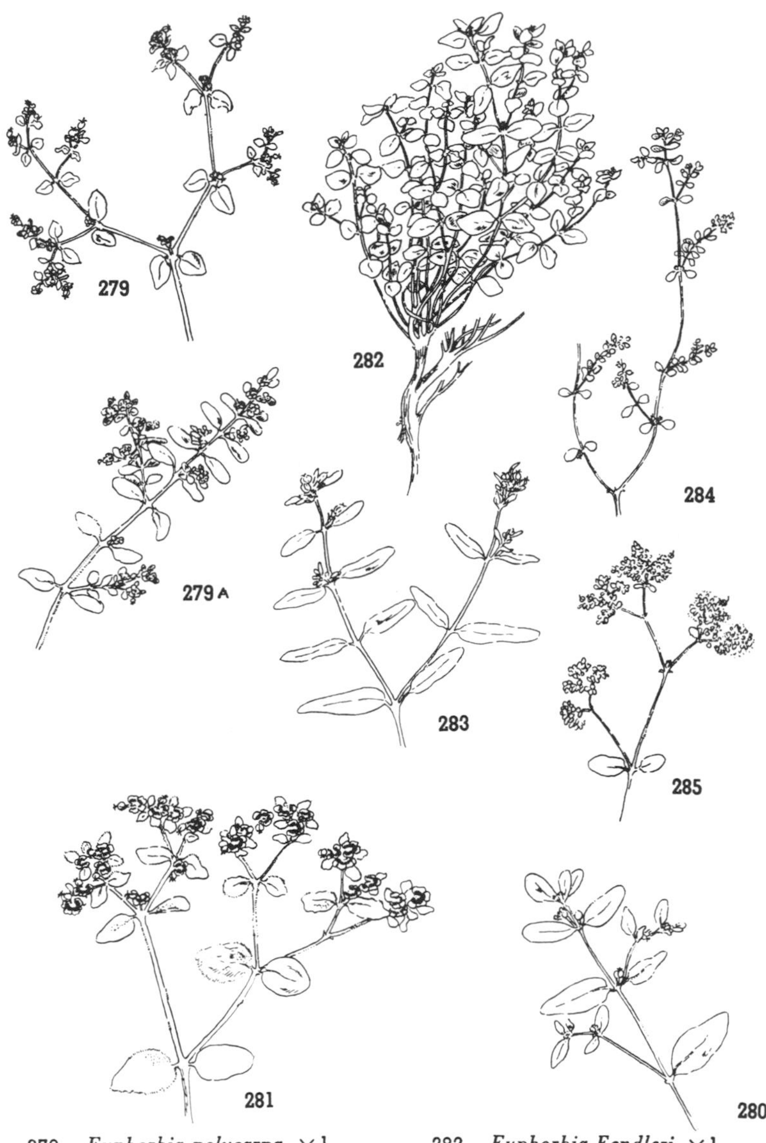

279. *Euphorbia polycarpa*, × 1
279A. *Euphorbia pediculifera*
280. *Euphorbia glyptosperma*, × 1
281. *Euphorbia vallis-mortae*, × 1
282. *Euphorbia Fendleri*, × 1
283. *Euphorbia neomexicana*, × 1
284. *Euphorbia micromera*, × 1
285. *Euphorbia setiloba*, × 1

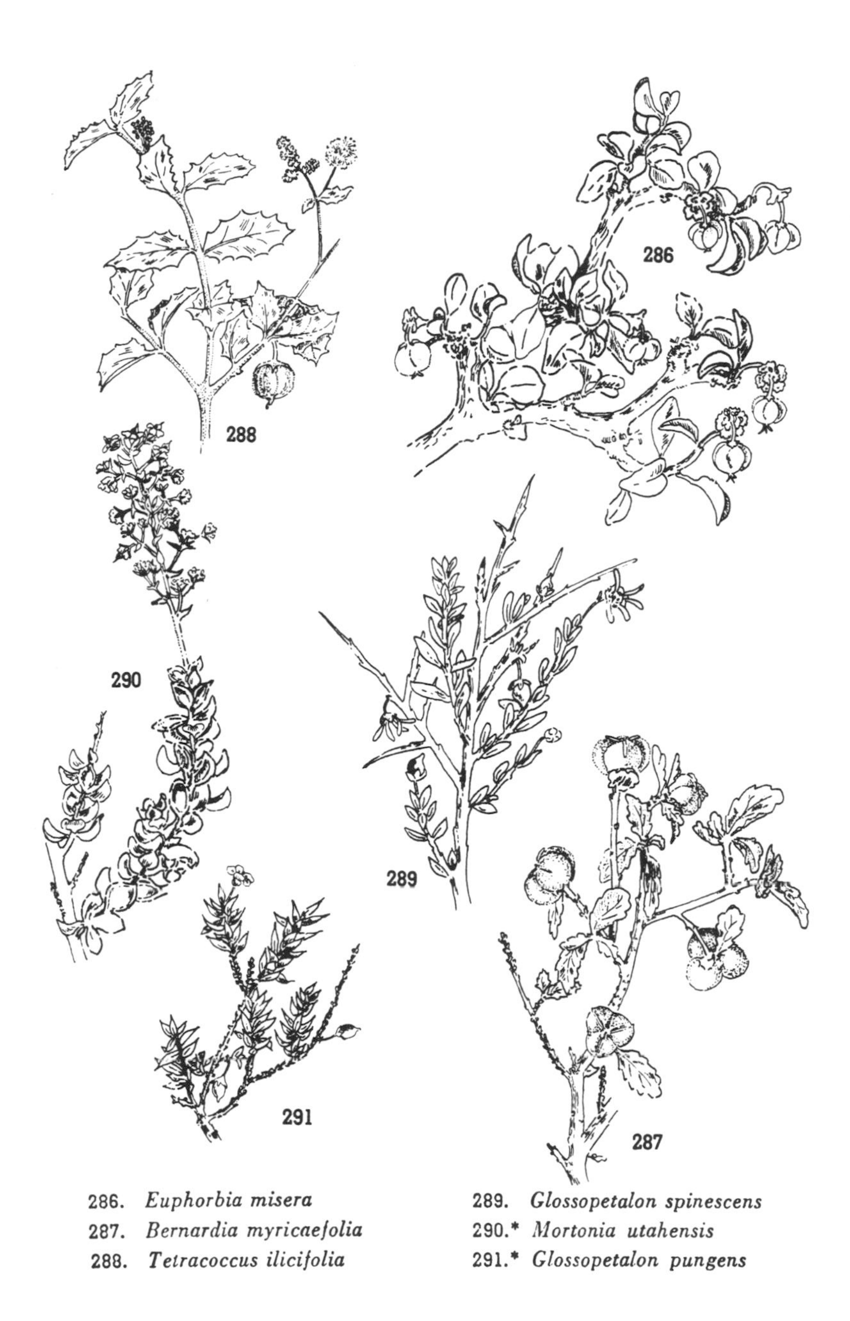

286. *Euphorbia misera*
287. *Bernardia myricaefolia*
288. *Tetracoccus ilicifolia*
289. *Glossopetalon spinescens*
290.* *Mortonia utahensis*
291.* *Glossopetalon pungens*

287. MOUSE-EYE. *Bernardia myricaefolia* (P. F. Bernard, 1749–1825, early French botanist; L., with *Myrica*-like leaves). **Fl.**: greenish. Stiff shrub, usually about 3 ft. high. The leaves are somewhat grayish, owing to a covering of fine, stellate hairs. Rocky canyons and slopes of the piñon-juniper belt of the arid ranges bordering the Salton Sink; to Tex., Baja Calif.

288. HOLLY-LEAF SPURGE. *Tetracoccus ilicifolius* (Gr., four-seeded; L., holly-leaved). **Fl.**: male flower sulphur-yellow, female flower dull reddish-green. A small, gray-barked shrub, recently discovered in mountains of the Death Valley Nat. Mon. The twigs of the season are purplish and the leaves leaden-green. The fruit is yellowish, being covered with a felt of small golden hairs. So rare is this shrub that only about fifteen specimens are known. The plants are 1–4 ft. high and grow in rock crevices. They have a short, enlarged basal stem, and from this the gray branches arise to form a low, squat bush sometimes as much as 8 ft. in diameter. "The existence of this new species in a single restricted locality in one of the severest of our deserts, the fewness of the individual plants, and the scarcity of fruit are evidence that the plants are in process of extinction through a still further increase in the aridity of Death Valley" (Coville).

CELASTRACEAE. Staff-Tree Family

289. SPINY-STEMMED TONGUE-FLOWER. *Glossopetalon spinescens* (Gr., "tongue petal," the petals being tongue-shaped; L., spiny). **Fl.**: white. Much-branched shrub, 1–2 ft. high. Eastern Mohave D.; north to eastern Ore. and east to Utah and Tex.

290. MORTONIA. *Mortonia utahensis* (Dr. Samuel G. Morton, 1799–1851, devotee of many branches of natural history, Professor of Anatomy at the Pennsylvania Medical College, and one-time President of the Philadelphia Academy of Natural Sciences. His world-famous collection of some 20,000 human crania is now housed in the academy museum. Of Utah). **Fl.**: white. Low, intricately branched shrub, with thick, leathery leaves. First discovered in the deserts of eastern Calif. by the author in 1934; to Utah and Ariz. Sometimes confused in the field with *Menodora spinescens.*

291. SPINY-TIPPED TONGUE-FLOWER. *Glossopetalon pungens* (L., prickly). Intricately branched, deciduous, weak-stemmed shrub. The twigs are brownish or olive, and the leaves have spiny tips. Rocky gulches of mountains of the piñon belt of the eastern Mohave D.; to the Sheep Mts. of Nev.

RUTACEAE. Rue Family

292. THAMNOSMA. *Thamnosma montana* (Gr., odorous shrub; L., montane). **Fl.**: purple. This is a low, aromatic, bushy shrub, of yellowish-green color. Except in the growing season its numerous gland-dotted stems are quite barren of leaves. In early April when in full flower and fruit it is a really handsome plant. It bears numerous, double, sac-like fruits, which finally turn bright yellowish-green, making them very conspicuous. These are about the size of peas and have gland-dotted skins like an orange; but at this we need not be surprised, for *Thamnosma* belongs to the same family that includes the important citrus fruits. The numerous blister-like glands yield an oil which when rubbed on the skin is a very powerful irritant. The crushed stems, at first giving off a rank odor, later yield a most pleasant, cocoanut-like scent. The Panamint Indians rubbed the crushed stems into open wounds and claimed this practice promoted rapid healing. Sometimes medicine men drank a tea made from it. "They soon went crazy like coyotes," said an old squaw, "but when they were that way they could find things long lost." Mohave and Colorado deserts, to Death Valley; Utah, N.Mex., Sonora, and Baja Calif.

ZYGOPHYLLACEAE. Caltrop Family

293. SMOOTH-STEMMED FAGONIA. *Fagonia chilensis laevis* (G. C. Fagon, seventeenth-century French botanist; L., of Chile; L., smooth). **Fl.**: purple to crimson. A low, spiny, slightly shrubby plant, with many intricately branching yellowish-green, angular, almost smooth stems; ¼–2 ft. high. On bare rocky soils of hottest hills and basins of the Colorado D.; to Baja Calif. The var. *glutinosa* has larger flowers, smooth, lacquer-bright leaves, and densely glandular stems.

294. CREOSOTE BUSH, COVILLEA. *Larrea divaricata* (John Anthony de Larrea, Spanish promoter of science; L., diverging). **Fl.**: yellow. Creosote bush is the most widespread, successful, and conspicuous xerophyte (dry-land plant) of the arid regions of North America. Its dark green leaves and blackish stems stand out so markedly against the background of light desert soils that it is easy to distinguish it from all other desert plants. It is a most important zone plant, for with its aid we can trace, even from a distance, the boundaries between the upper and lower divisions of the Lower Sonoran Life Zone. Over wide areas in the Mohave Desert it occurs in such pure stands as to constitute true larrea plains. In the bottom of Panamint Valley, where one reaches what is very near the absence of plant life, scarcely any other shrub is to be seen, and here it is so stunted that it is no more than 2 ft. high. The plant obtains its best development in the deep, pervious soils of those areas protected from wind, and there may reach a height of 10 ft. Its most common floral associate is the grayish burro-bush (*Franseria dumosa*) which generally occupies the open spaces between the seemingly methodically placed creosote bushes. If winter rains come, the creosote bush bursts into full flower in April and May, and soon after develops an abundant crop of fuzzy white "seed-balls," almost as spectacular as the flowers themselves. During late summer the leaves turn brownish, most of the fruits drop, and the plant remains in a condition known as drought dormancy. It is the strong, penetrating, resinous odor suggestive of creosote that gives *Larrea* its common name. Mexicans sometimes call it *hediondilla* ("the little bad smeller"). Col. Frémont spoke of it as "a rather graceful plant, its leaves exhaling a singular but very agreeable odor." This resiny, sweet scent is particularly noticeable when the desert is wetted by rains. Beans baked in molasses and in the presence of creosote-bush smoke take up the pleasant tang and are most delicious. In Mexico, where the plant is widely distributed, it is considered a medicinal shrub of high value and with many uses. A decoction of its leaves is employed for rheumatism and as an antiseptic for wounds and burns. Internally it is taken as a cure for tuberculosis and gastric complaints. The exudate of the lac scale, *Tachardiella larreae*, which forms resin-brown incrustations upon the stems, is used by the Indians in mending pottery and waterproofing their baskets.

The creosote gall midge, *Asphondylia auripila*, causes those conspicuous round, leafy balls, the size of small walnuts, which we so often see on the stems. A remarkable cigar-shaped case of silk, about 1¼ in. long, ornamented with fragments of the leaves, is built and hung in the branches by a moth larva known as the creosote bagworm (*Thyridopteryx meadii*). Strangely, the female insects never leave their cases, but lay their eggs and die there. After the eggs hatch, the larvae crawl out of the sac and immediately begin making their own tiny cases. "Distribution," says Dr. John A. Comstock, "must occur by means of wind transport as it seems unconceivable that the minute larvae can crawl over the desert from bush to bush carrying their cases with them. The larva never leaves its case. This is simply added to in length as it grows." Maturity is reached in May and June.

Successive lines of creosote and saltbushes mark the upper strands of many of the lakes which in bygone days occupied the floors of desert basins.

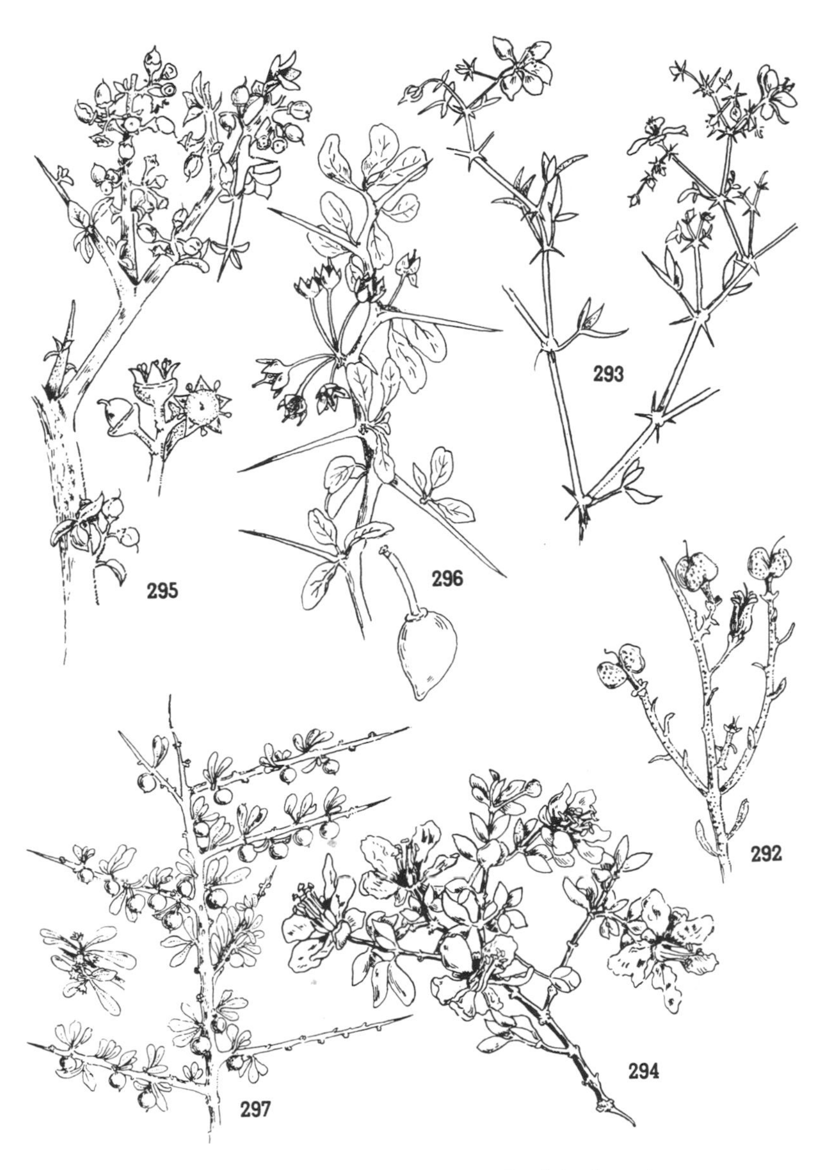

292. *Thamnosma montana*
293. *Fagonia chilensis laevis*, ×1
294. *Larrea divaricata*
295. *Condalia lycioides canescens*
296. *Condalia Parryi*
297. *Condalia spathulata*

RHAMNACEAE. Buckthorn Family

295. GRAY-LEAVED ABROJO. *Condalia lycioides canescens* (A. Condal, Spanish physician; L., like *Lycium;* L., grayish). **Fl.**: yellowish-green. Large, scraggly, grayish shrub, with many prominent, somewhat wand-like stems, the few ultimate branches of which are studded with a multitude of short, spiny branchlets, the whole somewhat resembling the green-stemmed crucifixion thorn (266). Common to the hot, sandy washes and narrow gullies of mountains between the Salton Sea and the Colorado R.; to Ariz. The dark blue, thin-skinned fruits contain hard, stony seeds.

296. PARRY ABROJO. *Condalia Parryi* (Dr. C. C. Parry—see **24**). A large, spiny shrub of the rough mountain slopes from Morongo Pass and the western edge of the Colorado D.; to Baja Calif. It was first collected in 1850 near San Felipe by Dr. C. C. Parry. The desert people call it "wild plum" and prize it much for fuel. The fruits ripen in mid-August, turning golden yellow. They then present a fine sight against the background of yellow-green leaves and black, thorny twigs. The fruit pulp is very bitter, but the Cahuilla Indians pounded it into a coarse meal.

297. SPINY ABROJO. *Condalia spathulata* (L., "spatulate," in reference to the leaf form). **Fl.**: greenish. Densely branched shrub, often tree-like in form, and 1–2 yds. high. The fruit is black or purplish and, though bitter, was formerly eaten by the Indians. When in full leaf this is a very handsome shrub. The bark of the twiggy, spinose stems is often made black by a smut fungus. It is quite abundant along the shallow, narrow washes of the Chuckawalla Bench and the eastern slope of the Chocolate Mts. in eastern Imperial Co., Calif.; to Tex. and Mex.

298. EVERGREEN BUCKTHORN. *Rhamnus crocea ilicifolia* (from *Rhamnos,* the old Gr. name used by Theophrastus for a species of crocus, from the stigmas of which a deep yellow dye is extracted; L., saffron; L., holly-leaved). **Fl.**: greenish. A shrub, 3–12 ft. high, with small, rather inconspicuous flowers. Occasional in the juniper-piñon belt of the Little San Bernardino Mts.; common in the Sierra Nevada foothills and coastal southern Calif.

299. MOHAVE BUCKBRUSH. *Ceanothus vestitus* (Gr., name applied by Theophrastus to a prickly plant; L., clothed, adorned). **Fl.**: white. A low, intricately branched shrub with grayish-green leaves. Mohave D., from 3,500 to 7,000 ft. altitude. Replaced in the Little San Bernardino Mts. by *C. perplexans,* a plant in many ways similar but with leaves yellowish-green.

300. CALIFORNIA SNAKE-BUSH. *Colubrina californica* (L., "like a serpent," in allusion to the twisted filaments of the stamens; of California). **Fl.**: yellowish. This is an intricately branched, dull-colored, somewhat spiny shrub, 3–6 ft. tall. Locally abundant in many of the sand washes and steep gullies of the rocky slopes on the north face of the Chuckawalla Mts. The stems are covered with dense, matted, wool-like hairs. The inconspicuous flowers, which occur in axillary clusters, give rise to dark purple fruits which often persist long after the leaves have fallen in late summer. First discovered in California by the author in 1927 in a rock-sheltered wash of the Eagle Mts. Southeastern Colorado D.; to Ariz. and Baja Calif. The type locality is Las Animas Bay, Baja Calif.

ANACARDIACEAE. Sumac Family

301. DESERT SQUAW-BUSH. *Rhus trilobata anisophylla* (Anc. Gr. name for sumac; L., three-lobed; Gr., "unequal-leaved," in reference to the lateral leaflets). **Fl.**: yellowish. A small, slender-stemmed shrub usually growing in clumps. The wand-like, somewhat arching stems were utilized in basketry by the aborigines. Mountain ranges of the Mohave and Colorado deserts.

BUXACEAE. Box Family

302. GOAT-NUT, JOJOBA. *Simmondsia californica* (T. W. Simmonds; of California). Goat-nut is a leathery-leaved shrub, 2–3 ft. high. It was first described from San Diego by Thos. Nuttall, the famous conchologist, botanist, and ornithologist. He named the genus in honor of Thos. William Simmonds, "ardent Botanist and Naturalist who accompanied Lord Seaforth to Barbadoes about the year 1804, and died soon after while engaged in exploring the island of Trinidad." The thickly set, gray-green leaves are cropped by browsing animals; and the oily nuts, which taste much like filberts, were long an important article of food among Indians and Mexicans. The seeds were eaten fresh or, when dried, ground, and roasted, were used in the making of a beverage. A sort of confection was made by grinding the seeds and combining them with sugar to form tablets. The little white-tailed antelope ground squirrels, nicknamed "ammos," store the fresh seeds and thus aid in dissemination. The male and female flowers are borne separately on the same plant, or on different plants.

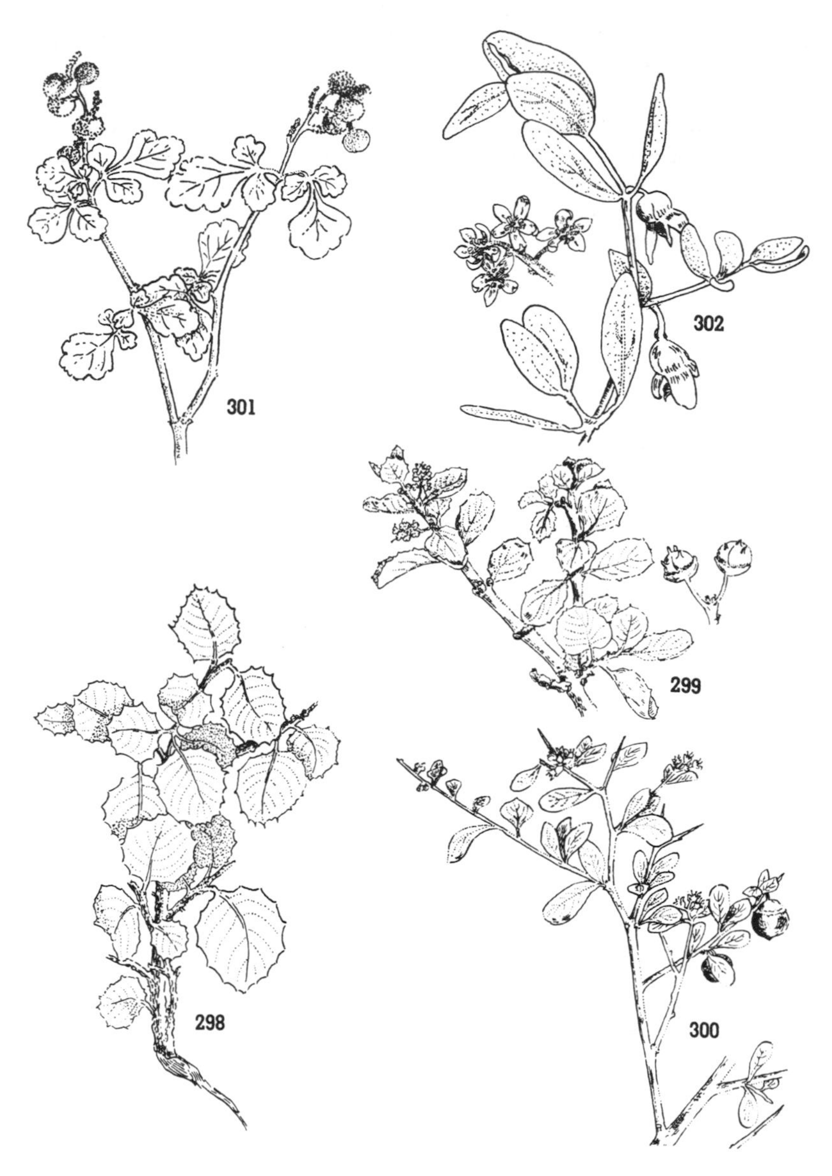

298. *Rhamnus crocea ilicifolia*
299. *Ceanothus vestitus*
300. *Colubrina californica*
301. *Rhus trilobata anisophylla*
302. *Simmondsia californica*

303. *Bursera microphylla*
304.* *Sphaeralcea pulchella*
305. *Sphaeralcea rosacea*
306. *Malvastrum rotundifolium*

BURSERACEAE. Torchwood Family

303. ELEPHANT TREE, TOROTE, DYE BARK. *Bursera microphylla* (Joachim Burser, 1593–1689, German botanist; Gr., small-leaved). A spicy-odored shrub or small tree, 4–10 ft. high, common on the dry hillsides and plains of Sonora, Baja Calif., and southern Ariz. A few elephant-tree colonies are known in California from the Vallecito Mountains of the Colorado Desert. The stems yield a resin which in Mexico is used as a cement or for the making of varnish. The bark exudes a red sap which is used in dyeing. The leaves are deciduous; the fruit is a bluish, one-seeded berry. This shrub should not be confused with that remarkable dropsical tree of the genus *Veatchia,* also called elephant tree, of the deserts of Baja Calif.

MALVACEAE. Mallow Family

304. MOHAVE DESERT MALLOW. *Sphaeralcea pulchella* (Gr., "globe," and *Alcea,* the marsh mallow, referring to the round-headed fruit; L., diminutive of beautiful, i.e., only fairly beautiful or perhaps "a little beauty"). **Fl.**: deep apricot. A particularly handsome species from the rocky slopes and canyons of the ranges paralleling the Death Valley trough. The dark brown stems and deep green leaves are covered with white hairs.

305. ROSE MALLOW. *Sphaeralcea rosacea* (L., rose-colored). **Fl.**: pink, drying rose-violet. Somewhat shrubby perennial of the rocky canyons, 1,000–3,500 ft., of the western Colorado D.; to Baja Calif., Ariz. Both stems (2–3 ft. long) and leaves are ashen because of a felt of short, branching hairs.

306. DESERT FIVE-SPOT, LANTERN FLOWER, CHINESE LANTERN. *Malvastrum rotundifolium* (a variant of Malva; L., round-leaved). **Fl.**: rose-purple with spots of carmine. This elegant species makes its best showing on black lava buttes, where it forms strong contrasts with the somber rocks. The globular corolla, with its fine red spots on the inner bases of the petals, suggested its local names of "lantern flower" and "five-spot." Confined to the Lower Sonoran Life Zone: both Calif. deserts; to Ariz., Nev., Baja Calif., and northwestern Sonora.

307. DESERT MALLOW. *Sphaeralcea ambigua* (Gr., round mallow; L., uncertain—as to affinities). **Fl.**: grenadine to peach-red. A somewhat shrubby perennial, so handsome when in flower that many desert folk think it the climax of floral beauty. A very variable species as to both herbage and inflorescence. Common to the Mohave and western Colorado deserts; to Ariz., Nev., and northern Baja Calif.

The sphaeralceas are locally known in Arizona as "sore-eye poppies" because the hairs on the plants are irritating to the eyes. In Baja Calif. the people for the same reason call them "plantas muy malas" (very bad plants).

308. FENDLER GLOBE MALLOW. *Sphaeralcea Fendleri californica* (August Fendler—see **538**; of California). **Fl.**: orange-scarlet. Erect perennial, 1–2½ ft. high, generally found in aggregations on clays and about sinks where water is ephemerally ponded. Colorado D.; to N.Mex.

309. WHITE MALLOW. *Malvastrum exile* (Gr., star mallow; L., meager, slender). **Fl.**: white or pinkish. Low, spreading or erect, green-leaved, winter annual, with stems 4–16 in. long. Sand and clay loams of desert flats. One of the least handsome but most common desert mallows. Both Calif. deserts, up to 5,000 ft.; to Ariz.

310. ALKALI PINK. *Sidalcea neomexicana parviflora* (*Sida* and *Alcea,* two malvaceous genera; of New Mexico; L., small-leaved). **Fl.**: purple to rose-pink. A several-stemmed perennial, ½–3 ft. tall, found about alkaline, marshy soils of the flat, western arm of the Mohave D. and across the Santa Ana Basin to Santa Monica.

311. ROCK HIBISCUS. *Hibiscus denudatus* (name used by Dioscorides for the marsh mallow; L., stripped). **Fl.**: white or pinkish, with red or purplish claw. Handsome tufted, openly branched perennial 1–2 ft. high, found mostly at low altitudes among rocks on slopes of mountains of the Colorado D.; east into Ariz., south into Sonora and Baja Calif. Both stems and leaves are yellow, with a dense covering of branching hairs.

312. SMALL-LEAVED ABUTILON. *Abutilon parvulum* (Arabic name for a plant analogous to the marsh mallow; L., very small). **Fl.**: brick red to pinkish. Tufted perennial, 6–18 in. tall, known from the dry, rocky slopes and gorges of the Providence Mts.; eastward to Colo., Tex., and Mex.

313. YELLOW FELT-PLANT. *Horsfordia Newberryi* (F. H. Horsford, New England botanical collector; Dr. J. S. Newberry, 1822–1892, geologist, paleontologist; naturalist on the Williamson Survey. He also collected under Lieutenant Ives along the Colorado River in 1857–58 as far north as the Grand Canyon). **Fl.**: yellow. A somewhat woody-stemmed perennial, with erect, wand-like stems 1–2 yds. high. The plants are covered with a dense, yellow felt of stellate hairs. Among rocks bordering washes and in lower canyons along the western side of the Colorado D.; to Baja Calif., Ariz., and Sonora.

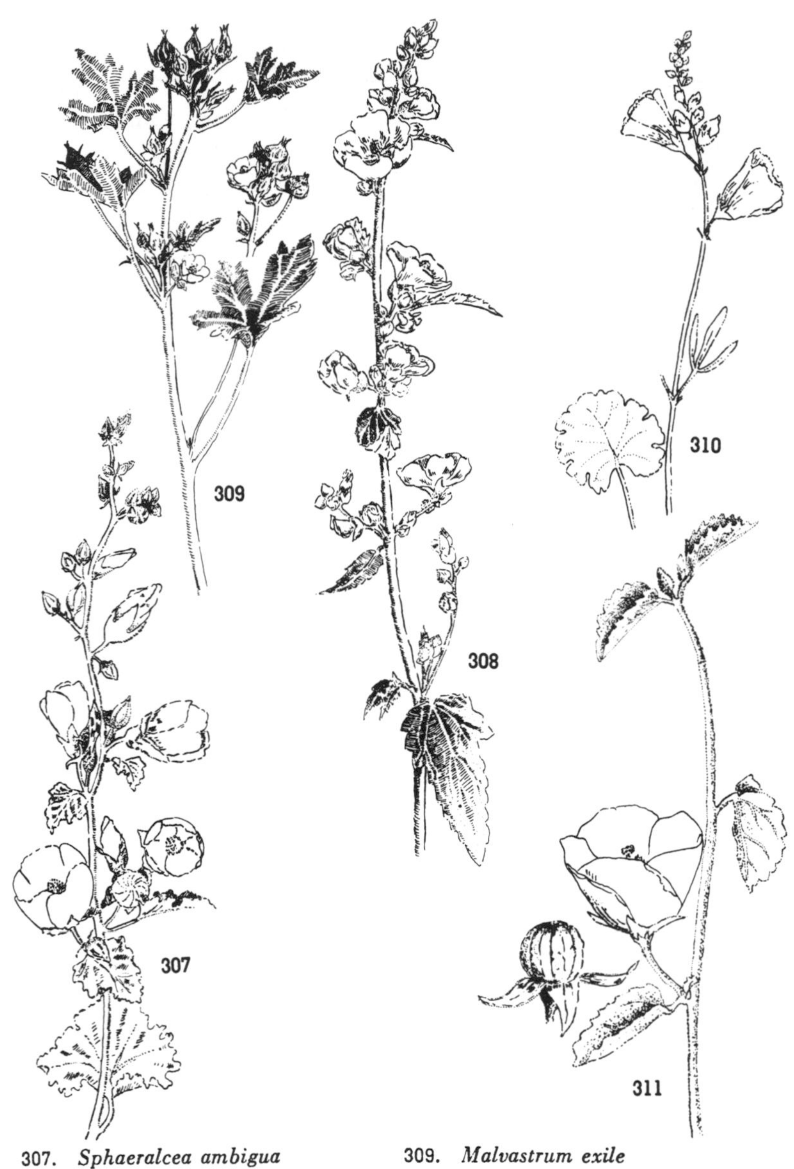

307. *Sphaeralcea ambigua* 309. *Malvastrum exile*
308. *Sphaeralcea Fendleri californica* 310. *Sidalcea neomexicana parviflora*
311. *Hibiscus denudatus*

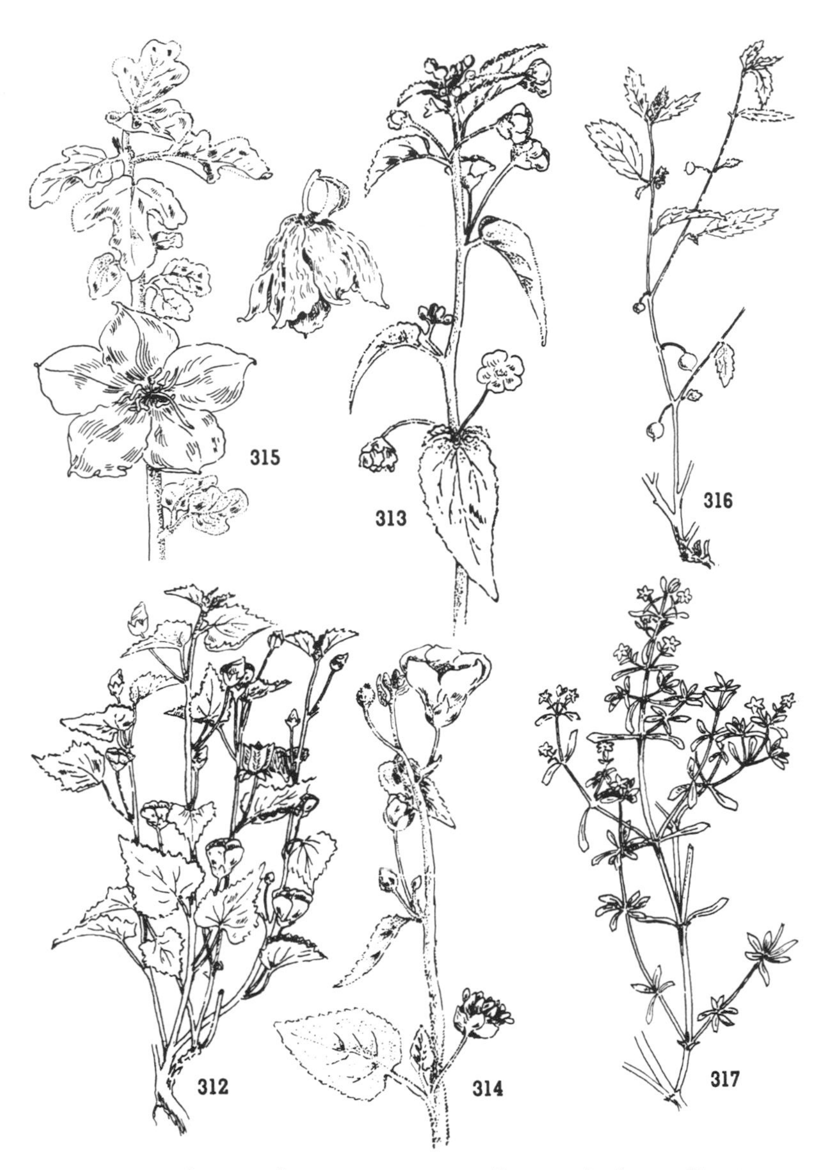

312.* *Abutilon parvulum*
313. *Horsfordia Newberryi*
314. *Horsfordia alata*
315. *Fremontodendron californica*
316. *Ayenia compacta*
317.* *Frankenia grandifolia campestris*

314. PINK FELT-PLANT. *Horsfordia alata* (L., winged, in reference to the long upper wings of the ovule). **Fl.**: pink. An erect, sparingly branched shrub, 1–3 yds. tall, with thickish, felty leaves. Often found growing up among shrubs in several canyons of the western side of the Colorado D.; it is probably a migrant from Baja Calif. or Sonora.

STERCULIACEAE. Cacao Family

315. CALIFORNIA SLIPPERY ELM. *Fremontodendron californicum* (John C. Frémont, early Western explorer and once presidential candidate; of California). **Fl.**: lemon-yellow. Called "flannel-bush" because of the felt-like surface of the under sides of the small, fig-like leaves. "A beautiful shrub usually 3–4 ft. high, but occasionally reaching a height of 10 feet and having very much the appearance of a fig tree." Such was Dr. Torrey's original description of this hardy plant of the lower arid slopes of the mountains along the western rim of the Mohave Desert. The specimens brought to Dr. Torrey were among the plants gathered by Frémont on his second expedition to California in 1843–44. "In this journey," says Dr. Torrey, "he made large collections in places never before visited by a botanist; but unfortunately a great portion of them was lost. In the gorges of the Sierra Nevada, a mule loaded with some bales of botanical specimens gathered in a thousand miles of travel, fell from a precipice into a deep chasm, from whence they could not be recovered. A large part of the remaining collection was destroyed on return of the Expedition by the great flood of the Kansas River. Some of the new interesting plants that were rescued from destruction were published in the Botanical Appendix to Col. Fremont's Report of the Second Expedition." The flowering season of the California slippery elm is May. Later the leaves turn leathery brown. This fine shrub is being seen increasingly in cultivation both in this country and in Europe. From the outer bark, rope was sometimes made by the Indians. (The illustration on the right is of a fruit with its covering of persistent calyx lobes.)

316. DESERT AYENIA. *Ayenia compacta* (Duc d'Ayen, 1739–1824; L., joined, i.e., of compact habit). **Fl.**: brownish. A rather inconspicuous, few-stemmed shrub, with small, 5-petaled, long-pediceled flowers and globose fruits. Among rocks on mountains of western side of Colorado D. Duc d'Ayen, after whom the genus was named by Peter Loefling, the Swedish traveler, was a Frenchman who forsook the profession of a soldier in order to devote himself to scientific studies, particularly in the field of chemistry. In 1777 he was elected to the French Academy of Sciences. His name as a citizen was Jean Paul Francois.

FRANKENIACEAE. Frankenia Family

317. ALKALI-FLAT FRANKENIA. *Frankenia grandifolia campestris* (John Frankenius, 1590–1661, Professor of Botany at Upsala, who first made a check-list of Swedish plants in his *Speculum Botanicon;* L., large-flowered; L., of the fields). **Fl.**: pinkish-purple. Low, herbaceous or slightly woody plant, of alkali flats such as are found at Saratoga Springs and in Panamint Basin. This *Frankenia* possesses little comeliness, but in Europe some of the other species are considered suitable for rock gardens and borders.

FOUQUIERIACEAE. Ocotillo Family

318. OCOTILLO, CANDLEWOOD (see p. 148). *Fouquieria splendens* (P. E. Fouquier, French physician; L., gleaming, vividly bright). **Fl.**: scarlet, "as brilliant as if dipped in vivid paint." Occasionally one sees a plant with white flowers. The ocotillo (Spanish diminutive for the Aztec term, *ocotle,* pine; Mex., *ocote* + Sp. *illa,* little) is one of the oddest shrubs of the deserts of southwestern United States and northern Mexico. Be it noted at the outset that ocotillo is not in any sense a cactus, despite its thorny armature. In California it is confined to rocky soils and arid hillsides of the Colorado and far eastern Mohave deserts. Much of the year the tall, radiating stems are mere "spiny canes," but whenever rains really wet the soil, bright green leaves spring forth above the sharp, stout spines and hide the thorny stems in foliage. Flowers soon tip the branches to make a striking show. The ocotillo loses its leaves very quickly when drought ensues. Since every summer rain produces a new set, the number of crops of leaves may be several a year. The roots are shallow, wide-spreading, and corky-sheathed, as in cacti and many other woody desert plants, enabling quick utilization of shallow-penetrating rains. The bark contains resin, gum, and wax, and burns with a fierce flame, giving off intensely black smoke. The Cahuilla Indians eat both flowers and seeds and make an agreeable beverage by soaking the blossoms in water. This shrub impresses one as a plant of extraordinary vigor. Seldom does one ever find a plant with even a single dead "cane," though there may be as many as a hundred or more in the bunch. About the only dead ocotillos one may see are those which have been prostrated by storm winds.

318. OCOTILLO is a wide-spread and characteristic shrub of the hot southern desert. It comes into flower in April and May and is rated by many as the desert's finest flowering shrub.

Photo by Avery Edwin Field

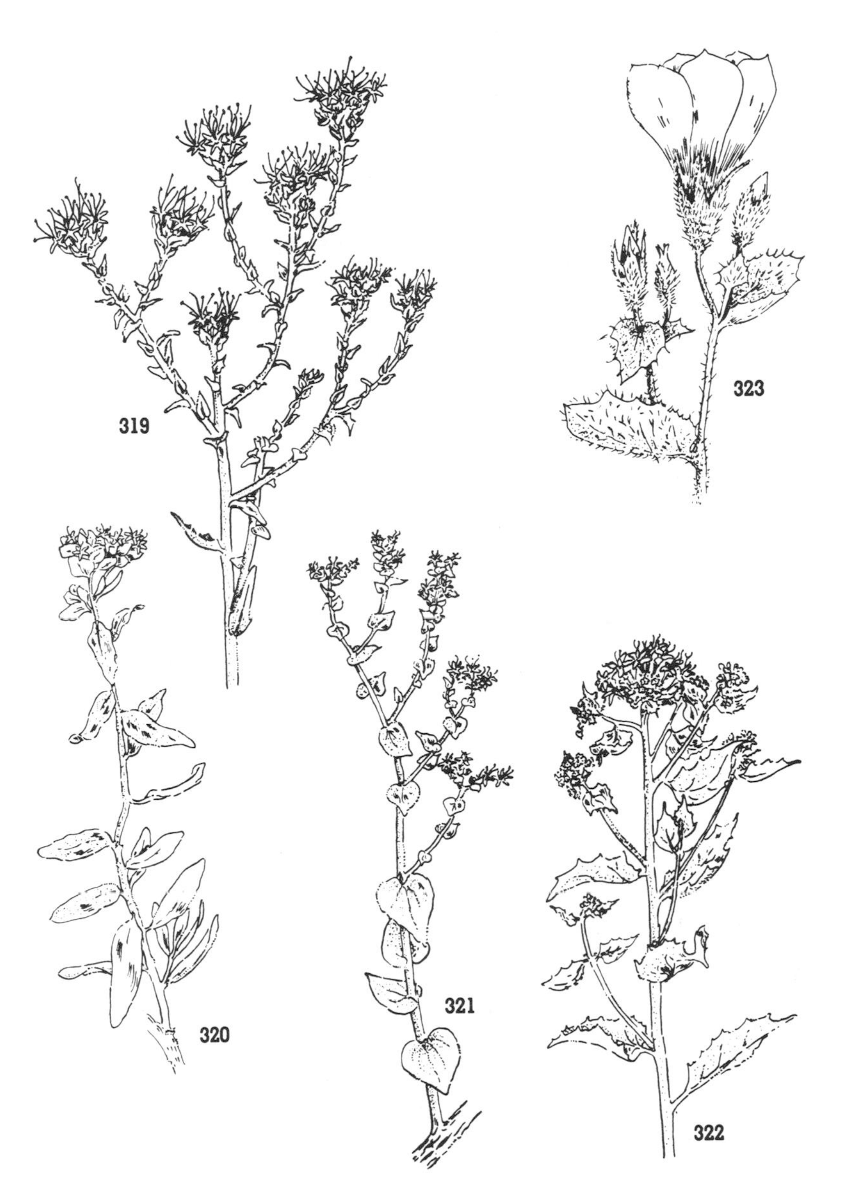

319. *Petalonyx Thurberi*
320. *Petalonyx linearis*
321.* *Petalonyx Gilmanii*
322. *Petalonyx nitidus*
323. *Eucnide urens*

LOASACEAE. Loasa Family

319. THURBER SANDPAPER - PLANT. *Petalonyx Thurberi.* (Gr., leaf + claw; Geo. Thurber, 1821–1890, "most accomplished horticulturist in America"; botanist of the Mexican Boundary Survey, 1850–1854. Nine plant species and one genus were named by Asa Gray to commemorate Thurber's extensive work on this journey. The new plants were described in *Plantae Novae Thurberiana.* It was Thurber who first collected adequate specimens of the desert smoke tree. Much of Thurber's collecting was done under great hardship, particularly that in western Sonora.) **Fl.**: white. A fragrant-flowered, somewhat woody-based perennial, blooming profusely in midsummer. The green stems and leaves have a covering of short, rigid, retrorsely appressed spines, which gives them the texture of sandpaper, and it is to this character that the plant's common name refers. Common on both Calif. deserts; to Nev. and Ariz.

320. LONG-LEAVED SANDPAPER-PLANT. *Petalonyx linearis* (L., linear [leaved]). **Fl.**: whitish. A small, rounded bush, 4–12 inches high. Occasional in rocky canyons of the Salton Sink south to Baja Calif. and Ariz.

321. GILMAN SANDPAPER-PLANT. *Petalonyx Gilmanii* (M. French Gilman—see **79**). **Fl.**: white. A low, twiggy shrub found in Ryan Wash near Death Valley. It is distinguished by heart-shaped leaves and the felty covering on its young stems. The bark of the lower, woody stems is often fissured.

322. SHINY-LEAVED SANDPAPER-PLANT. *Petalonyx nitidus* (L., shining). **Fl.**: whitish. A low, leafy shrub distinguished from *P. Thurberi* by its large, petiolate, usually shining leaves and papery floral bracts. Rare, at altitudes between 3,500 and 4,000 ft., western Mohave D., as in Cushenbury Canyon, to Inyo Co.; and Nev.

323. STING BUSH. *Eucnide urens* (Gr., true sea-nettle; L., "burning," "stinging," "the setae irritating the skin like nettles"). **Fl.**: pale yellow. This low, rounded bush, 1–2 ft. high, lodges its roots in crevices of rocks in the hot, arid desert ranges of eastern Calif., particularly in the Death Valley area; to southern Nev. and Utah. A number of instances are known where bats, emerging from cave entrances where plants were growing, have impaled themselves upon the barbed, stinging hairs.

324. VENUS BLAZING STAR. *Mentzelia nitens* (Christian Mentzel, 1622–1701, German botanist and philologist; L., "shining," with reference to the very white stems). **Fl.**: bright yellow. Annual or perennial, with loosely spreading, very white stems. The tuberculate seeds are sharply angled. Both deserts; to western Ariz.

325. INYO BLAZING STAR. *Mentzelia oreophila* (Gr., mountain-loving). **Fl.**: yellow. Branched perennial, with light green leaves. The calyx lobes become reflexed on the ripening fruit. The punctate, light brown seeds are broadly winged. Mountainous areas of Inyo Co.

326. YELLOW COMET. *Mentzelia affinis* (L., near, as to relationship). **Fl.**: bronzy-yellow. Annual, 8 in. to 2 ft. high, with stoutish, white, shining stems and numerous flowers. Both Calif. deserts below 3,000 ft., also interior southern Calif.; to southwestern Ariz.

327. PYGMY BLAZING STAR. *Mentzelia reflexa* (L., "turned back," because the seed capsules are often pendulous or reflexed). **Fl.**: silky, straw-colored, tinged with bronze. Annual, 5–8 in. high, with stout stems from a white, radish-like taproot. In barren soils strongly impregnated with salts. Northern Mohave D., principally of the Death Valley area, south to Barstow.

328. SPINY-HAIRED BLAZING STAR. *Mentzelia tricuspis* (L., "three-pointed," probably referring to the apex of the outer stamens). **Fl.**: straw-white. Coarse-stemmed, branching annual, 3–7 in. high. The narrowed floral bracts are green and leaf-like. The stamens, arranged in four series, number between 80 and 100. Stony slopes. Both deserts to eastern Nev.

329. ROUGH-STEMMED BLAZING STAR. *Mentzelia puberula* (L., minutely hairy). **Fl.**: yellow. Perennial, 4 in. to 1 ft. high. The slightly punctate seeds are light brown and broadly winged. Colorado D.; to Ariz.

330. PANAMINT BLAZING STAR. *Mentzelia longiloba* (L., "long-lobed," with reference to the calyx lobes, which are 9–10 mm. long, longer than in either *M. puberula* or *M. oreophila*). **Fl.**: light yellow. The flowers are subtended by linear bracts, as in *M. puberula.* Panamint Mts. to northern Colorado D.

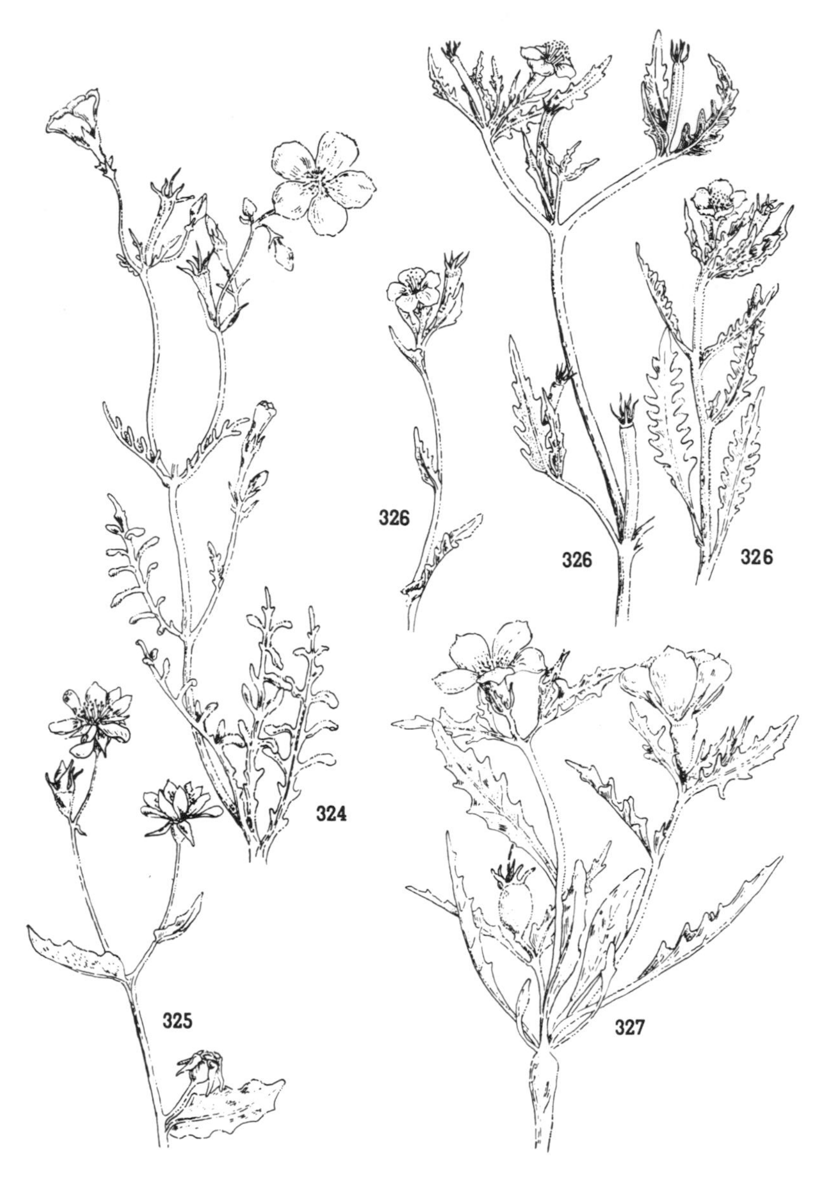

324. *Mentzelia nitens*
325.* *Mentzelia oreophila*
326. *Mentzelia affinis*
327. *Mentzelia reflexa*

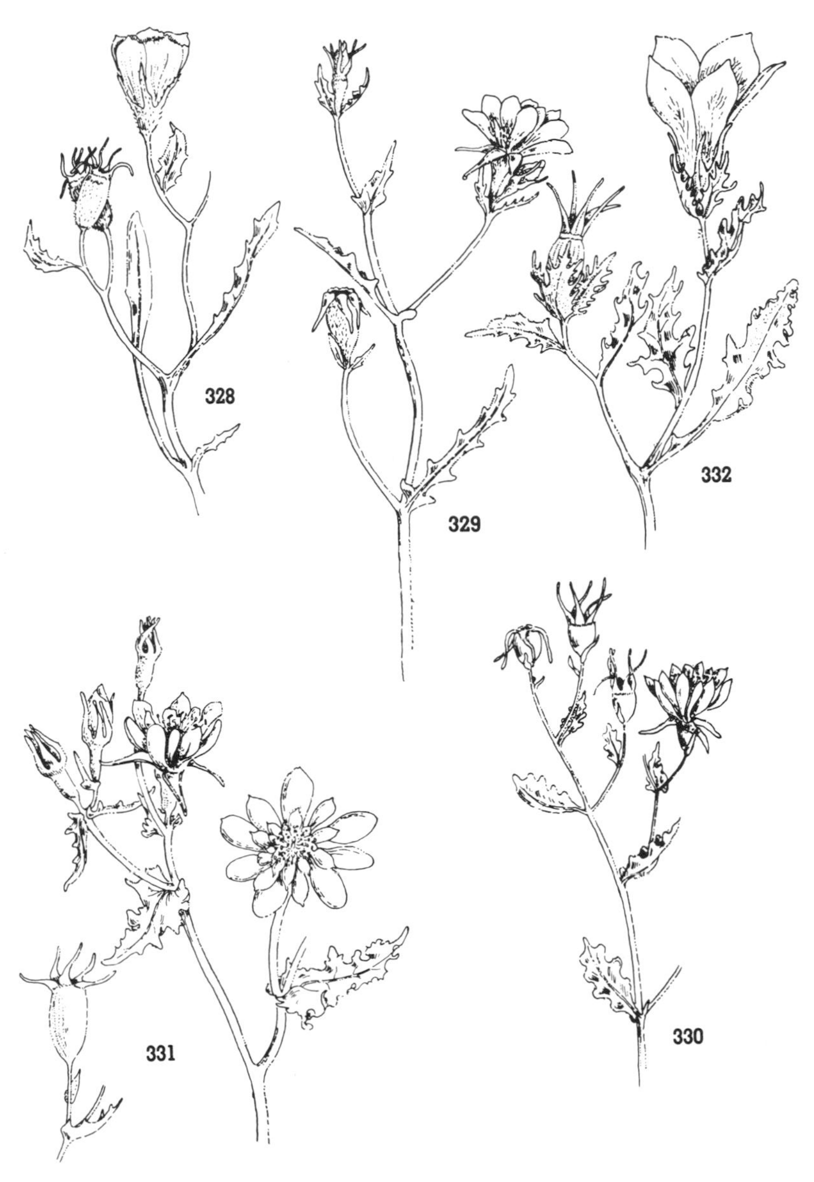

328. *Mentzelia tricuspis*
329. *Mentzelia puberula*
330. *Mentzelia longiloba*
331. *Mentzelia multiflora*
332. *Mentzelia involucrata*

331. ADONIS BLAZING STAR. *Mentzelia multiflora* (L., many-flowered). **Fl.**: light yellow. Perennial, 15 in. to a yard high, from a stout taproot. Leaves 1–3 in. long. The flattish, white seeds are bordered with a broad ring. A plant of much character and worthy of a place in gardens. The flowers open in late afternoon. Sandy areas of Colorado and eastern Mohave deserts; Wyo. south to Tex. and Mex.

332. SAND BLAZING STAR. *Mentzelia involucrata* (L., with an involucre). **Fl.**: pale cream, with crimson-tinged center. An annual, branching from the base, 4–16 in. high; the herbage covered with stiff hairs. Low, hot washes and canyon sides. Colorado and eastern Mohave deserts; western Ariz. and Sonora. The variety *megalantha* (Gr., "large flower") "differs from the typical species in the larger, more conspicuous flowers of brighter yellow."

333. SMALL-FLOWERED BLAZING STAR. *Mentzelia albicaulis* (L., white-stemmed). **Fl.**: bright yellow. A very variable species that needs further study, especially by field students and geneticists. Seeds should be planted and several years of intensive experimental work should be done on the variants. Plants growing in the open are often quite different from those growing in the partial shade of shrubs. Annual, 4–16 in. high, branched at the base, with slender, white, shining stems. Petals 2–4 mm. long. Dry sandy or gravelly soils of both deserts; eastward to Neb.

334. APOLLO BLAZING STAR. *Mentzelia albicaulis heliophila* (Gr., sun-loving). **Fl.**: bright yellow. A compact, low-growing variety, affecting open sunny situations. Both Calif. deserts.

335. MOHAVE COMET. *Mentzelia albicaulis inaequiloba* (L., unequal-lobed). **Fl.**: bright yellow. A larger-flowered, coarser-stemmed plant than either of the preceding. Mohave D.

336. PAIUTE BLAZING STAR. *Mentzelia albicaulis gracilis* (L., slender, graceful). **Fl.**: yellow. The leaves are all pinnately cleft. Both deserts; to Ariz., Wash., and Wyo.

337. FLOR DE LA PIEDRA. *Sympetaleia rupestris* (Gr., joined petals; L., of the rocks). **Fl.**: yellow. A Mexican species which has been collected in canyons of the western Colorado D. near the boundary. The low plants are brittle-stemmed, and the leaves appear to be covered with a sort of thin varnish.

CACTACEAE. Cactus Family

Cacti are abundant only in places where water supplies are seasonally plentiful. Such conditions exist in the higher desert ranges and on many of the encircling, detrital fans whose washes carry the rapid run-off of summer cloudbursts and winter rains. Loose gravels or rock crevices where percolation is adequate are very necessary for prolific growth. This accounts for the peculiar distribution of cacti and their widespread absence over large areas in the bottoms of desert basins. Cacti are much more abundant in the Colorado than in the Mohave Desert.

The long, corky-barked roots, many of them yards in length, are purposely laid close to the surface of the soil where they can quickly take up the water of either shallow or penetrating rains. This water, which is stored in the fat stems, is given up very reluctantly even during the hottest days. The ingress of carbon dioxide is also checked at the same time, and this accounts for the slow growth of cacti.

338. BIGELOW CHOLLA (see p. 157). *Opuntia Bigelovii.* (Old Latin name used by Pliny, derived from the city of Opus; John M. Bigelow, M.D., 1804–1878, who collected in the West under Whipple in the Pacific Railroad Survey, 1853–1854, and for whom not only the *Opuntia* but many other California plants were named. In the Report of the Botany of the Expedition (Pac. R.R. Rept. Vol. IV, 1856) he is the author of the general description and also of the division, Forest Trees). **Fl.**: yellowish. The spiniest but most handsome of all our cacti; usually gregarious, often extensively so on detrital fans and benches. The yellowish, retrorsely barbed spines covering the short, turgid joints are very penetrating and often cause painful sores. They crowd the upper joints so thickly that they give the plant a brilliant yellowish-white color, noticeable even from a distance. The dark brown color of the lower part of the plant is due to the discolored spines of the lower, dead joints and of the trunk below. Curiously, pack rats (*Neotoma*) transport the prickly joints and use them to protect their runways and nests, while cactus wrens lodge their nests in the thickets of spiny branches. Because the seeds are usually sterile, propagation is principally effected by the detached, fleshy joints. The fruits are small and quite spineless. Colorado D.; to Nev., Ariz., Baja Calif., and Sonora.

333. *Mentzelia albicaulis*
334. *Mentzelia albicaulis heliophila*
335. *Mentzelia albicaulis inaequiloba*
336. *Mentzelia albicaulis gracilis*
337. *Sympetaleia rupestris*

338. BIGELOW CHOLLA occupies many of the alluvial fans of the southern deserts. With it is seen barrel cactus and Engelmann calico cactus (*right*). The shrub occupying the intervening spaces is brittle bush (*Encelia farmosa*).

Photo by Avery Edwin Field

339. THORNY-FRUITED CACTUS (see p. 161). *Opuntia echinocarpa* (Gr., "hedgehog-fruited," i.e., spiny-fruited). **Fl.**: yellow, tinged with red. A low shrub, sometimes sprawling but occasionally compact, with few ascending stems and many short, spreading branches. The numerous straw-colored spines, borne on the short tubercles, are covered with thin, papery sheaths, which, when wetted, taste like witch hazel. The fruit is dry and very spiny, especially the upper part. The woody, reticulate, tubular stem skeletons, appropriately called "ventilated wood" by desert campers, make most excellent fuel. Common throughout the southern deserts; to Utah, Ariz., and Baja Calif. The plants are solitary or approximate, but seldom form thickets.

340. DEER-HORN CACTUS (see pp. 160 and 161). *Opuntia acanthocarpa* (Gr., thorny-fruited). **Fl.**: burnished-yellow. A much-branched, long-stemmed cane-cactus, somewhat larger than thorny-fruited cactus. Useful marks of identification are: the prominent, much-elongated, laterally flattened tubercles, each bearing near the upper end 8–25 yellow spines, 1 in. or less in length; and the dry fruits armed with bunches of 8–12 long, stout spines. Occasional in the western Colorado D. but forming dense stands in the mountains of the eastern Mohave D., where it is closely associated with Mohave yucca and juniper. The reddish tinge of the flower is due to fine lines of scarlet against the yellow background. More abundant in southern Nev., Ariz.; to Utah, Sonora.

341. DARNING-NEEDLE CACTUS, DIAMOND CACTUS (see pp. 160 and 163). *Opuntia ramosissima* (L., very much branched). **Fl.**: greenish-yellow fading to salmon. A branching shrub, 2 to 5 ft. high, with very slender stems. The flowers are small and occur infrequently; they are seldom noticed, and many botanists confess they never have seen one. The handsome fruits are about 1 in. long and covered with tawny, stout spines and axillary wool. The very slender, woody joints are usually marked with diamond-, quadrilateral-, or heart-shaped figures, from the center of which spring the long, solitary, yellow-sheathed spines. The old stems of late summer are often practically spineless and reddish. Common in basins and valleys of both Calif. deserts; to Nev., Ariz.

342. MUNZ CHOLLA (see p. 113). *Opuntia Munzii* (Dr. Philip A. Munz, botanist, of Pomona College, specialist on the genus *Oenothera*). **Fl.**: yellowish-green, tinged with red. First recognized as a possible new species by the author in 1922. It was then brought to the attention of botanists but was not described until 1938. It is believed to be a hybrid between *O. Bigelovii* and *O. acanthocarpa.* This is a rapidly growing species, and next to *Cereus giganteus* is the tallest of California cacti. The plants when young are quite trim and symmetrical but in age grow 6–12 ft. high and become shaggy, giving a particularly wild appearance to the rocky canyons and plains where they grow. They are often intimately associated with Bigelow cholla, ironwood, and palo verde trees. Munz opuntia is known only from canyons intersecting the Chocolate Mts. from near Iris Pass south of the Chuckawalla Bench to the upper reaches of the Arroyo Seco in eastern Imperial Co.

343. DEAD CACTUS (see p. 160). *Opuntia Parishii* (S. B. Parish —see **603**). **Fl.**: greenish-yellow, fading to salmon. This is a ground-covering species with much-flattened, grayish-brown woody spines. Though much alive, the joints appear most of the year to be wholly dead. The rapidly growing young shoots present an array of brilliant red spines contrasting with the dull color of the older growth. The creeping stems, which root from the under side, form dense patches on sand and gravel flats in high areas of the Joshua Tree Nat. Mon. The small, dry, club-shaped fruits, which show a deep umbilicus at the summit, are very handsome, with their numerous radially arranged yellow glochids. The blossoms remain open only about two hours. Little San Bernardino Mts., Ivanpah Mts.; southern Nev.

344. BEAVERTAIL CACTUS (see p. 163). *Opuntia basilaris* (Gr., regal). **Fl.**: brilliant magenta or sometimes white. A low-spreading species, with flat joints, 4–12 in. long, occurring in small clumps. The bluish-green stems bear no spines, but in their places are cushions of very short, yellowish-brown or reddish spicules set in slight depressions of wrinkles. The young fruits, which are borne on the margins of the flattish stems, having been divested of their spicules by being rolled in sand, are cooked with meat by the Cahuilla Indians. The Panamint Indians dry the joints, later boil them with salt, and eat them. In its native habitat this species is subject to the attacks of two cottony cochineal scales (*Dactylopius tomentosus* and *D. confusus*), and the white, cotton-like wax made by the red insects sometimes almost covers the plant. Colorado and eastern Mohave deserts; to Ariz. and Sonora. A variety *brachyclada* (Gr., short shoot), with small, thick, reddish joints, is found at Cajon Pass and in western Mohave D. above 3,500 ft.

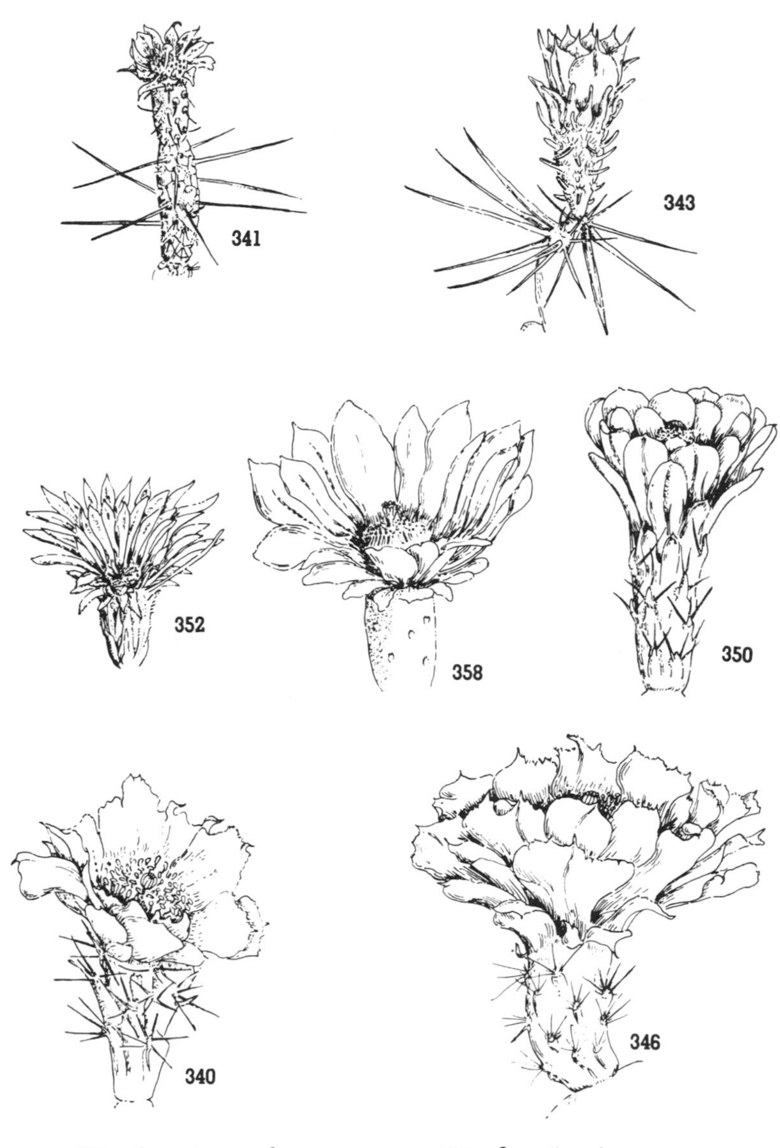

340. *Opuntia acanthocarpa*
341. *Opuntia ramosissima*
343. *Opuntia Parishii*
346. *Opuntia erinacea*
350. *Echinocereus mohavensis*
352. *Mamillaria deserti*
358. *Echinocactus polyancistrus*

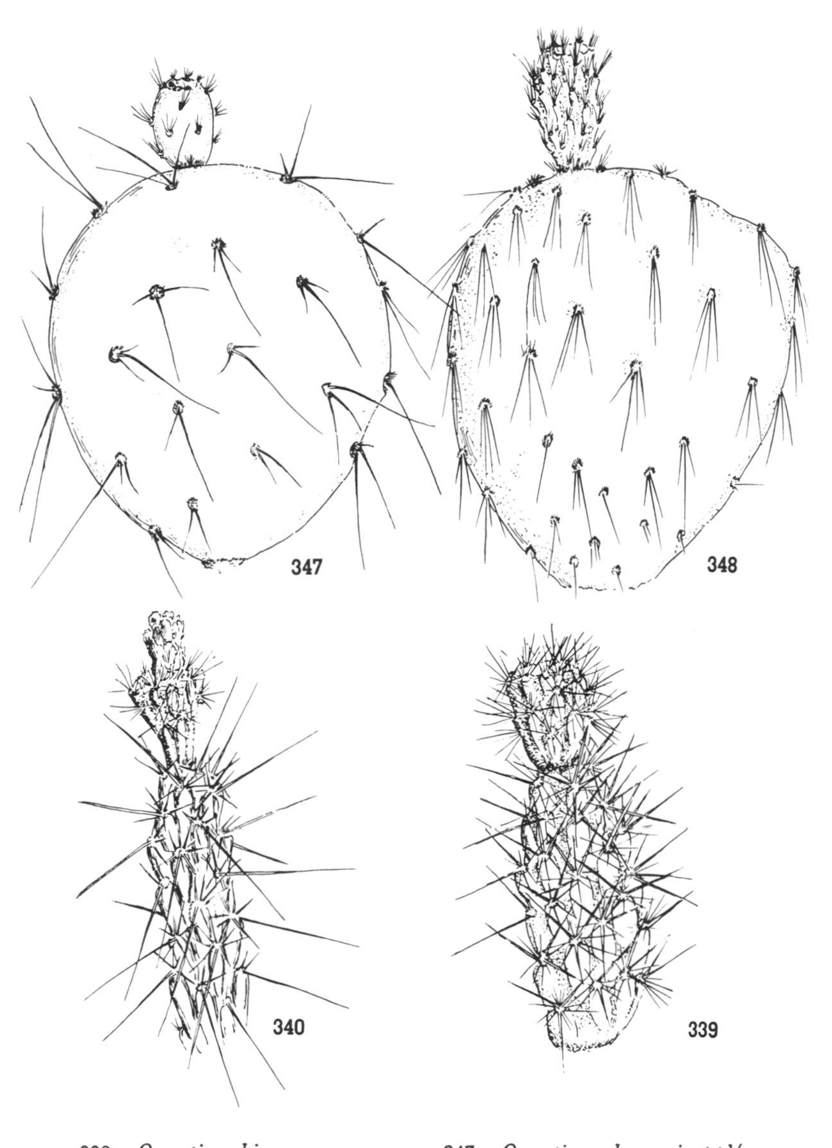

339. *Opuntia echinocarpa*
340. *Opuntia acanthocarpa*
347. *Opuntia mohavensis*, × ¼
348. *Opuntia chlorotica*, × ¼

345. GRIZZLY-BEAR CACTUS (not illustrated). *Opuntia ursina* (L., pertaining to a bear). **Fl.**: yellow. This species is best known from the Ord Mountains of the Mohave Desert. Its most outstanding character is the concealing covering of flexible ashy-gray spines ("like bear's hair"), sometimes up to 8 in. long. The plants generally occur in clumps or in large colonies. Most unfortunately, some of the finest of these colonies have been destroyed by commercial "cactus hogs." The fruit is dry and very spiny.

346. HEDGEHOG CACTUS (see pp. 160 and 163). *Opuntia erinacea* (L., hedgehog-like). **Fl.**: lemon-yellow, turning bronzy. About fifteen per cent of these plants have reddish-purple flowers. The ovate to oblong joints, 2–5 in. long, bear long, shaggy, flexible brown or white spines; the fruit also is covered with slender spines. Locally abundant on high gravelly or stony slopes of the mountains of the Mohave D.; to Nev., Utah, and Ariz. Closely allied to and perhaps but a geographical variety of the grizzly-bear cactus.

347. MOHAVE PRICKLY PEAR (see p. 161). *Opuntia mohavensis* (of Mohave [Desert]). **Fl.**: sulphur-yellow. The large pancake-like joints of this plant, 8–12 in. in diameter, bear numerous spicules and somewhat flattened, yellow-tipped, reddish-based spines of unequal length (up to 2 in.); the long, central spine is often twisted at the middle. This rather rare, prostrate-stemmed *Opuntia* is confined to the higher mountains bordering the Mohave D. on the west and to high mountain slopes of the eastern and northern Mohave D.

348. PANCAKE CACTUS (see p. 161). *Opuntia chlorotica* (Gr., pale yellowish-green). **Fl.**: yellow; 2–3 in. in diameter. This arborescent species, which sometimes grows to a height of 7 or 8 ft., has a definite trunk crowned with great, flat, disk-like joints, shaped "like pancakes," 3–8 in. in diameter. These joints are light green in color and armed with many close-set groups of handsome, down-pointing, yellow spines and smaller glochids. Though purple without, the fruit is green within and edible. The very erect plants usually occur singly in rough, mountainous country and immediately attract attention. Both Calif deserts; to Nev., Ariz., and Mex.

349. CALICO CACTUS (see p. 163). *Echinocereus Engelmannii* (Dr. G. Engelmann, 1809–1884, intimate friend of Asa Gray and one of the greatest botanists of his time. He was long head of the Missouri Botanical Gardens and author of that magnificent work, *The Cactacea of the Mexican Boundary*). **Fl.**: showy, purplish-red. Named calico cactus because of its many-colored spines. The numer-

341 and 356. DARNING-NEEDLE CACTUS (*Opuntia ramosissima*) and MOHAVE NIGGERHEADS (*Echinocactus polycephalus*) in foreground

344. BEAVERTAIL CACTUS (*Opuntia basilaris*)
Photo by Mary Damerel

346. HEDGEHOG CACTUS (*Opuntia erinacea*)
Photo by Avery Edwin Field

349. CALICO CACTUS (*Echinocereus Engelmannii*)

ous, stout, cylindric stems, each from 6–12 in. high, have 11–13 interrupted ribs, bearing radial spines in clusters of 10–12. The 4–6 central spines are long and stout, often curved and twisted, and yellowish or brown. A common species of both Calif. deserts; to Utah, Ariz., Sonora, and Baja Calif. Considerable variation in the size, length, and coloration of the spines is seen even in individual plants.

350. MOHAVE MOUND CACTUS (see p. 160). *Echinocereus mohavensis* (of the Mohave [River]). **Fl.**: spectrum-red to scarlet. The many small, globose to oblong stems, 6 to 600 in number, are huddled together in cushion-like mounds up to a yard across; and when the plant is in full flower the color effect is most startling. The strongly tuberculate ribs, 8–13 in number, bear groups of 3–10 grayish, curved, radiating, interlocking spines clustered about 1–3 centrals. The spatula-shaped "petals" of the flower are peculiarly fleshy. In crevices of rocks or on rocky slopes of mountains of the Mohave D., to Inyo Co.; Nev. and Ariz.

351. SAHUARO, GIANT CACTUS, PITAHAYA (not illustrated). *Cereus giganteus* (L., of the giants, gigantic). **Fl.**: cream-white. The sahuaro is the state flower of Arizona. In California a few of these great columnar cacti occur in picturesque groups in the rocky hills bordering the Colorado River, north of Yuma. The little elf owl finds nesting sites in the gourd-shaped pockets made in the stems by Gila woodpeckers. Dried ribs from the stems are used by Indians for lances and for framework for their huts. In May the numerous flowers appear, and by the end of June the famed sahuaro fruit is ripe. The fruit is eaten raw or cooked, or rolled into balls and dried to make a conserve. Syrup from the fruit is fermented to make an intoxicant, and the oily seeds are ground into a paste to spread like butter upon tortillas. Abundant in southern Ariz. and Sonora.

352. FRINGE-FLOWERED NIPPLE CACTUS (see p. 160). *Mamillaria deserti* (L., little nipple; of the desert). **Fl.**: sometimes straw-colored but generally brilliant pink. A small *Mamillaria*, single- or several-stemmed. From each grooved nipple 3–4 short, straight, brown- or tan-tipped spines emerge, surrounded by 15–20 slender, white, interlocking radials, which quite conceal the surface of the stem. The small flowers which appear at the summit of the stem have their outer segments fringed with fine, silky hairs. To be found in the mountains of the eastern Mohave D.; to Nev. The generic name, *Mamillaria*, is here spelled with but one *m* as did Haworth, who first described the genus.

353. ALVERSON NIPPLE CACTUS, FOX-TAIL CACTUS (see p. 166). *Mamillaria Alversonii.* (Andrew Halstead Alverson, 1845–1916, jeweler, mineralogist, cactus and succulent dealer of San Bernardino. In company with Frank Coffey, prospector of Dos Palmas, he made several prospecting trips into the Eagle and Chuckawalla mountains, and collected specimens of this cactus for the trade. Alverson obtained the type near Twenty-nine Palms.) **Fl.**: light purple. Another of the straight-spined *Mamillaria;* but in this one the outer segments of the flower are *margined* with hairs. The body is densely covered with stout, straight, black- or purple-tipped spines and slender, white, interlocking radials. The fruit is grayish-green and is edible. Chuckawalla, Eagle, and Little San Bernardino mountains.

354. CORKSEED CACTUS (see p. 166). *Mamillaria tetrancistra* (Gr., four fishhooks). **Fl.**: pink. A fine little *Mamillaria,* usually simple-stemmed; 1–4 purplish, central, hooked spines on each nipple; these are surrounded by 30–60 white radials, quite obscuring the surface. The strawberry-red, finger-shaped fruits were eaten by the aborigines. The seeds have a peculiar, corky, brown appendage, which suggested its common name. Colorado D. and southern and eastern Mohave D.; to Utah and Nev.

355. GRAHAM NIPPLE CACTUS (not illustrated). *Mamillaria microcarpa* (L., small-fruited). **Fl.**: amaranth-pink. A small cactus, with but one purple, hooked spine to the nipple; this is surrounded by 12–30 straight, nearly equal-length, white radials. The small flowers, each about ¾ in. long, give rise to scarlet, club-shaped fruits ½–1 in. long. Dry hills of the Colorado and Mohave deserts to Ariz. and Sonora.

356. NIGGERHEADS, COTTONTOP CACTUS (see p. 163). *Echinocactus polycephalus* (Gr., "hedgehog cactus," a most appropriate name for this genus of spiny cacti; L., many-headed). **Fl.**: light picric-yellow. The very fragrant flowers appear in August, making this the latest-flowering of any of our cacti. Most commonly called "niggerheads," because of the general dark color of the melon-shaped heads. Sometimes the compact, rounded clumps are made up of as many as 60 heads. The name "cottontop" refers to the generous tufts of cottony hairs enveloping the flower base and fruit. This woolliness of the fruits is natural and not due to injury made by rodents, as has been said in a recent government publication on Death Valley. The Indians extract the seeds for food. In rocky areas of the Mohave and northern Colorado deserts; to Utah and Ariz.

353. ALVERSON NIPPLE CACTUS (*Mamillaria Alversonii*)
Photo by Nicholas N. Kosloff

354. CORKSEED CACTUS (*Mamillaria tetrancistra*) in fruit
Photo by Avery Edwin Field

359. BISNAGA OR BARREL CACTUS in flower

Photo by Eugene N. Kozloff

357. BEEHIVE CACTUS, JOHNSON CACTUS (not illustrated). *Echinocactus Johnsonii octocentrus* (Joseph Ellis Johnson, 1817–1882, amateur Mormon naturalist of St. George, Utah, known "for his zeal for the development of the natural history and resources of his region" and once awarded a gold medal for having the best garden in the state of Utah; Gr., "eight-pointed"). **Fl.**: pink. Stems 4–9 in. high, the narrow ribs (18–20) almost concealed by the ashy-gray, sometimes reddish, awl-shaped, basally enlarged spines. Resting Springs, Inyo Co. The variety *lutescens* (L., yellowish) with lemon-yellow flowers, which are chocolate-brown at the base, is quite common in the vicinity of Searchlight, Nev. The flowers are said to remain open in sunshine about five days.

358. MOHAVE FISHHOOK, PINEAPPLE CACTUS (see p. 160). *Echinocactus polyancistrus* (Gr., many fishhooks). **Fl.**: pink, with darker midrib. Base of petals greenish. A simple-stemmed cactus, 8–12 in. high, shaped much like a pineapple, and almost hidden in spines. The long, stout, central, white spine of each nipple is flattened above and often somewhat twisted. This is surrounded by 5–7 brown, hooked, and 15–20 slender, short, white, straight, radial ones. The 3–5 large, showy flowers appear at the summit. The fruit is dry, pear-shaped, and nearly naked. This is a very difficult plant to subject to cultivation. On gravelly flats of mountainous areas of the Mohave D. north of Victorville; to southern Nev. Sometimes called hermit cactus "because it is rare to find more than one in a place."

359. BARREL CACTUS, BISNAGA (see p. 157). *Echinocactus acanthodes* (Gr., "thorn," plus the ending *-odes*, "form," i.e., "of the form of"). **Fl.**: yellow. The bisnaga, at first globular in form, eventually becomes cylindrical and may grow to a height of 5 or 6 ft. This is strictly a spring-flowering species, whereas the related barrel cactus (*E. Wislezenii*) of Arizona blooms all through the summer. The seedling plants usually begin life in the shade cast by rocks or protecting shrubs. Since there is a tendency to grow toward the most intense light, they often establish a leaning habit which persists in the adult plants. The annulated spines, more or less curved, and white-, red-, or rose-colored, are borne on stout ribs 18–28 in number. The slimy alkaline juice, so often reputed as a wonderful thirst-quencher, might in emergency save life but is for no other purpose fit to drink. Thirsty jack rabbits, mountain sheep, and wild burros occasionally eat the flesh. The largest plants are probably 20–30 years old. Colorado and eastern Mohave deserts; to Utah, Ariz., and Baja Calif. A variety, *Rostii,* with slenderer stems and straw-colored spines so closely set as to almost hide the stem, is found on the southwestern borders of the Colorado D.; to Baja Calif.

ONAGRACEAE. Evening Primrose Family

360. PYGMY PRIMROSE.* *Oenothera pterosperma* (Gr., "wine pursuit," a name given by Dioscorides to some plant whose roots when eaten were said to incite a desire for wine; L., "wing-seeded," the flattened seeds having a revolute, wing-like margin). **Fl.**: white, fading to rose. Tiny annual, 1–4 in. high, from mountainous areas of the northern Mohave D.; to Nev. and Utah.

361. BIG TOOTH-LEAVED PRIMROSE. *Oenothera dentata Johnstonii* (L., "toothed," in reference to the leaf margins; Dr. Ivan M. Johnston, 1898– , specialist on borages, taxonomist at Gray Herbarium, Harvard University). **Fl.**: bright yellow. A very handsome, free-flowering, wiry-stemmed primrose, forming small, often rounded plants 2–8 in. high. Found in sandy areas of the Mohave D.; to Nev. The flowers, which open near sundown, close about eight o'clock next morning. Plants of the eastern Mohave D. have the largest flowers. The var. *Parishii*, of the western Mohave, has narrower, smooth-margined leaves and smaller flowers.

362. GILMAN TOOTH-LEAVED PRIMROSE. *Oenothera dentata Gilmanii* (M. F. Gilman—see **79**). **Fl.**: yellow. Divaricately branching, the branches zigzag, with minute leaves at the angles. The stems are densely glandular. Bradbury Wash, Death Valley Nat. Mon.

363. BOOTH PRIMROSE. *Oenothera Boothii* (named by David Douglas, the Scotch botanist, "in compliment to M. B. Booth," of whom little is known). **Fl.**: white, fading to pink. A rare, slender-stemmed, tiny-flowered, spreading annual with central stem 4–16 in. high. The bark of the stems splits and peels as the plants mature. Found in small colonies in sandy washes near Hesperia on the Mohave D.; to Wash., Ida., and Utah.

364. DWARF CONTORTED PRIMROSE. *Oenothera contorta flexuosa* (L., twisted; L., bending). **Fl.**: yellow. Low, erect plants, generally less than 6 in. high, with long, unbeaked capsules. Occasional in mountainous areas of the western Mohave D.; to Ore. and Wyo.

365. WOODY BOTTLE-WASHER. *Oenothera decorticans desertorum* (L., deprived of bark; L., of deserts). **Fl.**: white, fading to pink. Stout, almost woody-stemmed annual with white, shining bark that soon cracks and peels. The plants, as shown in the illustration,

* The name primrose, so commonly applied to these plants, is really a misnomer, since the true primroses belong to the genus *Primula*.

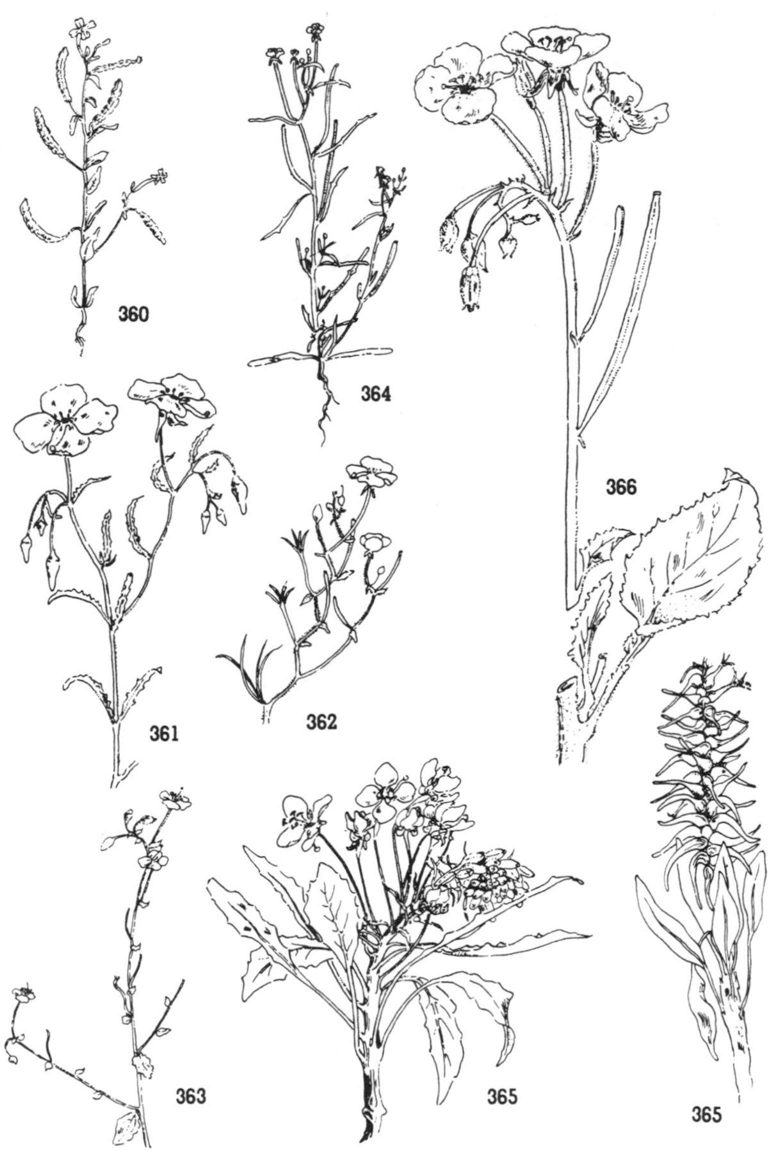

360. *Oenothera pterosperma*
361. *Oenothera dentata Johnstonii*
362.* *Oenothera dentata Gilmanii*
363.* *Oenothera Boothii*
364. *Oenothera contorta flexuosa*
365. *Oenothera decorticans desertorum*
366. *Oenothera brevipes*

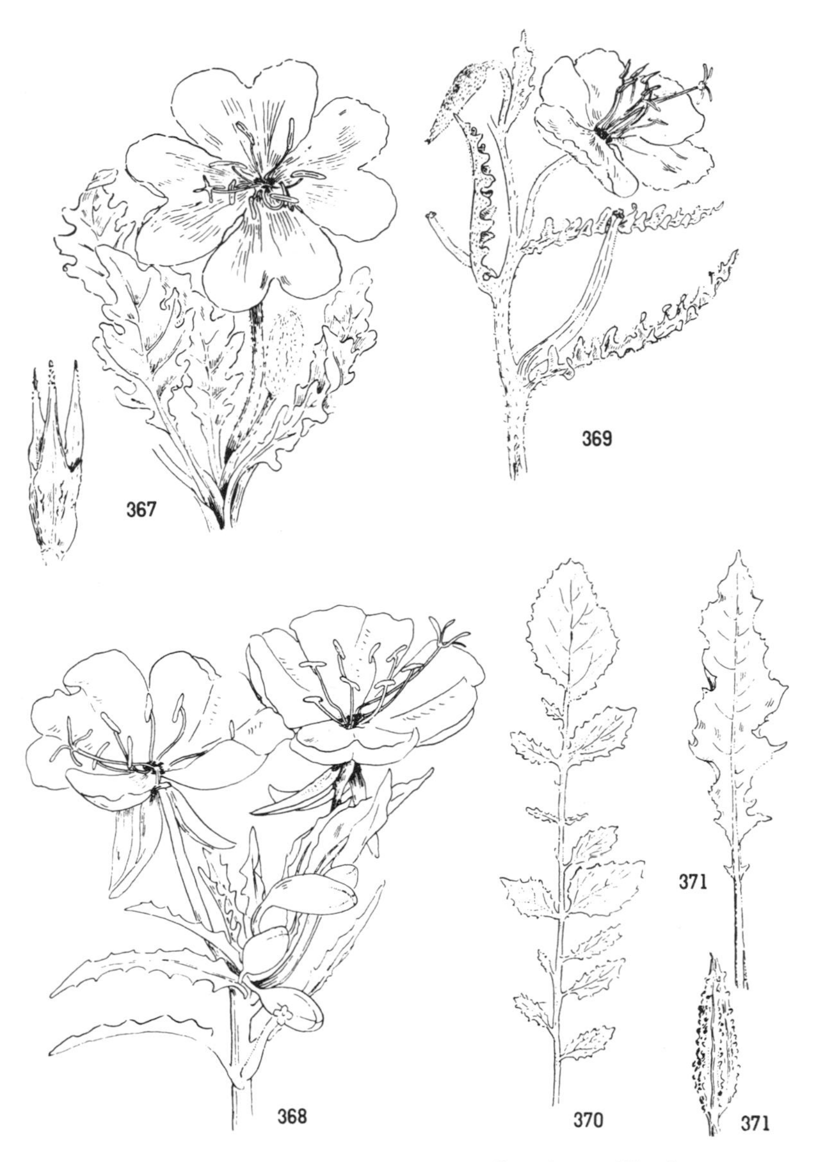

367. *Oenothera primiveris* 369. *Oenothera californica*
368. *Oenothera deltoides* 370. *Oenothera multijuga parviflora*
371. *Oenothera caespitosa marginata*

appear heavy and nodding with bloom when very young. The spikes become erect and lengthen if the season is favorable, until sometimes a foot or more in height. When the plants dry, there remains, as shown in the illustration on the right, a long-persistent, woody core, adorned with numerous splitting capsules. The var. *condensata,* of the eastern Mohave and Colorado deserts, has a thicker stem and very woody, thick-based, squarish capsules.

366. YELLOW CUPS. *Oenothera brevipes* (L., "short-footed," with reference to the pedicels, which are very short in comparison with the length of the fruit). **Fl.**: brilliant yellow. A bright, showy, free-flowering annual, with 1–6 reddish stems up to 15 in. high. The leaves are largely in a basal rosette and often markedly red-veined beneath. Common to the higher open deserts and on stony hills.

367. LARGE YELLOW DESERT PRIMROSE. *Oenothera primiveris* (L., "first-spring," that is, "early-blooming"). **Fl.**: lemon-yellow, aging to orange-red. A low, showy, long-flowering annual, many times found in aggregations on flat, sandy areas of both deserts. It bursts into flower in the evening and the blossoms remain open most of the following day. It occurs as far east as Utah and Texas.

368. DUNE PRIMROSE. *Oenothera deltoides* (Gr., for the Gr. letter Δ, delta, in reference to the leaf form). **Fl.**: white, turning to pink. This is the large-flowered, sweet-scented primrose so commonly seen adorning the sandy areas near Indio and the great dunes on the Mohave near Kelso. The outer stems of old plants have a peculiar way of turning inward and upward so that they form, when dried, what desert people often call "baskets" or "bird cages." The larvae of the two-lined sphinx moth feed greedily upon both leaves and flowers and sometimes occur in such numbers in spring that they destroy most of the plants. When plentiful, these larvae were utilized as food by the Cahuilla Indians. Watch for the beautiful yellow-green crab spiders hidden among the webby anthers; lying in wait for their prey, they simulate the flower color so closely that they are almost invisible.

369. CALIFORNIA PRIMROSE. *Oenothera californica* (of California). **Fl.**: white. A cismontane, gray, hairy-leaved species which usually occurs in colonies on sandy areas of the western end of our deserts. The leaves vary considerably in the degree of their dentation.

370. FROST-STEMMED PRIMROSE. *Oenothera multijuga parviflora* (L., "many-jointed," in reference to the many lobes of the pinnate leaf; L., "small-flowered"). **Fl.**: yellow. Annual, 8–30 in. tall, with medium-sized flowers and long, prominently veined, mostly basal leaves, one of which is shown in the illustration. The stems often appear covered with frost because of numerous white hairs. The tall stems, which are almost devoid of leaves, the calyces, and other parts of the herbage are at times quite reddish. Washes and rocky slopes of the Death Valley region.

371. LARGE WHITE DESERT PRIMROSE. *Oenothera caespitosa marginata* (L., turfy; L., margined). **Fl.**: white, turning pink. A widespread perennial species of mountainous areas of the Mohave Desert, with large flowers quite resembling in size and form those of the large yellow desert primrose. Only a drawing of the leaf and fruit is shown. The young plants closely hug the ground. They continue blooming over a long period, and the short, thick stems eventually reach a length of 6 in. or more. The large, sweet-scented flowers open in the afternoon and remain expanded until morning. Young plants transplanted to gardens will take root and bloom over a long period.

372. NARROW-LEAVED PRIMROSE. *Oenothera refracta* (L., "bent, turned," referring to the linear capsules). **Fl.**: white. An annual, 4–10 in. high, common to open, sandy areas of both deserts. Especially plentiful from Baker northward on the Mohave D.

373. PURPLE PRIMROSE. *Oenothera heterochroma* (L., different-colored). **Fl.**: purple. A Nevada species which enters the mountains of the Death Valley Nat. Mon. The spreading plants are 8–15 in. high and have herbage which is both hairy and glandular.

374. PALE YELLOW PRIMROSE. *Oenothera pallidula* (L., somewhat or a little pale). **Fl.**: pale yellow. A species intermediate between *O. brevipes* and *O. clavaeformis.* It lacks the large flowers and spreading hairs on the stems of the former and has a flower more nearly like the latter species. It ranges from the arid portions of southwestern Utah to the deserts of eastern Calif., including the Death Valley Sink.

375. BROWN-EYED PRIMROSE. *Oenothera clavaeformis* (L., "club-shaped," with reference to the capsules). **Fl.**: white. A simple, dentate-leaved annual, common in the Death Valley area and on the almost level plains dominating the western half of the Mohave D. The var. *aurantiaca,* with mostly pinnate leaves, is common about Palm Springs but is also found on the Mohave D. The var. *Piersonii,* with pale yellow flowers and with densely hairy stems, is seen occasionally on the southern Colorado D.: to Baja Calif.

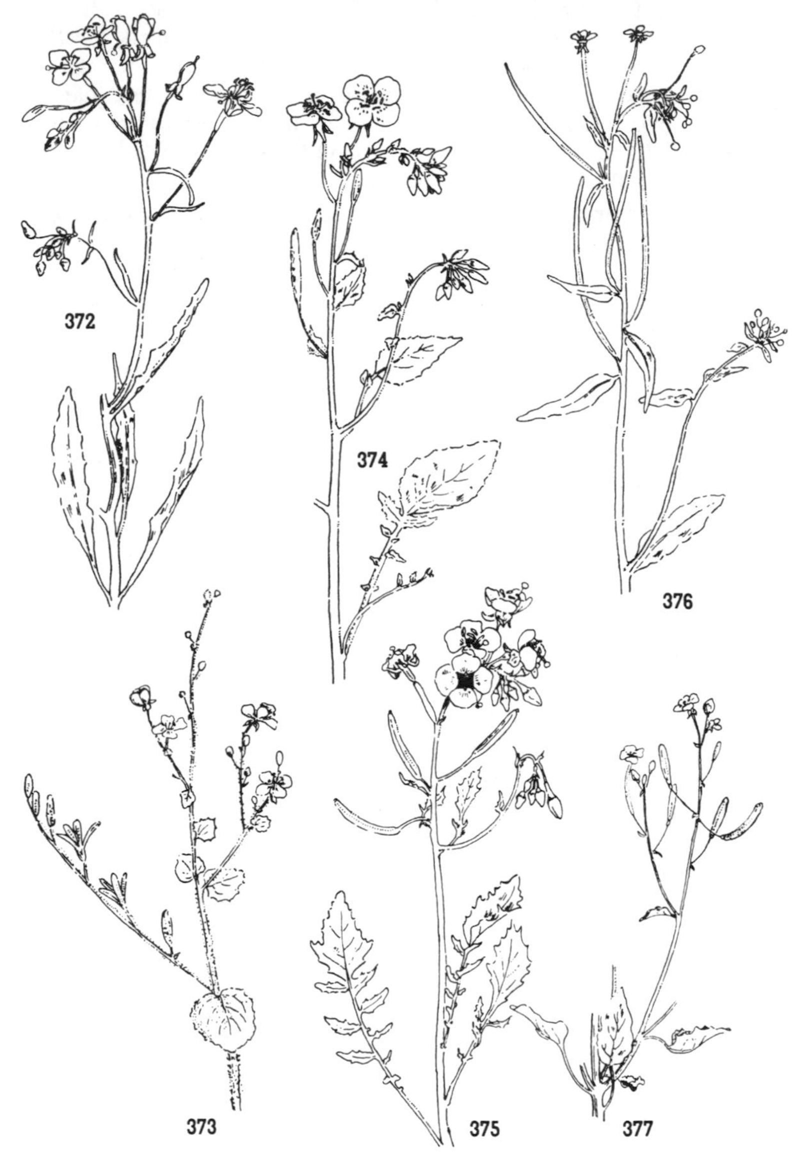

372. *Oenothera refracta*
373.* *Oenothera heterochroma*
374. *Oenothera pallidula*
375. *Oenothera clavaeformis*
376. *Oenothera chamaenerioides*
377.* *Oenothera scapoidea seorsa*

378. *Oenothera cardiophylla*
379. *Oenothera cardiophylla splendens*
380. *Oenothera Palmeri*
381. *Oenothera micrantha exfoliata*

376. LONG-CAPSULED PRIMROSE. *Oenothera chamaenerioides* (Gr., "like *Chamaenerion*," a genus related to *Epilobium*). **Fl.:** white. Slender annual, with the general appearance of an *Epilobium*, 10–15 in. tall, and usually branching. It is characterized by its very long capsules, narrow leaves, and small flowers. At times common in sandy washes and open places of both deserts, more so on the Colorado D.: to Utah and Tex.

377. PAIUTE PRIMROSE. *Oenothera scapoidea seorsa* (L., scape-like; L., separate or apart). **Fl.:** yellow. Many-branched annual, 8–14 in. high, known from the Death Valley Sink.

378. HEART-LEAVED PRIMROSE. *Oenothera cardiophylla* (L., heart-leaved). **Fl.:** yellow, drying red. A handsome, erect species, with dark or pale green foliage, growing in sheltered, rocky canyons. It is an annual, or more often a somewhat woody, freely branched perennial, 4–20 in. tall. Most parts of the herbage are covered with soft hairs, giving the plants a somewhat grayish appearance. Particularly abundant in canyons along the southwestern side of the Salton Sink.

379. LONG-TUBED PRIMROSE. *Oenothera cardiophylla splendens* (L., splendid). **Fl.:** yellow. A tall and robust plant found from the Salton Sea region east to the Colorado R.

380. PALMER PRIMROSE. *Oenothera Palmeri* (Edward Palmer —see **474**). **Fl.:** golden. Dwarf, tufted, gray-green annual, strong-rooted, and with short, stiff branches having a peeling bark. The first or outer leaves dry very early, giving the plant a singularly coarse, xeric appearance. The snugly fitting, quadrangular capsules are uniquely winged and unexpectedly large and woody in comparison with the small flower. Flat, open spaces of the Mohave D.; to Ore. and Ariz.

381. SPENCER PRIMROSE. *Oenothera micrantha exfoliata* (Gr., "small-flowered"; L., "stripped of leaves," i.e., coming off in scales, in reference to the splitting bark). **Fl.:** yellow. Single- to several-stemmed, erect or reclining annual, up to 10 in. high. It has straight or somewhat curled fruits, and gray herbage, owing to numerous white hairs. The papery bark splits and peels off in age. Named in honor of Mary F. Spencer, of San Diego, zealous lifelong collector, who even in her eightieth year continued her botanical visits to the deserts and mountains. Large collections of her plants are in the leading herbaria of the world. Both deserts; to Ariz.

UMBELLIFERAE. Carrot Family

382. PARISH WILD PARSLEY. *Lomatium nevadense Parishii* (Gr., "border," referring to the seed wings; of Nevada; S. B. Parish —see **603**). **Fl.**: white. Rocky slopes of the mountains of the western and northern Mohave D. The two illustrations show the plant in fruit and in flower.

383. PARRY WILD PARSLEY. *Lomatium Parryi* (C. C. Parry—see **24**). **Fl.**: white. Between 4,000 and 8,000 ft. on dry, rocky slopes of mountains of the northern and eastern Mohave D.; to Utah.

384. UTAH CYMOPTERUS. *Cymopterus utahensis* (Gr., "wavy wings," in reference to the fruit; of Utah). **Fl.**: purple. Creosote-bush, sagebrush, and piñon belts from Ida. to N.Mex., Nev., and eastern Mohave D. of Calif.

385. DESERT UMBEL. *Ammoselinum occidentale* (*Ammi* + *Silinum*, two umbelliferous genera; L., western). **Fl.**: white. Low, diffuse annual, mostly of the eastern Colorado D.

386. PANAMINT CYMOPTERUS. *Cymopterus panamintensis* (of the Panamint [Mountains]). **Fl.**: greenish-yellow. Mohave D. The var. *actifolius* (L., sharp-pointed leaf) is a large, spreading perennial found about rocks on the higher mountains of both Calif. deserts.

387. PURPLE CYMOPTERUS. *Cymopterus multinervatus* (L., many-nerved). **Fl.**: purplish. A plant known in Calif. from a few collections made on the Mohave D.; to Ariz., Utah, and Tex.

388. DESERT CYMOPTERUS. *Cymopterus deserticolus* (L., desert-dwelling). **Fl.**: purple. Rare on Mohave D., near Victorville.

389. WHITE CYMOPTERUS. *Cymopterus aboriginum* (L., "of the original inhabitants," here referring to Indian Springs, Nev., rather than to the Indians themselves). **Fl.**: white. Rare, northern Mohave D.; to Mono Co. and Nev.

390. GILMAN PARSLEY. *Cymopterus Gilmani* (M. French Gilman—see **79**). **Fl.**: maroon. Recently discovered in the mountains of the Death Valley Nat. Mon.

391. MOHAVE WILD PARSLEY. *Lomatium mohavense* (of the Mohave [Desert]). **Fl.**: deep maroon. The leaves, stems, and fruit are grayish because of a dense covering of hairs. Mohave D. and adjacent Nev., 2,500–6,000 ft. altitude.

382. *Lomatium nevadense Parishii*
383. *Lomatium Parryi*
384. *Cymopterus utahensis*
385. *Ammoselinum occidentale*

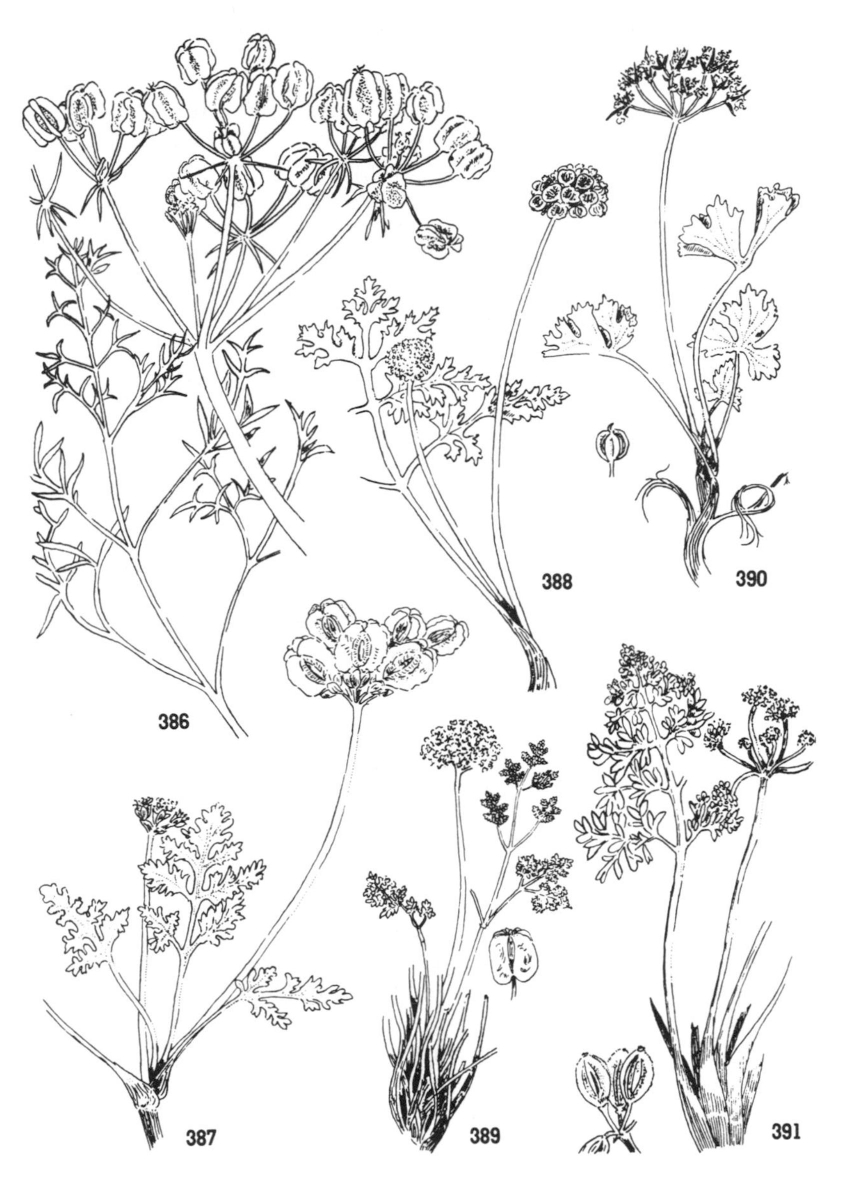

386. *Cymopterus panamintensis*
387.* *Cymopterus multinervatus*
388.* *Cymopterus deserticolus*
389.* *Cymopterus aboriginum*
390.* *Cymopterus Gilmani*
391. *Lomatium mohavense*

GARRYACEAE. Silk Tassel Family

392. QUININE-BUSH, YELLOW-LEAF SILK TASSEL. *Garrya flavescens* (Nicholas Garry, Deputy Governor of the Hudson Bay Co.; David Douglas named the genus in his honor because of his helpfulness in facilitating Douglas' botanical work in northwestern America. Dr. Lindley considered *Garrya* "the greatest botanical curiosity in all his [Douglas'] collection." L., becoming yellow). **Fl.**: lemon-chrome. Though the young, leathery leaves of this hardy evergreen shrub are grayish-green, the older leaves are distinctly yellowish. All parts of the plant, including the pendulous catkins, are intensely bitter and are unpalatable to most browsing animals. Dry canyons and slopes at the lower edge of the juniper belt in the Mohave D.; to Utah; south to San Diego Co.

LENNOACEAE. Lennoa Family

393. SCALY-STEMMED SAND PLANT. *Pholisma arenarium* (Gr., "scaly," in allusion to the stem; L., sandy). **Fl.**: smoky-purple, with white border. We are always agreeably surprised when on our desert walks we come upon these odd plants of the sandy washes. They generally occur in clumps with only the purple and brown, club-shaped heads above the sand. When dug up they show basal connections to the roots of such shrubs as cheese-weed and rabbitbrush which they parasitize. Blooms in May; on both deserts. Pholismas collected along the seashore are now referred to the species *panniculatum.*

394. SAND-FOOD, SAND-SPONGE, BIATATK. *Ammobroma sonorae* (Gr., sand food; of Sonora). **Fl.**: purple. Following a wet winter the parasitic sand-sponge is plentiful on the Algodones sand hills west of Yuma and at the head of the Gulf of California. The gray flower disks (1½–5 in. in diameter) lie flat like big buttons on the sand and are easily overlooked by the average observer. It is only by digging that we see the long, brittle stems which extend 2–5 ft. below the surface to contact the roots of such host plants as *Coldenia Palmeri, C. plicata,* and *Eriogonum deserticola,* from which they gain their sustenance. The number of plants branching from the host root is usually 4 to 8, and their combined weight may be many times that of their host. *Ammobroma* was once an important and much-prized article of food for the Gulf Indians. The Papagos called it "biatatk" (*bia,* sand or sandhills, + *tatk,* root). The stems are eaten raw, cooked, or roasted. When roasted they resemble, in flavor, well-browned yams.

ERICACEAE. Heath Family

395. DESERT MANZANITA. *Arctostaphylos glauca eremicola* (Gr., "bear-grape"; L., "bluish-gray," in reference to the leaf; L., "desert-dweller"). **Fl.**: pink. Many of the manzanitas have berries containing several nutlets, but this one has fruits with single, solid stones. Moreover its berries are not round as in other species but elliptic in outline, and the leaves are peculiarly purple-veined. Confined to the piñon-covered mountains along the southwestern edge of the Joshua Tree Nat. Mon. The coastal manzanita with rounded fruit, *A. canescens,* also enters the Monument from the west.

OLEACEAE. Olive Family

396. BROOM TWINBERRY. *Menodora scoparia* (Gr., enforced gift; L., "broom," which the many stems somewhat resemble). **Fl.**: yellow. A branching, green bush coming into flower in June and continuing through midsummer. Only careful examination of the twinned capsules and floral parts would lead one to know that this species is in the same genus with the shrubby twinfruits. Very common among rocks at Jacumba. Mountainous areas of the Colorado and eastern Mohave deserts; to Ariz. and Baja Calif.

397. SPINY MONODORA. *Menodora spinescens* (L., spiney). **Fl.**: white tinged with brownish purple without. Widely scattered over the Mohave D. on dry flats and rocky slopes; to Nev.

398. TANGLEBRUSH. *Forestiera neomexicana* (M. Forestier, French physician; of New Mexico, where Fendler took the type specimens). We have come upon colonies of this odd, opposite-leaved shrub in most unexpected places. It usually occurs in mountainous areas where ground waters are near the surface. Gavilan Hills, San Jacinto Mts., Joshua Tree Nat. Mon.; eastward to Colo. and Tex.

399. LEATHER-LEAVED ASH. *Fraxinus velutina coriacea* (L., name for the ash; L., covered densely with silky hairs; L., leathery). Tree, 12–30 ft. high, generally found in aggregations. Both Calif. deserts (Owens Valley, Victorville, Keyes' Ranch), below 5,000 ft.; to Utah.

400. SINGLE-LEAVED ASH. *Fraxinus anomala* (Gr., "not regular," with reference to the singular leaf). A queer dwarf tree or low-spreading shrub with rounded leaves, which often turn yellow and drop early in summer. Generally in canyons of the piñon-juniper belt of mountains of the northern and eastern Mohave D.; to Colo. and N.Mex.

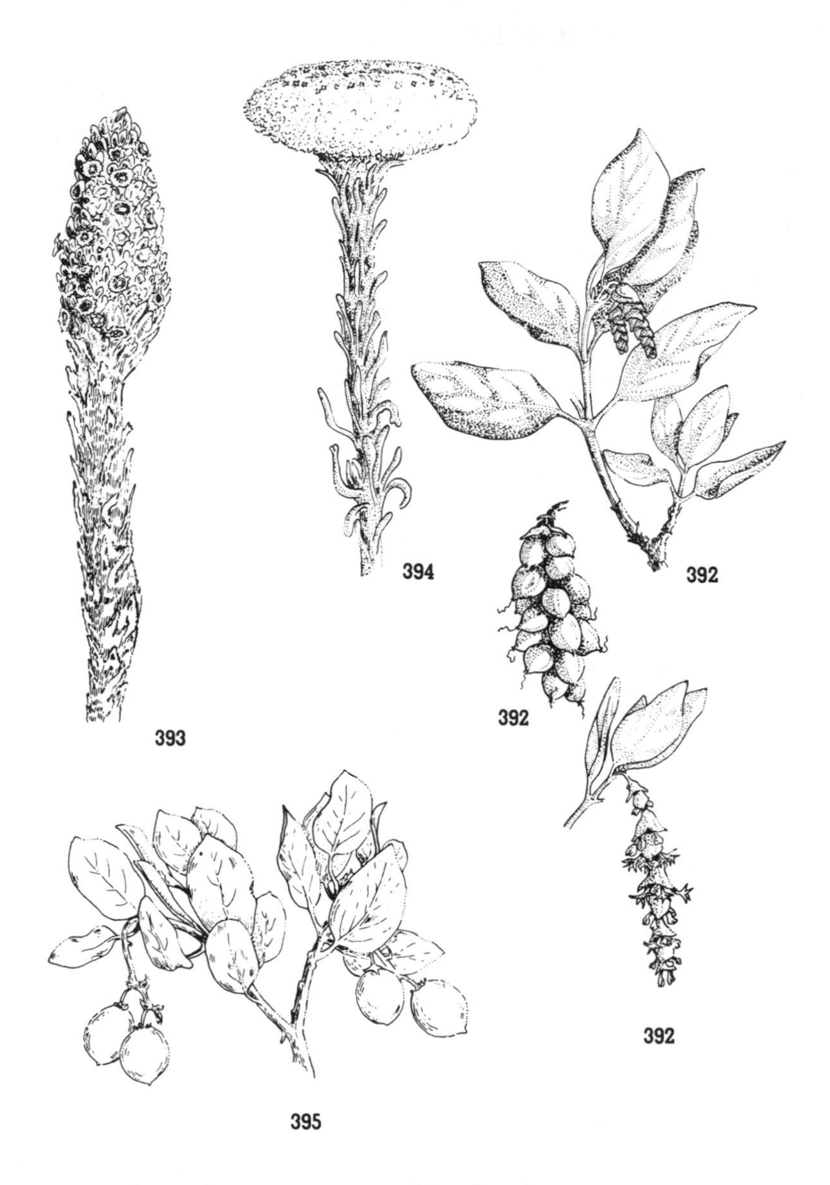

392. *Garrya flavescens*
393. *Pholisma arenarium*
394. *Ammobroma sonorae*
395. *Arctostaphylos glauca eremicola*

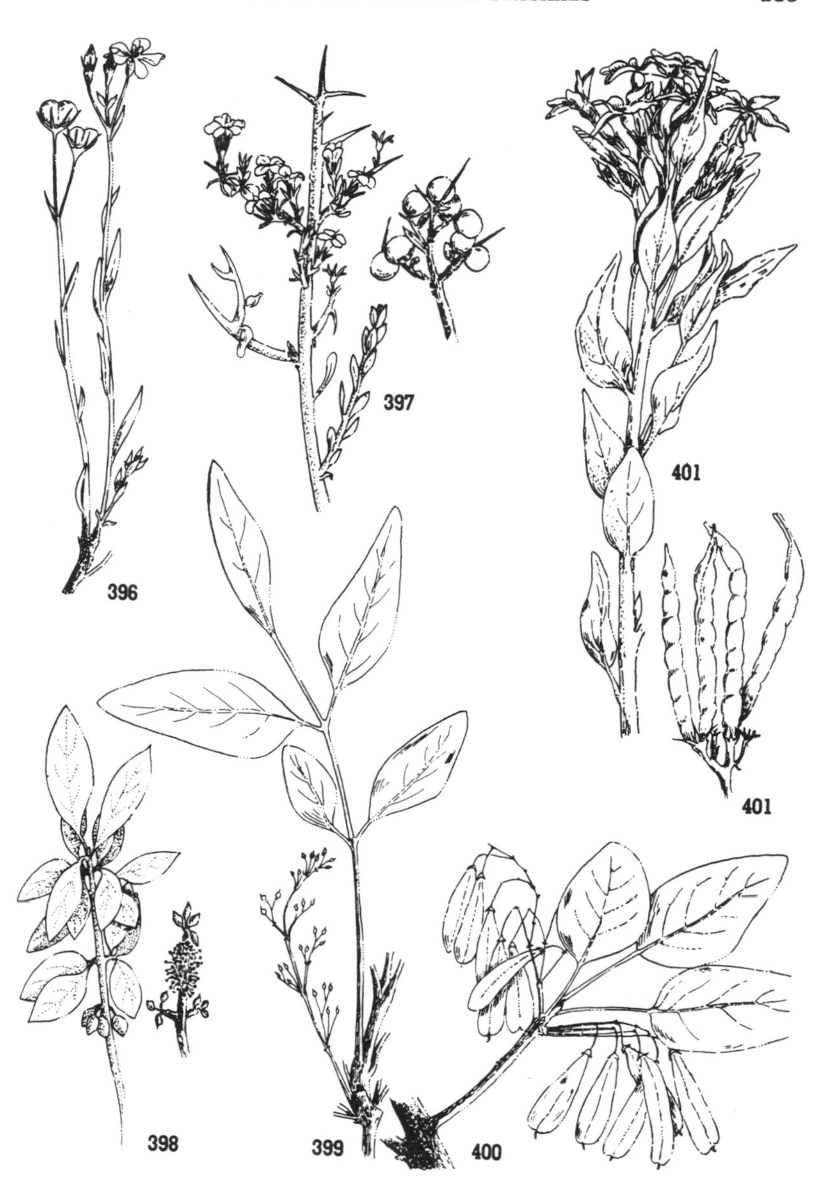

396. *Menodora scoparia*
397. *Menodora spinescens*
398. *Forestiera neomexicana*
399. *Fraxinus velutina coriacea*
400. *Fraxinus anomala*
401. *Amsonia brevifolia*

APOCYNACEAE. Dogbane Family

401. SMALL-LEAVED AMSONIA. *Amsonia brevifolia* (Dr. Charles Amson, Colonial Virginia physician of the eighteenth century; L., short-leaved). **Fl.**: sky-blue. A many-stemmed, perennial herb, with milky juice; stems 8–15 in. tall. It is of quite local distribution on dry, gravelly soils bordering washes in the higher portions of the Mohave and northern Colorado deserts. Abundant near Quail Springs in the Joshua Tree Nat. Mon.; to Utah and northwestern N.Mex. With the green-leaved species occurs the more handsome variety, *tomentosa,* having herbage of gray-green color.

ASCLEPIADACEAE. Milkweed Family

402. DEBOLTIA. *Astephanus utahensis* (Gr., "crownless," with reference to the flowers; of Utah). **Fl.**: dull yellow. A thick-rooted perennial, with slender, twining stems 4–16 in. long. The lobes of the shallowly cleft, bell-shaped corolla turn inward, almost closing the throat. Occasional in the dry sands from widely scattered localities of both Calif. deserts; to Ariz. and Utah. Named in honor of Frank Debolt, locally known as "Dutch Frank," who on a number of occasions acted as guide, using his burros to pack bedding and provisions for scientific explorers on the Colorado D.

403. SPEAR-LEAF, TALAYOTE. *Vincetoxicum hastulatum* (L., "poison conqueror," because of its supposed antidotal powers; L., spear-shaped). **Fl.**: greenish. An exceedingly rare species of the Colorado D.; first collected in Baja Calif. In Mexico, young fruits of some of the "talayotes" are eaten raw or cooked. Sweetmeats are made by boiling the green pods in syrup.

404. AJAMETE. *Asclepias subulata* (Gr., name for Æsculapius, the father of medicine; L., "awl-shaped," referring to the leaves). **Fl.**: greenish-yellow. Many to several hundred gray-green, rush-like stems, 2–5 ft. tall, from a perennial root. The latex has been analyzed and found to yield rubber. In Mexico, the milky juice is used as an emetic and purgative. Often seen along the edge of washes in the low, hot portions of the Colorado and eastern Mohave deserts; to Mex.

405. DESERT MILKWEED. *Asclepias erosa* (L., "gnawed at the edges," referring to the ruffled, saw-edged leaf margins). **Fl.**: greenish-white, hoods yellowish. The tall stems, 2–4½ ft. high, are fine-woolly, and remain standing long after the whitish leaves and flowers have withered and fallen. Confined almost wholly to sandy washes from the low deserts to the juniper belt. The pollen is not suitable for use by bees, but they visit the blossoms for nectar. *Pepsis,* the large tarantula wasp, is a constant flower visitor. The Indians rubbed the milky juice on warts. Both Calif. deserts; east to Utah.

406. PURPLE CLIMBING MILKWEED. *Funastrum heterophyllum* (L., "rope star," with reference to stems and flower form; Gr., different-leaved). **Fl.:** purplish. A climbing perennial, with rank, milky juice; often growing among and over shrubs and trees of the low, hot deserts. Frequently 8 or 10 of the long twining stems twist themselves into veritable ropes which extend to the very tops of wash-bordering trees.

407. RAMBLING MILKWEED. *Funastrum hirtellum* (L., minutely stiff-haired). **Fl.:** greenish-yellow. Much like the purple climbing milkweed but with smaller flowers and more hairy herbage. The long stems ramble widely through shrubs or over the ground. In sandy washes of the Colorado D. and northeastern Mohave D.; to Ariz. and Nev.

408. WHITE-STEMMED MILKWEED. *Asclepias albicans* (L., whitish). **Fl.:** greenish-white, with some brown, and the hoods yellowish. If you see several long, gray-green stems, appearing like stalks of old-fashioned buggy whips, arising from some niche in the rocks of a low desert hill or canyon-side, you may be quite sure you have found the white-stemmed milkweed. The pliant, woody stems are very strong and in the days when prospectors traveled with donkeys they were often employed as whips. Colorado D.; to Baja Calif. and Sinaloa.

409. FOUR-O'CLOCK MILKWEED. *Asclepias nyctaginifolia* (*Nyctago* [genus name for the four-o'clock] + leaved). **Fl.:** greenish-white. Perennial herb, with several ascending green stems, 4–8 in. long. The leaves are ovate and somewhat resemble those of the garden plants called four-o'clocks. Mountains of the eastern Mohave D.; to Ariz.

410. ADORNED MILKWEED. *Asclepias vestita* (L., clothed, adorned). **Fl.:** purplish, with yellow hoods. Several-stemmed, somewhat ascending perennial, 8–20 in. high. The umbels appearing in May number 1 to 4, the lateral ones being without a stalk. In dry sands along the north side of the San Bernardino Mts.; north to Inyo Co.

411. ANTELOPE HORNS. *Asclepiodora decumbens* (Gr., gift of Æsculapius; L., prostrate). **Fl.:** greenish-white, with purplish hoods. A spreading perennial, of the piñon-juniper belt, with several stems 1–2 ft. long, each producing large balls of flowers. Eastern Mohave D.; to Ark. and Tex. The common name is derived from an appropriate Navaho name referring to the form of the green pods.

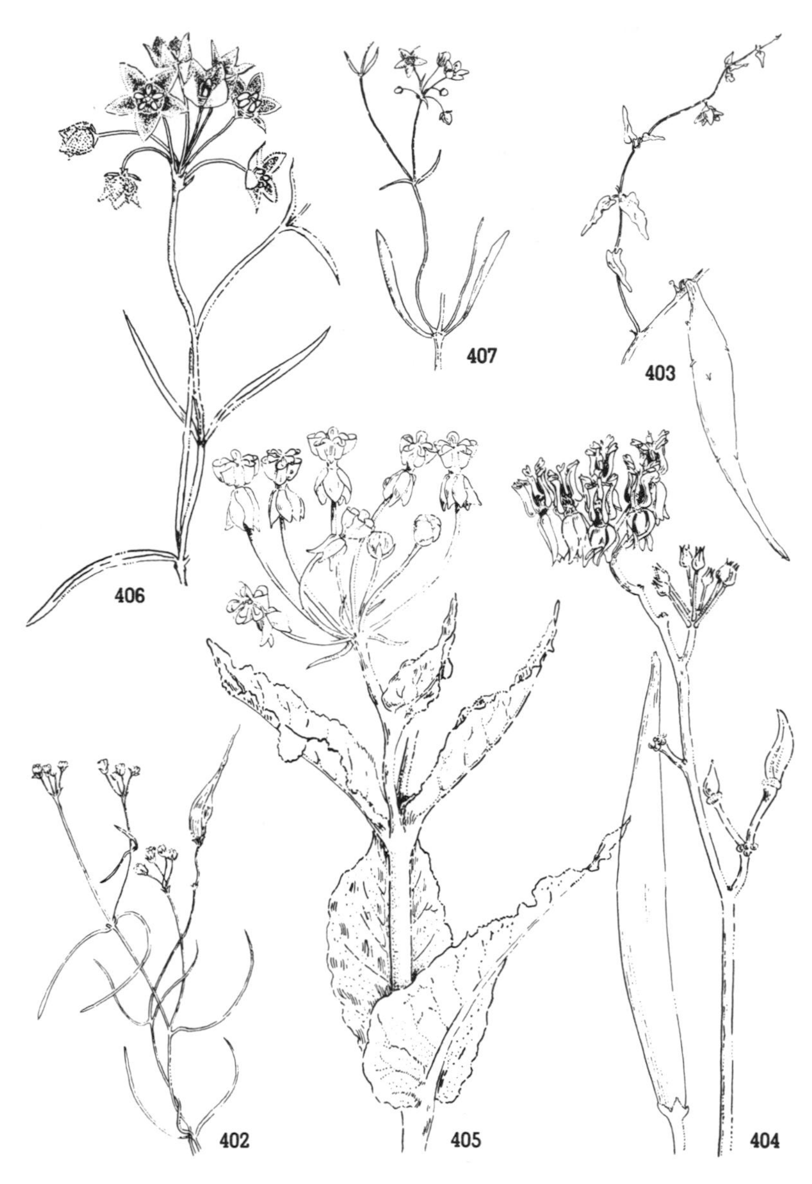

402. *Astephanus utahensis*
403.* *Vincetoxicum hastulatum*
404. *Asclepias subulata*
405. *Asclepias erosa*
406. *Funastrum heterophyllum*, × 1
407.* *Funastrum hirtellum*

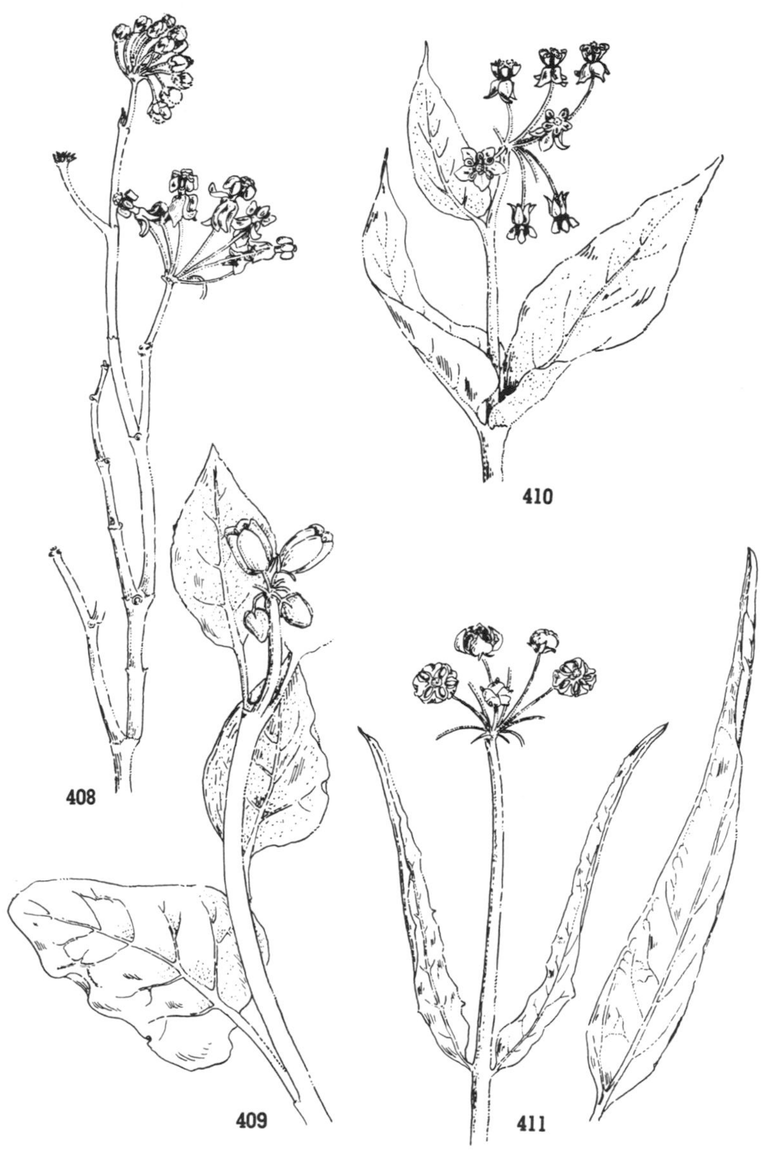

408. *Asclepias albicans*
409.* *Asclepias nyctaginifolia*
410. *Asclepias vestita*
411. *Asclepiodora decumbens*

CONVOLVULACEAE. Morning-Glory Family

Dodder, a relative of the sweet potato and morning-glory, is a parasite on a number of desert shrubs, including burro-bush and creosote bush. It can readily be recognized by the tangled bunches of coarse, brownish-orange or yellowish, thread-like stems which often almost hide the plant which is parasitized. Two species are common.

412. TOOTHED DODDER (not illustrated). *Cuscuta denticulata* (Arabic, *Kush-kūt;* L., small-toothed). This species, confined to the Mohave Desert, has pale yellow stems and very small, urn-shaped flowers, ovoid capsules, and toothed calyx lobes.

413. CALIFORNIA DODDER (not illustrated). *Cuscuta californica* (of California). This species and its variety *papillosa* (L., covered with small protuberances) have brownish-orange stems and somewhat larger, whiter flowers than *C. denticulata.* A cismontane plant occasionally reaching the Mohave and Colorado deserts.

POLEMONIACEAE. Phlox Family

414. STANSBURY PHLOX. *Phlox Stansburyi* (Gr., "flame," in allusion to the brilliancy of the flowers; Capt. H. Stansbury—see **184**). **Fl.**: pinkish-red. A woody-based perennial of the piñon-juniper belt, ranging in California from the Argus and Inyo mountains to the New York and Providence mountains and beyond to Ida., Utah, and N.Mex. Flowering begins in late May and may continue until July.

415. GOLDEN GILIA. *Gilia aurea* (Felippo Luigo Gilii, eighteenth-century Italian botanist and astronomer; L., golden). **Fl.**: yellow, with brownish throat. Fairly frequent in open places on both the Colorado and Mohave deserts, where it forms brilliant patches of color in early spring. Diurnal. The form *decora* (L., fitting, decorous) has white-to-violet flowers and is often equally plentiful on sandy flats of both deserts.

416. BLUE MANTLE. *Gilia densifolia austromontana* (L., dense-leaved; L., of the southern mountains). **Fl.**: lively blue. The many stems, which spring from a perennial root, are compactly clustered and in May almost "smothered" in flowers. Mountains of the coastal area, reaching the western edge of the Mohave D. in the Joshua Tree Nat. Mon.

417. EVENING SNOW. *Gilia dichotoma* (Gr., "twice-cut," i.e., "two-forked," with reference to the habit of branching). **Fl.**: white, with a brownish-pink area along one side of the back of each petal. The brown stems blend so perfectly with the soil that it is not until evening, when the large flowers open, that the presence of this dainty plant is disclosed. When the plants occur in numbers, wide areas are in a few moments of waning daylight changed to a floral blanket of white. Closely related and growing in similar situations is *G. Bigelovii,* with longer, wider leaves and a smaller, more tubular corolla.

418. LONG-TUBED GILIA. *Gilia brevicula* (L., "rather short," in allusion to the low stature, it being "but a span high"). **Fl.**: white to pink or bluish-pink, with purple throat. A dainty annual, which wanders from the slopes of the bordering desert mountains to the desert floor. To be sought in the western Mohave from near Old Woman Springs to northern Los Angeles Co.

419. PRICKLY GILIA. *Gilia pungens tenuiloba* (L., "prickly"; L., "slender-lobed," in reference to the corolla lobe). **Fl.**: whitish. Low, woody-based perennial, found on rocky slopes of the piñon-juniper belt. Blooms profusely at evening time. San Jacinto, San Bernardino, Inyo, Panamint, and Providence mountains; to Ariz.

420. PARRY GILIA. *Gilia Parryae* (Mrs. Parry, wife of Dr. C. C. Parry—see **24**). **Fl.**: white to bluish-purple. Much-branched little annual, blooming vigorously all during the sunlit hours. Common to sands and clays of the higher desert, ranging north as far as Kern and Inyo counties. In the juniper-yucca area between Cajon Pass and Victorville the ground is at times almost as white with these flowers as if covered with snow.

421. DAVY GILIA. *Gilia Davyi* (J. B. Davy, botanist, formerly of the University of California, now of Oxford University). **Fl.**: violet to pink, with yellow throat. A splendid, showy species, coming to perfection in the deep sandy soils of the western Mohave D.; to southern Nev.

422. HUMBLE GILIA. *Gilia demissa* (L., "humble, lowly," in reference to its habit of growth). **Fl.**: white, with maroon spots. Leafy annual, hairy-glandular throughout. The petals of the fragrant flowers are often twisted like the sails of a windmill. Rather common on desert expanses and in washes of the Mohave D.; to Ariz. and Utah.

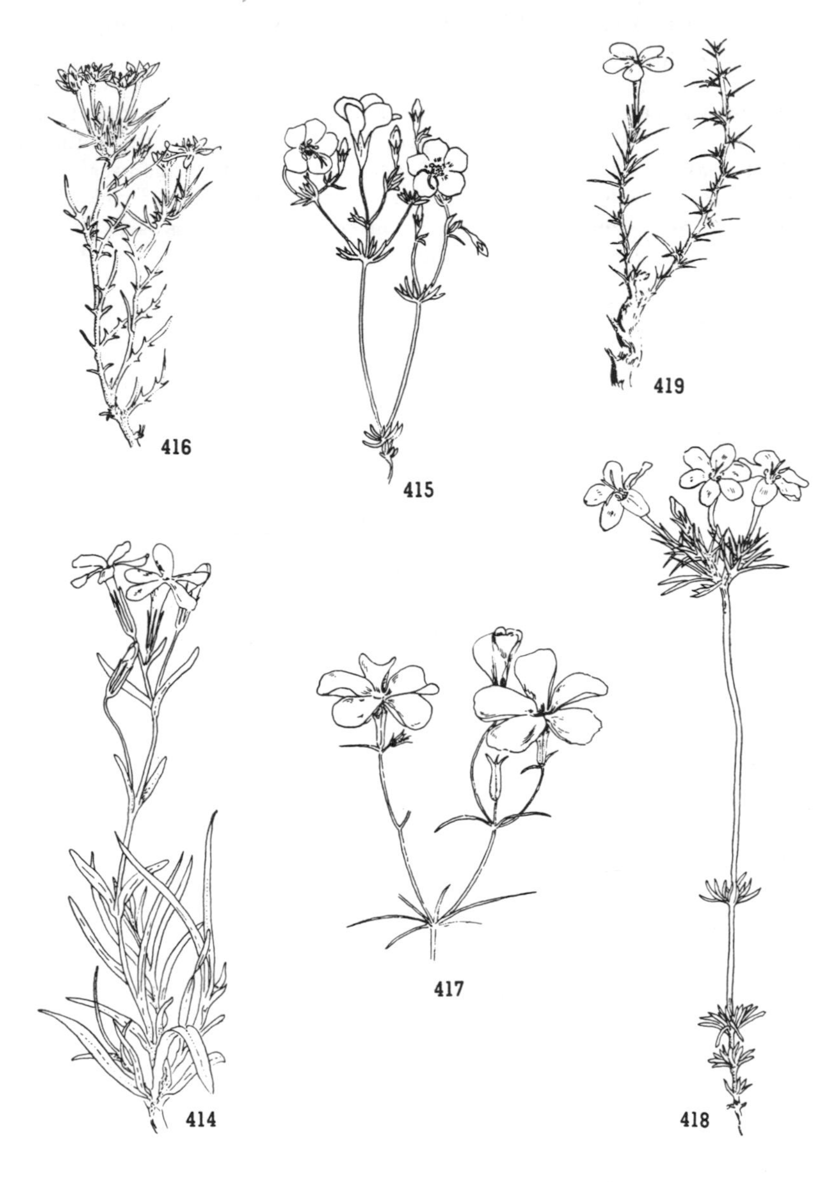

414. *Phlox Stansburyi*
415. *Gilia aurea*
416. *Gilia densifolia austromontana*
417. *Gilia dichotoma*
418. *Gilia brevicula*
419. *Gilia pungens tenuiloba*

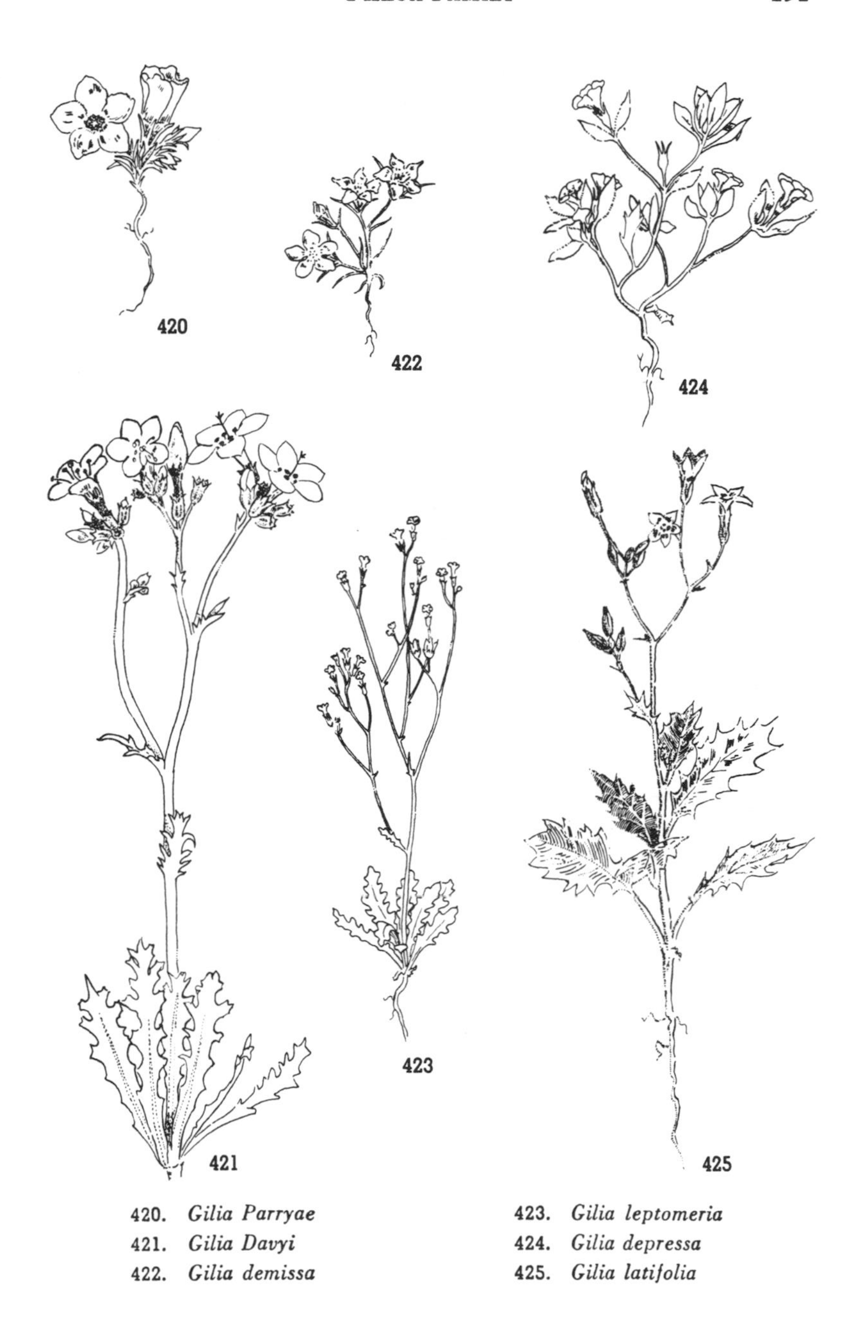

420. *Gilia Parryae*
421. *Gilia Davyi*
422. *Gilia demissa*
423. *Gilia leptomeria*
424. *Gilia depressa*
425. *Gilia latifolia*

423. TOOTH-LEAVED GILIA. *Gilia leptomeria* (Gr., slender-divisioned). **Fl.**: pale white or pinkish. A dainty-flowered annual, of the higher desert basins, blooming prolifically in April and May in the lower elevations and in May and June in the higher valleys of Inyo Co.; Nev. and Utah. Often several of the glandular-haired stems arise from the basal rosette of leaves and run upward to a height of 8 or 9 in.

424. ARGUS GILIA. *Gilia depressa* (L., pressed-down). **Fl.**: white. A northern Mohavean species; nowhere common, but widely scattered; eastward to Nev. and Utah.

425. BROAD-LEAVED GILIA. *Gilia latifolia* (L., broad-leaved). **Fl.**: Tyrian pink. A handsome plant, which keeps consistently to the low, sandy areas. Both deserts, below 2,000 ft.

426. SPOTTED GILIA, LILAC SUNBONNET. *Gilia punctata* (L., dotted). **Fl.**: violet, dotted with purple. A low, tufted, profusely blooming annual, confined to the Mohave D.; to Nev.

427. SCHOTT GILIA. *Gilia Schottii* (A. Schott—see **734**). **Fl.**: pale pink to tan, with purple spots. The illustration shows a single-stemmed plant; but there are often in addition a number of spreading branches 1–3 in. long. Both deserts.

428. PINK-SPOTTED GILIA, PYGMY PINK-SPOT. *Gilia maculata* (L., spotted). **Fl.**: corolla white, with pink spot on each lobe. Rare in sandy washes of the Morongo Pass area; also near Palm Springs.

429. DESERT CALICO. *Gilia Matthewsii* (Dr. W. Matthews, of the U.S. Army, stationed in Owens Valley in 1875). **Fl.**: whitish to pink. A freely flowering annual, with a short central stem from which several horizontal branches spring to make a plant 4 or 5 in. in diameter. When it occurs in dense aggregations, as it often does in sandy flats, its elaborately patterned flowers make a conspicuous show. It is very drought-resistant and persists in bloom late in the season.

430. BRISTLY GILIA. *Gilia setosissima* (L., "very bristly," because of the many stiffish, slender hairs found on the leaves). **Fl.**: light violet. Low, tufted, freely flowering annual, locally frequent, particularly on the sands of washes and mesas of the Colorado and eastern Mohave deserts. Its range extends east and north to Utah and Ida.

431. DESERT GILIA. *Gilia eremica* (Gr., lonely, desert). **Fl.**: pale lavender. The flowers are strongly 2-lipped. Very widespread and common in sand. Both Calif. deserts; to Nev. Another gilia, with pale blue flowers and bracts embedded in tufts of white wool, is *Gilia filifolia diffusa.* Its flowers are smaller and quite regular and the stems are reddish. It too is common in sandy areas of both deserts. The var. *Harwoodii,* with similar small, regular flowers but with marked pale, woolly herbage, is abundant in May in the sand hills near Blythe and Kelso.

432. THREAD-STEMMED GILIA. *Gilia filiformis* (L., formed like a thread, i.e., thread-like). **Fl.**: yellow. In aggregations among rocks of low hills of the Mohave D.; to Utah.

433. BROAD-FLOWERED GILIA. *Gilia latiflora* (L., broad-flowered). **Fl.**: violet to pink, with yellow throat variously spotted. Sandy areas western Mohave; to western Nev., 2,000–5,000 ft.

434. SPREADING GILIA. *Gilia polycladon* (Gr., many-stemmed). **Fl.**: white. Many-stemmed annual, with reddish-brown stems and with flowers in leafy terminal tufts. Argus Mts., mid-Mohave D.; Deep Spring Valley; to Utah, western Tex. Rare in Calif.

435. ROCK GILIA. *Gilia scopulorum* (L., of the rocks). **Fl.**: rose-lavender, with yellow throat. Both deserts on dry washes and rocky slopes. A very variable species, 4–12 in. high, with thin, sticky leaves.

436. LESSER GILIA. *Gilia inconspicua* (L., "inconspicuous," with reference to the small flowers). **Fl.**: light blue to cream. A variable, freely branching annual, from 4 to 20 in. tall. Common to dry and sandy soils of both Calif. deserts; to Wyo. and N.Mex.

437. MOHAVE GILIA. *Gilia densifolia mohavensis* (L., dense leaved; of the Mohave). **Fl.**: pale blue. A perennial species of rocky areas of the juniper belt. Joshua Tree Nat. Mon., western Mohave D.

HYDROPHYLLACEAE. Waterleaf Family

438. PARRY PHACELIA. *Phacelia Parryi* (Dr. C. C. Parry—see **24**). (Gr., "a cluster" or "fascicle," referring to the crowded flowers). **Fl.**: royal purple or deep violet. An erect annual, with handsome, shallow-cupped flowers and hairy-glandular herbage. Western edge of the Colorado D., thence westward to the coast. The type specimens were collected by Dr. Parry in 1850 somewhere in the mountains east of San Diego.

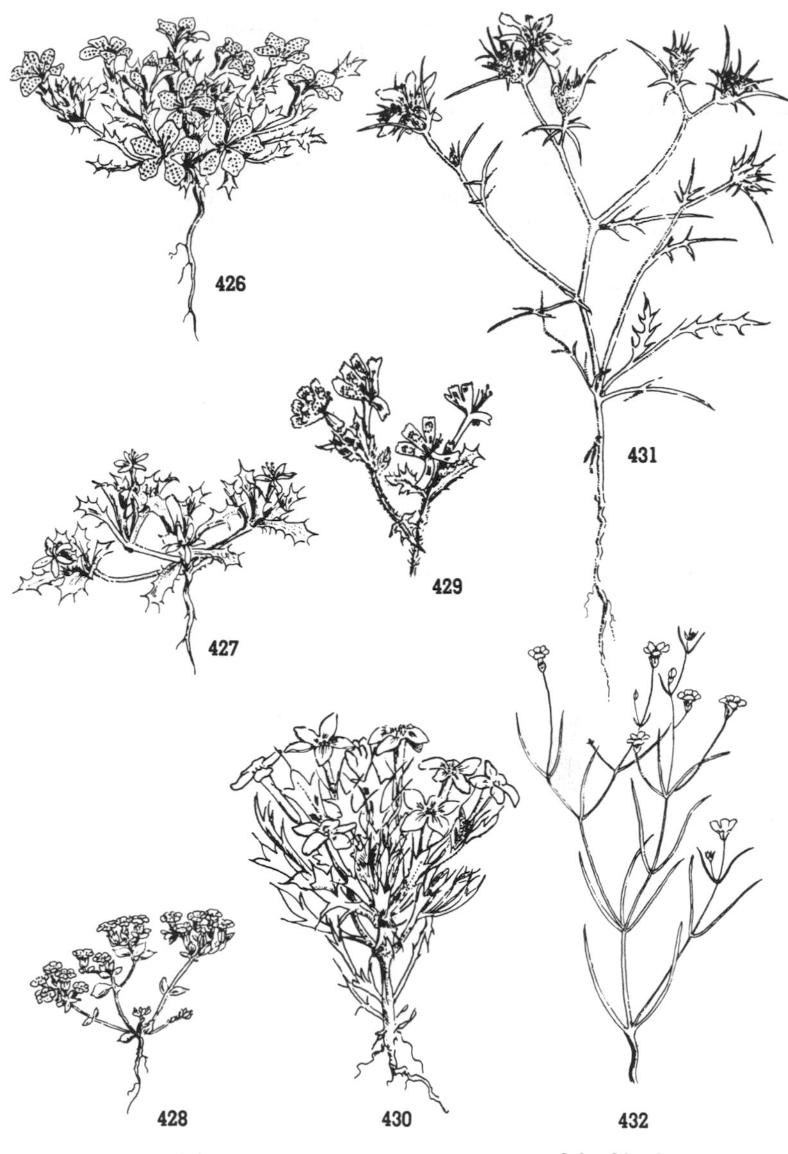

426. *Gilia punctata*
427. *Gilia Schottii*
428. *Gilia maculata*, ×1
429. *Gilia Matthewsii*
430. *Gilia setosissima*
431. *Gilia eremica*, ×1
432. *Gilia filiformis*

433. *Gilia latiflora*
434. *Gilia polycladon*
435. *Gilia scopulorum*
436. *Gilia inconspicua*
437. *Gilia densifolia mohavensis*

439. DEATH VALLEY PHACELIA. *Phacelia vallis-mortae.* (Of Death Valley). **Fl.**: pale purplish-lavender. A weak annual, often with purplish stems. The herbage, especially the calyx, covered with bristly hairs. The stamens are scarcely as long as the broad, funnelform corolla. Generally growing among shrubs or in the higher desert ranges along with *Chenopodium hybridum* in the partial shade of nut-pines. Mohave D. from Barstow to Death Valley and eastward to Nev.

440. FAT-LEAF PHACELIA. *Phacelia distans australis* (L., distant; L., southern). **Fl.**: light purple. A variety of *P. distans* (see **450**), distinguished by its almost succulent, less finely divided leaves. Infrequent, under, or mingling with low shrubs of the western Mohave D.

441. WEASEL PHACELIA. *Phacelia mustelina* (L., "pertaining to a weasel," because of its odor). **Fl.**: white. Low, rank-odored annual of mountains of the Death Valley area.

442. SPECTER PHACELIA. *Phacelia pedicellata* (L., "with a pedicel," because of thread-like stalks of the flower). **Fl.**: white or pale blue. An erect, heavy-stemmed, foul-scented, large-leaved annual, 4–20 in. tall, blooming in April and May in open valleys and canyons (500–4,000 ft.) of both deserts; to Baja Calif.

443. PARISH PHACELIA. *Phacelia Parishii* (S. B. Parish—see **603**). **Fl.**: bluish. A rare annual, with somewhat reclining stems 4–6 in. long. Known only from very scattered stations of the Mohave D. and the Santa Rosa Mts.

444. LEMMON PHACELIA. *Phacelia Lemmonii* (J. G. Lemmon —see **464**). **Fl.**: white. Rare on moist, sandy places of desert canyons (4,000–5,000 ft.) from the Little San Bernardino Mts. to mountains about Death Valley; to Nev. and Ariz.

445. PEARL-O'ROCK. *Phacelia perityloides* (like *Perityle*, which has somewhat similar leaves). **Fl.**: white, with purplish tube. Panamint Mts., in canyons at 3,000–4,500 ft.

446. JAEGER PHACELIA. *Phacelia perityloides Jaegeri* (Edmund C. Jaeger). **Fl.**: white. Clark Mt.; Sheep Mt., Nev.

447. ROUND-LEAVED PHACELIA. *Phacelia rotundifolia* (L., round-leaved). **Fl.**: white. On shady slopes and in shaded rock crevices of canyons of the Mohave D.; to Nev. and Utah.

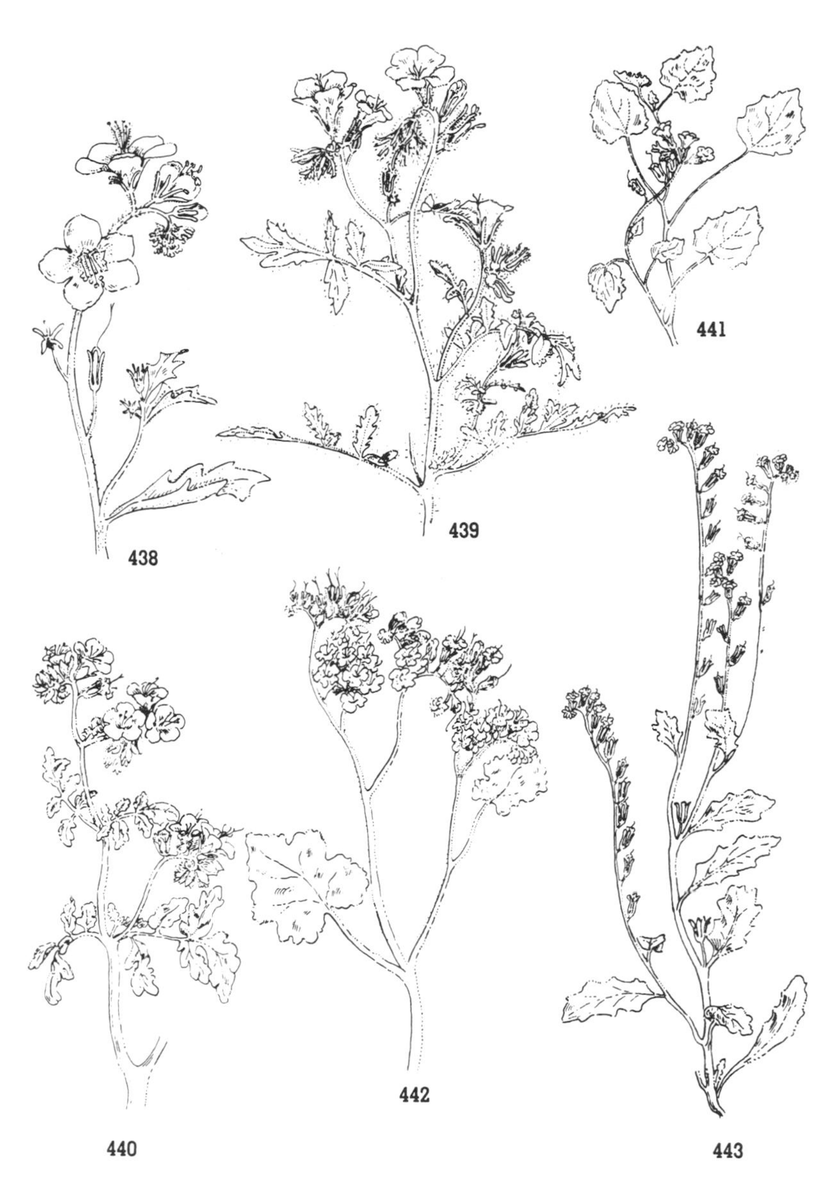

438. *Phacelia Parryi*
439. *Phacelia vallis-mortae*
440. *Phacelia distans australis*
441.* *Phacelia mustelina*
442. *Phacelia pedicellata*
443. *Phacelia Parishii*

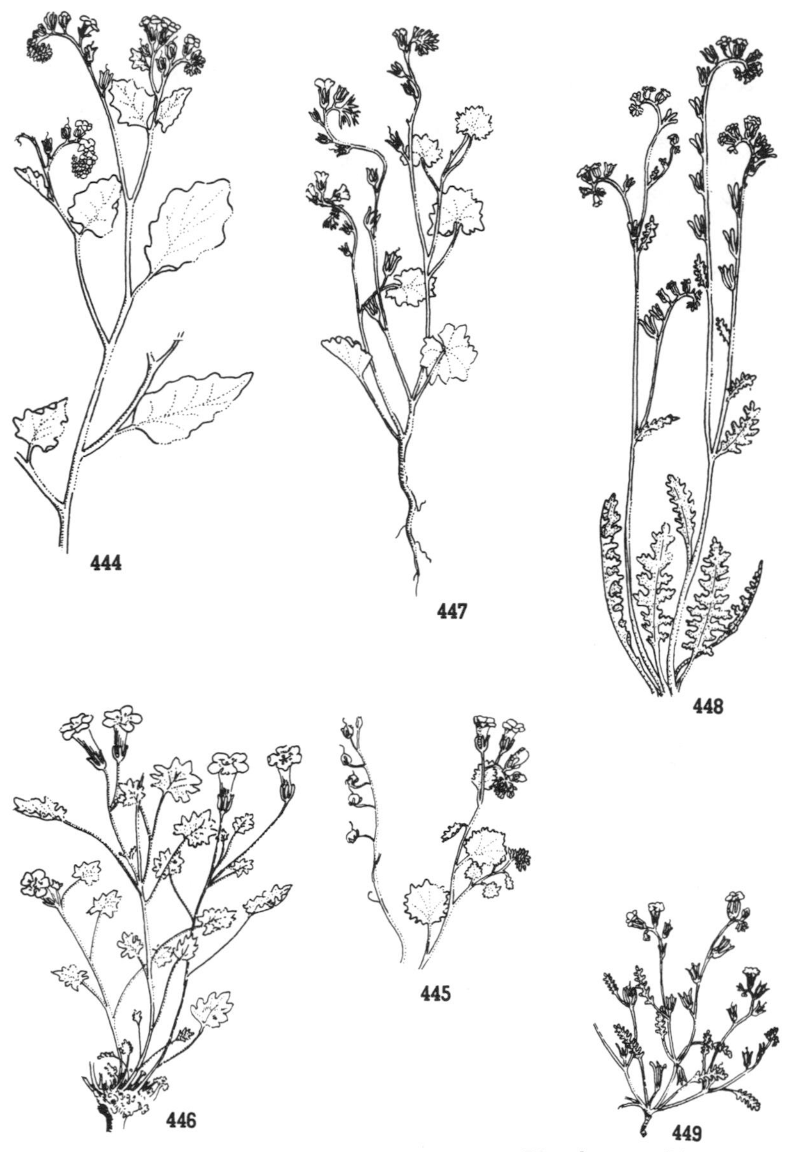

444.* *Phacelia Lemmonii*
445. *Phacelia perityloides*
446. *Phacelia perityloides Jaegeri*
447. *Phacelia rotundifolia*
448. *Phacelia affinis*
449. *Phacelia Ivesiana*

448. PURPLE-BELL PHACELIA. *Phacelia affinis* (L., adjacent, i.e., nearly related to *P. Orcuttiana* and *P. Fremontii*). **Fl.**: purplish. Occasional, 2,000–4,000 ft., Little San Bernardino Mts., Jacumba; to Baja Calif.

449. IVES PHACELIA. *Phacelia Ivesiana* (Lieut. Joseph C. Ives, in charge of a government party making a survey of the "Colorado River of the West" in 1857–58; Dr. J. S. Newberry was in charge of the natural history and geological investigations). **Fl.**: white or purplish. A dwarf plant, 2–5 in. high, growing at low elevations on both Calif. deserts; to Wyo. and Utah.

450. LACY PHACELIA. *Phacelia tanacetifolia* (*Tanacetum,* a genus which includes tansy, and L., leaf). **Fl.**: lavender or bluish. A stout-stemmed species, with long-exserted stamens, often growing about the bases of creosote or cat's-claw bushes. Western Mohave D. to Great Valley of Calif. Much like *tanacetifolia* is the purple-flowered *P. distans* (L., distant), with scarcely exserted stamens and shorter calyx lobes. Common in April on Colorado D. under cat's claw and ironwood.

451. HALL PHACELIA. *Phacelia Hallii* (H. M. Hall—see **275**). **Fl.**: bluish-purple, with paler tube. Annual, with several stems, 4–8 in. long. Dry slopes 3,500–7,000 ft., Little San Bernardino Mts. to Mt. Piños.

452. THICK-LEAVED PHACELIA. *Phacelia pachyphylla* (Gr., thick-leaved). **Fl.**: bluish-purple. Dwarf annual, 1½–4 in. high; the herbage often viscid. Both deserts; to Baja Calif. and Ariz., below 2,000 ft. Rare; on alkaline soils or on gravels.

453. CALTHA-LEAVED PHACELIA. *Phacelia calthifolia* (L., like the leaf of *Caltha,* the marsh marigold). **Fl.**: purple. Coarse-stemmed, erect annual, 4 in. to 1 ft. high, with sticky glands on the herbage. In drying it stains the herbarium sheets a rich reddish-brown. Dry washes from Death Valley south to Barstow.

454. SMALL-FLOWERED PHACELIA. *Phacelia cryptantha* (Gr., "hidden flower," because the flower is so small). **Fl.**: pale lavender. Panamint Mts. to the western edge of Colorado D.; and Ariz.

455. CAMPANULATE PHACELIA. *Phacelia campanularia* (like *Campanula,* the bellflower). **Fl.**: blue violet with white spot at base of each sinus. Annual, 4 in. to 2 ft. high, with flaring, somewhat vase-shaped flowers; often in aggregations in the shelter of

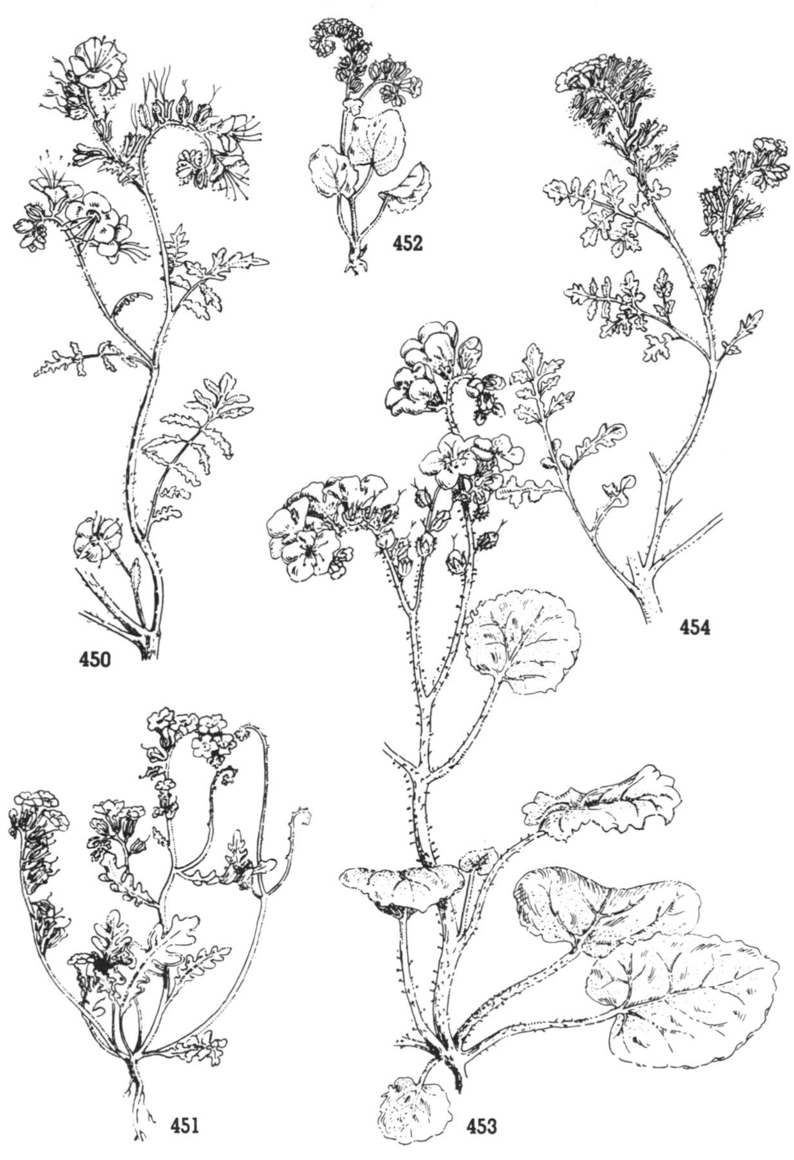

450. *Phacelia tanacetifolia*
451.* *Phacelia Hallii*
452.* *Phacelia pachyphylla*
453. *Phacelia calthifolia*
454. *Phacelia cryptantha*

455. *Phacelia Campanularia*
456. *Phacelia crenulata ambigua*
457. *Phacelia bicolor*
458. *Phacelia Fremontii*, ×1
459. *Tricardia Watsonii*

rocks, where it makes an almost startling show of color. Where it grows in the open, it is a smaller plant. Mohave and western Colorado deserts, below 4,000 ft. It was one of the plants found by Dr. C. C. Parry on the Whitewater Bench. Seeds were sent to England, and there it is still a favorite garden flower.

456. NOTCH-LEAVED PHACELIA. *Phacelia crenulata ambigua* (L., "minutely notched [leaf margins]"; L., "doubtful," expressing a state of the describer's mind). **Fl.**: deep violet or bluish-purple. Green-stemmed, strong-scented annual, 4–16 in. tall. The stamens are at least 1½ times the length of the corolla. Eastern Mohave and Colorado deserts. The var. *funerea* (known from the Death Valley area north to the White Mts.) has reddish stems, and the entire plant is covered with dark, stalked glands.

457. TWO-COLORED PHACELIA. *Phacelia bicolor* (L., "two-colored," with reference to the flower). **Fl.**: purple, with yellow tube. Annual, with stems 4–8 in. long. Joshua Tree Nat. Mon., western Mohave D. to Lassen Co.; Nev.

458. FRÉMONT PHACELIA. *Phacelia Fremontii* (Capt. John C. Frémont—see **315**). **Fl.**: lavender-violet, with yellow throat. A common annual of sandy stretches and dry streamways of mountains of the western and northern Mohave D.; to Nev. and Ariz. It comes into flower early in the season and persists in blooming over a long period. The flowers have a strong mephitic odor, that is, one like the scent of a skunk.

459. THREE-HEARTS. *Tricardia Watsonii* (Gr., "three hearts," in allusion to the sepals; Sereno Watson, 1826–1892, Asa Gray's assistant and successor, botanist of King's expedition and principal author of two volumes on the botany of California, 1876–1880). **Fl.**: purplish. An herb, 5–10 in. high, certain to attract attention because of the racemes of large, heart-shaped, pinkish-green, papery sepals. The several stems spring from a short, perennial rootstalk. Occasional among junipers and piñons from the Inyo Mts.; to Utah.

460. SMALL-LEAVED NAMA. *Nama pusillum* (Gr., "a spring of water or stream," in allusion to the natural place of growth; L., small). **Fl.**: whitish. Plants prostrate to slightly ascending. Stony flats and gentle slopes of the northern Mohave D. to the Colorado R.

461. PURPLE MAT. *Nama demissum* (L., humble, lowly). **Fl.**: deep pink to purple. The species and its varieties, *Covillei* (of the Death Valley area) and *deserti* (common to both deserts), are lovers of flat clay and sandy soils, and often in favorable years the plants spread enough to form broad, colorful mats. In dry years plants hurry on to maturity without spending much time making leaves, and under these circumstances a plant may bear but a single flower.

462. HISPID NAMA. *Nama hispidum spathulatum* (L., hairy with bristles; L., spatulate [leaf]). **Fl.**: purplish-red. Rare annual of both Calif. deserts; most common in eastern portions near the Colorado R.

463. NARROW-LEAVED NAMA. *Nama depressum* (L., pressed-down). **Fl.**: whitish. Low, yellow-green plant, of sandy areas of the western half of the Mohave D., north to Owens Valley.

464. LEMMONIA. *Lemmonia californica.* (John Gill Lemmon, 1832–1908, botanical explorer, writer on Southwestern plants, and California's first scientific forester. For many years he carried on a spirited correspondence with Asa Gray and sent him many plants which proved to be new to science. He made a number of collecting trips on the deserts in company with Dr. C. C. Parry [see **24**] and other well-known botanists of his time. Of California.) **Fl.**: white. Small, depressed annual found in dry areas of mountains bordering the western Mohave D. The corolla is open, bell-shaped, and the two-valved capsule contains four seeds.

465. WHITE FIESTA-FLOWER. *Pholistoma membranaceum* (Gr., "scaly mouth," because of the scales in the throat of the flower; L., "membranous or skin-like," referring to the leaves). **Fl.**: white, sometimes with lanceolate purplish spots. A weak annual growing in the shade of rocks or other shelter. Stems 8–20 in. long, with minute retrorse prickles on the angles; these are very evident when the plant is handled. The calyx lobes bear bristles on their edges. Often the leaves are but once pinnately parted. Both deserts to central Calif., at low altitudes. The Indians used the young plants for greens.

466. SMALL-FLOWERED EUCRYPTA. *Eucrypta micrantha* (Gr., "well-hidden," in allusion to the concealed cells of the capsule; Gr., small-flowered). **Fl.**: purplish-blue or white, with yellow tube. A weak, erect, slender branching annual, 4–10 in. tall, and of variable habit. Well distributed in sheltered places over eastern Mohave and northern Colorado deserts; to southern Nev., Utah, and Tex. A more erect, robust form found in sunny situations among piñons in the southern Inyo range has been recently collected by the author. It has much the appearance of a tiny *Phacelia.*

467. TORREY EUCRYPTA. *Eucrypta chrysanthemifolia bipinnatifida* (L., chrysanthemum leaf; L., twice pinnately cleft). **Fl.**: white or bluish. A weak-stemmed, few-flowered annual, growing in

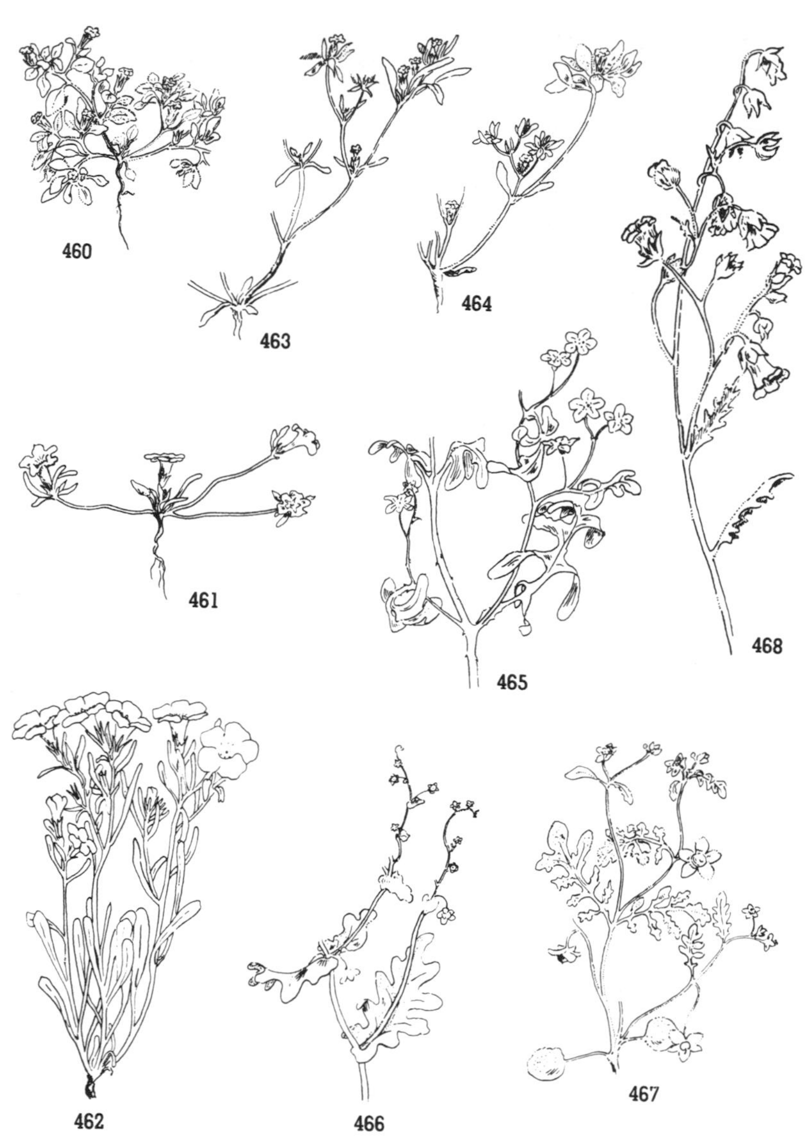

460. *Nama pusillum*
461. *Nama demissum*
462. *Nama hispidum spathulatum*
463.* *Nama depressum*
464* *Lemmonia californica*
465. *Pholistoma membranaceum*
466. *Eucrypta micrantha*
467. *Eucrypta chrysanthemifolia bipinnatifida*
468. *Emmenanthe penduliflora*

469. *Nemophila Menziesii*
470.* *Nemophila rotata annulata*
471. *Eriodictyon trichocalyx lanatum*
472. *Eriodictyon crassifolium*
473.* *Eriodictyon angustifolium*

the shelter of rocks or bushes in desert areas of Calif; to Baja Calif., southern Nev., and Ariz. The tiny corolla about equals or slightly exceeds the lobes of the calyx. The leaves are peculiarly blotched with white.

468. CALIFORNIA YELLOW BELLS, WHISPERING-BELLS. *Emmenanthe penduliflora* (Gr., "abiding flower," in allusion to the persistent corolla. In the nearly related *Phacelias* the corolla falls away in age. L., hanging-flowered). **Fl.**: pale yellow. A yellowish-green annual, 4–20 in. high; the herbage covered with soft, short hairs. Below 5,000 ft., from southern to central Calif., occasional on both deserts; to Utah and Ariz.

469. BABY BLUE-EYES. *Nemophila Menziesii* (Gr., groves + to love; Archibald Menzies, 1754–1842, Scottish botanist, born in Perthshire. He was surgeon on the ship "Discovery" under Captain Vancouver and visited California on numerous occasions—southern California in 1794). **Fl.**: dark to light blue, and often dotted or veined with purple. Slender, succulent annual of cismontane California reaching the desert's western edge.

470. PALE BABY BLUE-EYES. *Nemophila rotata annulata* (L., wheel-shaped; L., ringed). **Fl.**: pale blue. The short-haired, oblong scales or appendages of the corolla are attached along one side only. A cismontane species entering the western edge of the Mohave D.

471. WOOLLY YERBA SANTA. *Eriodictyon trichocalyx lanatum* (Gr., "wool network," in reference to the under surface of the leaves; Gr., hairy-calyxed; L., woolly). **Fl.**: pale lilac or white. Erect shrub, 1½–4 ft. tall, with leaves dark green and sticky above but white and felty beneath. The young twigs are covered with minute hairs. A tea made from the leaves is used as a remedy for coughs. It is rather "peppery" but is pleasant to the taste. Physicians use the sweetened fluid extract as a vehicle for other drugs. Canyons of the western Colorado D., from the Santa Rosa Mts. south.

472. FELT-LEAVED YERBA SANTA. *Eriodictyon crassifolium* (L., thick-leaved). **Fl.**: lavender. An open, erect shrub, with white-felty leaves and twigs. About canyons among rocks. Below 6,000 ft. from Tehachapi, south to Palm Springs; Baja Calif.

473. NARROW-LEAVED YERBA SANTA. *Eriodictyon angustifolium* (L., narrow-leaved). **Fl.**: white. Lower piñon belt in scattered patches. Plentiful in southern Nev. and reaching Calif. in mountains of the eastern end of the Mohave D.

BORAGINACEAE. Borage Family

474. PALMER COLDENIA. *Coldenia Palmeri.* (Dr. Cadwallader Colden, Lieutenant Governor of New York, Colonial botanist and correspondent of Linnaeus; Dr. Edward Palmer, 1831–1911, ethnobotanist, botanical explorer in southwestern United States and Mexico, and one of the plant collectors for the California State Geological Survey in 1861. After collecting *Canbya candida*—see **142**—on his way from Fort Mohave to San Bernardino for supplies, Dr. Palmer made a trip to San Gorgonio Peak in May 1876. On the way he fell from his horse, severely injuring his spine. Obliged to lie on an improvised bed for a day until he could be carried home, a report was published in San Bernardino that the Doctor had been left "on the mountain without grass or water with a man to look after him." "Wherever I went for sometime afterwards," said Dr. Palmer, "I was pointed out as the man who had been left on Grayback* without grass or water and sometimes jocularly addressed, 'Hello, old grass and water! How's your back?' ") **Fl.**: bluish-white. A rather coarse, hemispherical perennial, with trailing stems often spreading to form a plant 2½ ft. across. The older branches are white, with splitting, papery bark. After a wet winter the plant early comes into full flower and is most appealing. In dry seasons it presents a very ragged appearance and there is little growth. Confined to sands, silts, and clays of the southern Colorado D.; southwestern Ariz. and Baja Calif.

475. SHRUBBY COLDENIA. *Coldenia canescens* (L., grayish with hairs, alluding to the herbage of the species). **Fl.**: deep lavender, with white throat. A woody shrub, 4–6 in. high, with gnarled, spreading branches covered much of the year with dry, gray, outrolled, bristly leaves. After rains new green leaves and, later, flowers appear near and at the ends of the branchlets. The regular period of flowering is late March and early April. Rocky slopes and benches of the Eagle, Chuckawalla, and Chocolate mountains; Ariz., Tex., and Mex. A large-flowered form from the Chocolate Mts. recently described by Dr. Ivan M. Johnston is given the varietal name *pulchella.*

476. NUTTALL COLDENIA. *Coldenia Nuttallii* (Thomas Nuttall—see **87**). **Fl.**: pink or white. A Mohavean species, which spreads eastward to Utah and Wyo.; known also from the high Andes

* In Civil War days the designation for the "cootie" was "grayback" and the early Californians are said to have applied this wretched name to the mountain peak because of some fancied resemblance in its form to the lowly body louse.

of Argentina! It is a prostrate annual, with somewhat ashen-colored herbage and revolute-margined leaves.

477. PLICATE COLDENIA. *Coldenia plicata* (L., "plaited," because of the appearance of the leaves). **Fl.**: light blue. The stems lie flat to the earth, forming dark gray-green mats 4–16 in. across. Common to the lower, sandy flats of both deserts; to western Ariz., southern Nev., and Baja Calif. A perennial species, flowering late in March.

478. FALSE MORNING-GLORY. *Heliotropium convolvulaceum californicum* (Gr., "sun-turning," in reference to the summer solstice, when the first-described species were supposed to bloom; L., like *Convolvulus,* the morning-glory; of California). **Fl.**: pure white. A low-growing, rough-hairy annual, with handsome, sweet-scented, morning-glory-like flowers, which open at sunset. It can be readily grown in gardens from seed. Sandy mid-Mohave D.; to Ariz. and Sonora.

479. CHINESE PUSLEY. *Heliotropium Curassavicum oculatum* (adjective for Curaçao, Dutch West Indies, i.e., "Curaçaoan," one of the first collections having been made there; L., "having eyes," with reference to the spots in the flower throat). **Fl.**: white, with yellow spots in throat, commonly becoming purplish about the center. A fleshy-stemmed, spreading plant, from a perennial root, particularly common about irrigation canals in the Imperial Valley. The flowers open as the spike uncoils. Very susceptible to frost. Saline and alkaline soils over southern Calif. up to 6,700 ft.; to Ariz., southwestern Utah, and central Nev. The common name, "pusley," is probably a corruption of "purslane," which in turn is derived from the L. *Portulaca,* a fleshy annual used in Europe as a salad and potherb.

480. ARCHED - NUTTED COMB - BUR. *Pectocarya recurvata* (Gr., comb nut; L., "bent back," in allusion to the reflexed body of the nutlets). **Fl.**: white. Stems mostly prostrate. Mountains of both deserts; to Baja Calif., Sonora.

481. BROAD - NUTTED COMB - BUR. *Pectocarya platycarpa* (Gr., wide-fruited). **Fl.**: white. Spreading annual herb, embedding its roots in dry gravel and stony soils of both deserts; south to Sonora. The nutlets have very broad-toothed margins.

482. HAIRY-LEAVED COMB-BUR. *Pectocarya penicillata heterocarpa* (L., with brush of "soft hairs," the upper margin of the nutlets being "pectinate with uncinate hairs"; Gr., different-fruited). **Fl.**: white. Spreading annual, 4–8 in. across. The nutlets have hooked bristles only at the end. A neat little plant which keeps consistently to gravelly or sandy slopes and valleys. Both deserts; to Baja Calif.

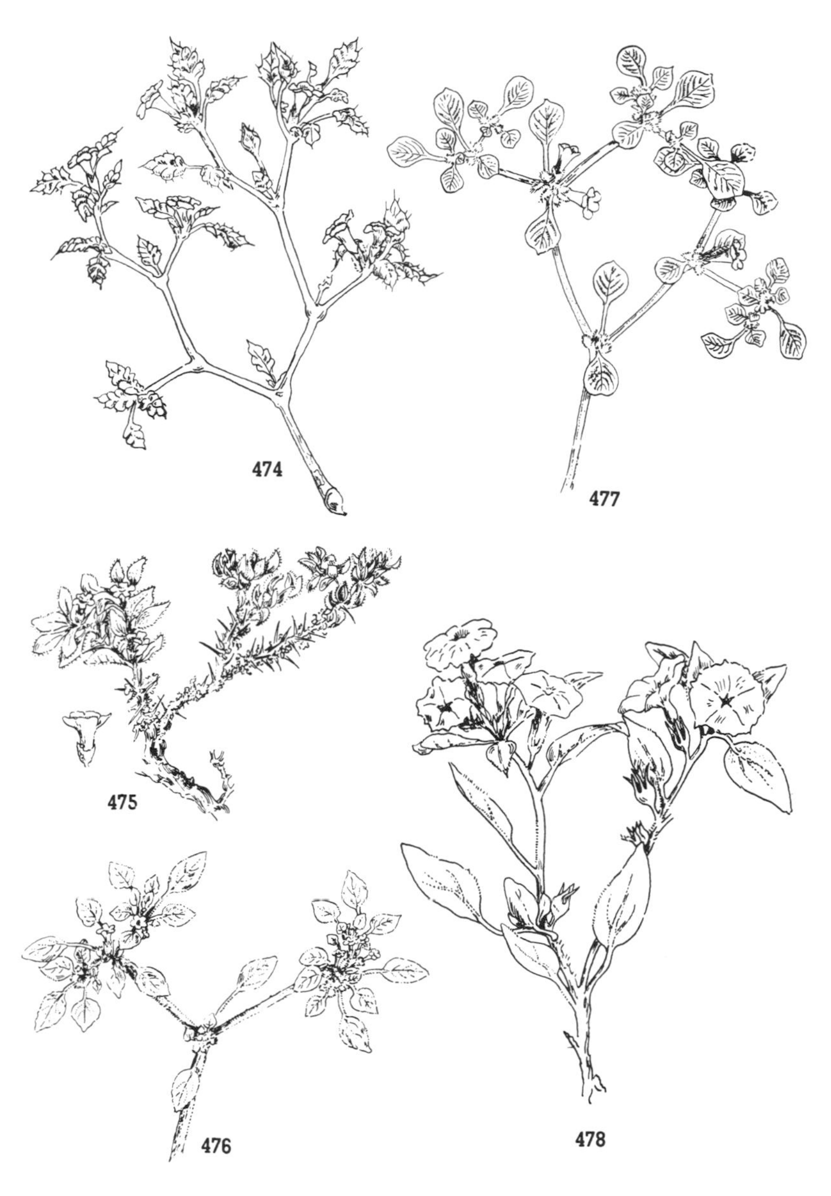

474. *Coldenia Palmeri*, ×1
475. *Coldenia canescens*
476. *Coldenia Nuttallii*
477. *Coldenia plicata*, ×1
478. *Heliotropium convolvulaceum californicum*

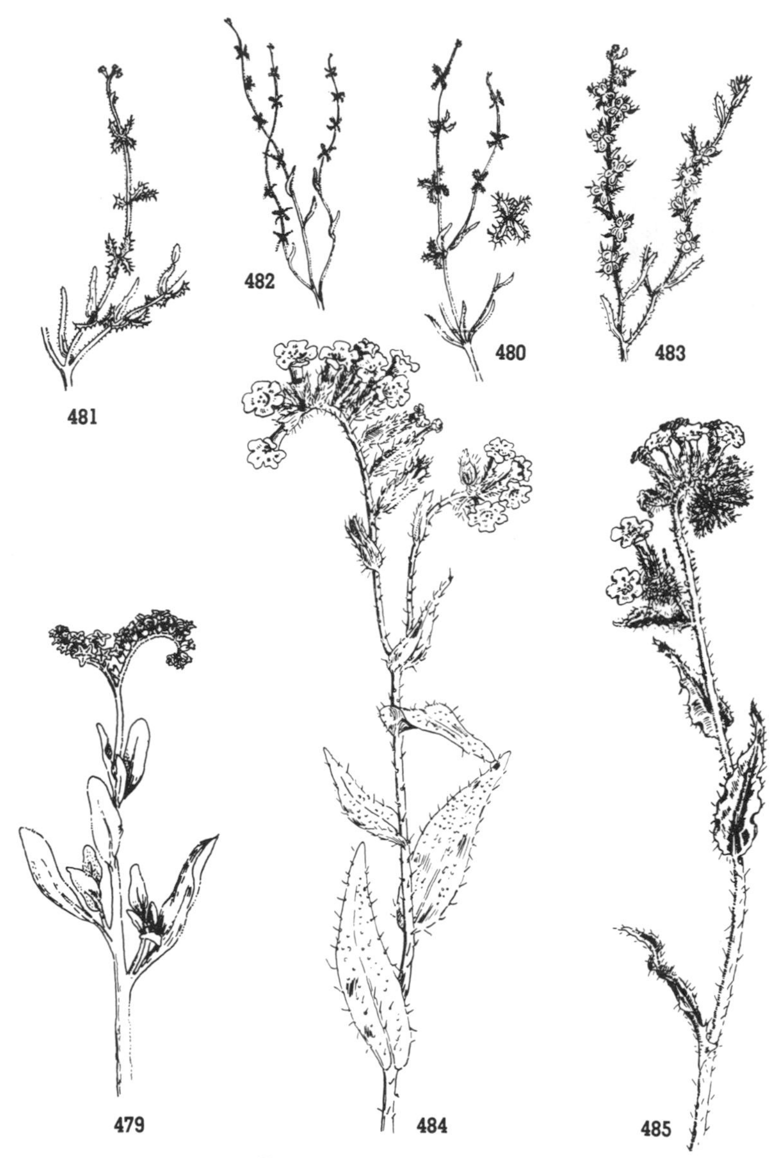

479. *Heliotropium Curassavicum oculatum*

480. *Pectocarya recurvata*
481. *Pectocarya platycarpa*
482.* *Pectocarya penicillata heterocarpa*
483. *Pectocarya setosa*
484. *Amsinckia vernicosa*
485. *Amsinckia tessellata*

483. STIFF - STEMMED COMB - BUR. *Pectocarya setosa* (L., "bristly," with reference to the herbage). **Fl.**: white. Stems erect or ascending, 3–8 in. long, armed––as are also the leaves—with stiffish hairs. Both deserts, on dry gravel or sand slopes and benches; to eastern Wash., Utah, and Ariz.

484. SHINY - SEEDED FIDDLENECK. *Amsinckia vernicosa* (Wilh. Amsinck, Bürgermeister and botanic patron of Hamburg; L., "brilliantly polished," because of the smooth, shiny nutlets). **Fl.**: golden yellow. A fleshy-leaved borage found on mountains of the northern Mohave region. The stout stems, 7–18 in. high, are for the most part without hairs, especially below. The pustulate disks on the upper sides of the dry leaves are very noticeable. Mohave D., also Coast Ranges of San Luis Obispo and Kern counties.

485. CHECKER FIDDLENECK. *Amsinckia tessellata* (L., "checkered," in allusion to the arrangement of warts on the backs of the nutlets). **Fl.**: orange-yellow. This leafy borage, unlike *A. vernicosa,* is armed throughout with numerous stiff hairs. It is very abundant and has considerable aesthetic appeal, especially on the Mohave Desert, where it sometimes forms vast fields. Both deserts, San Joaquin Valley; to Utah, northern Baja Calif., eastern Wash., Ida., and also northern Patagonia in South America!

486. WHITE-HAIRED FORGET-ME-NOT. *Cryptantha maritima* (Gr., "hidden flower," the flowers being small; L., "of the sea," it often being found along the coast). **Fl.**: white. A widespread, often reddish-stemmed annual, of sands and gravelly areas of both deserts as far north as Death Valley; also coastal Calif., adjacent Ariz., and Baja Calif.

487. SCENTED FORGET-ME-NOT. *Cryptantha utahensis* (of Utah). **Fl.**: white. In this species the stems are erectly branched, the base of the calyx is densely covered with appressed, silky hairs, and the flowers are very fragrantly scented, like carnations. Of broad altitudinal range (4,000–6,000 ft.), in California desert mountains and basins as far north as Inyo Co.; east to Utah and Ariz.

488. GRAVEL FORGET-ME-NOT. *Cryptantha decipiens* (L., "deceiving," with respect to identity). **Fl.**: white. Slender-stemmed annual, 4–11 in. tall, with flower spikes usually in pairs. Sand and gravelly stretches of the creosote-bush belt of the eastern Mohave D.; adjacent Ariz. and Nev.

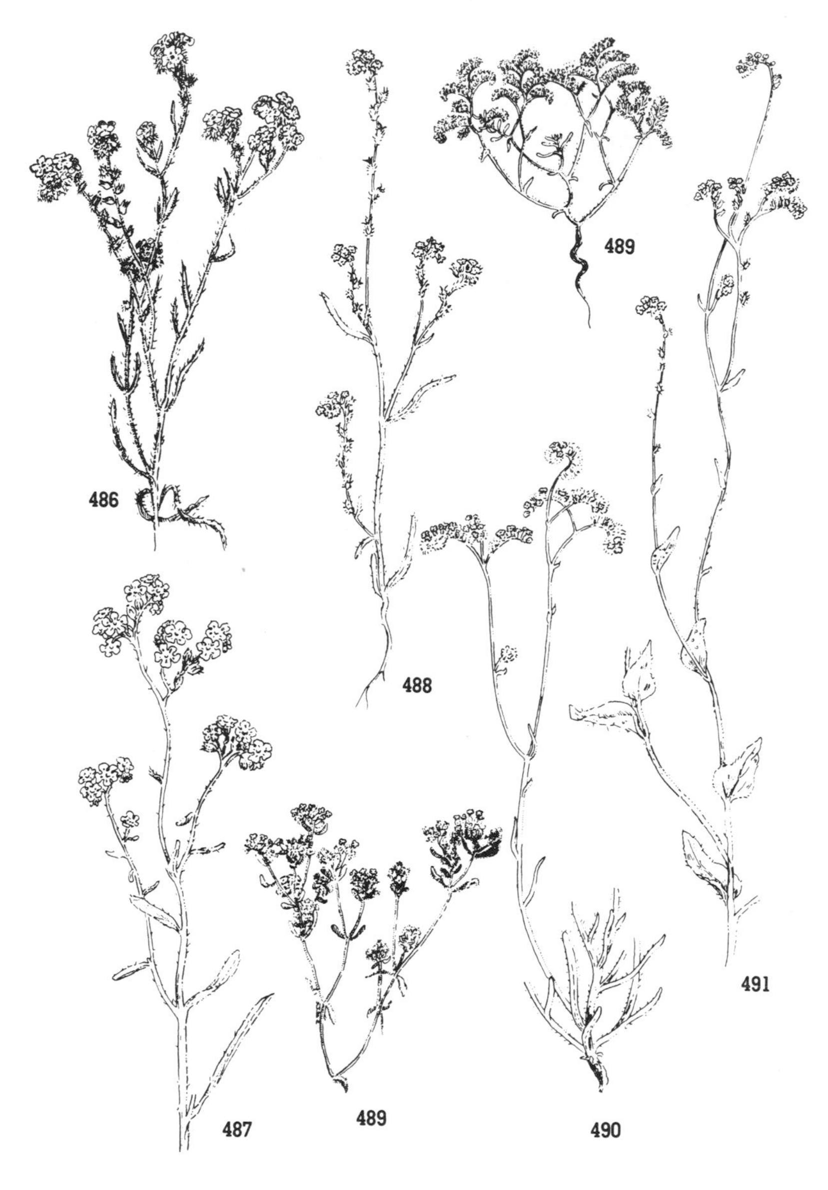

486. *Cryptantha maritima*
487. *Cryptantha utahensis*
488. *Cryptantha decipiens*
489. *Cryptantha micrantha*
490. *Cryptantha gracilis*
491. *Cryptantha dumetorum*

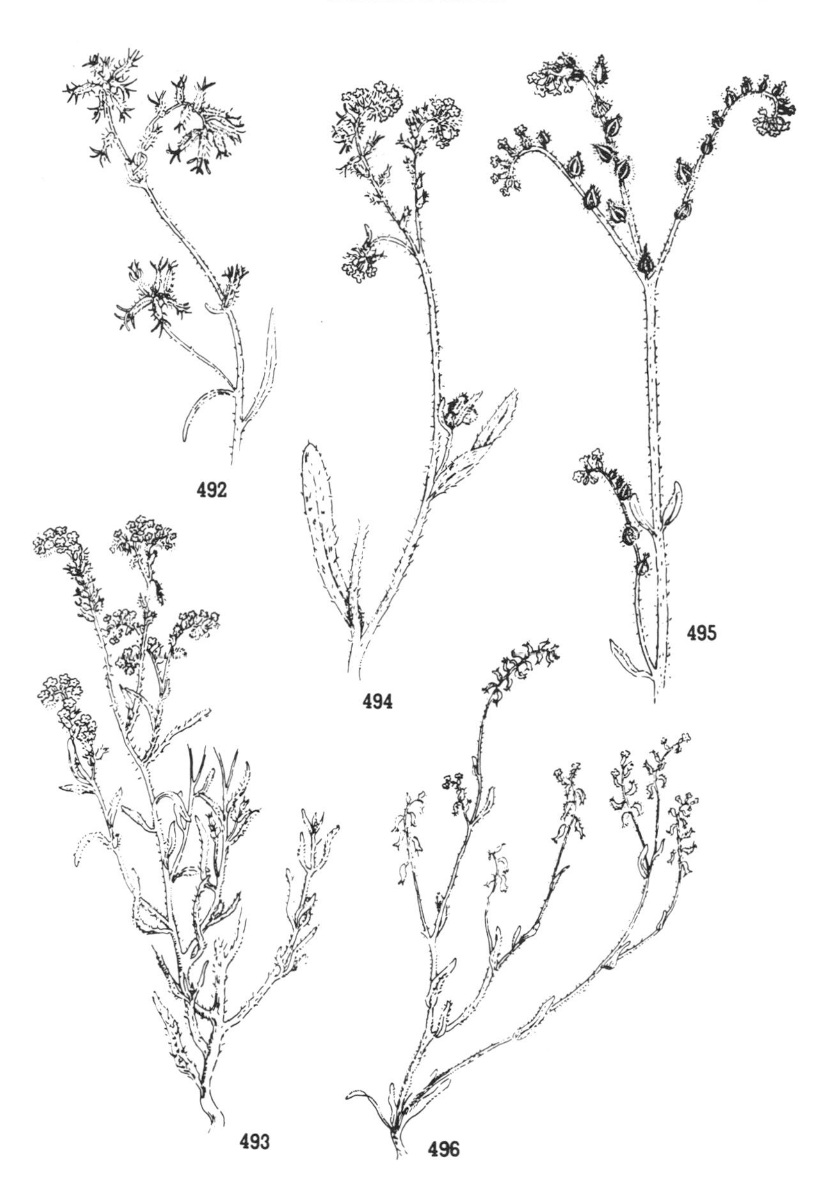

492. *Cryptantha nevadensis*
493. *Cryptantha inaequata*
494. *Cryptantha barbigera*
495. *Cryptantha pterocarya*, ×1
496. *Cryptantha recurvata*

489. PURPLE-ROOTED FORGET-ME-NOT. *Cryptantha micrantha* (Gr., small-flowered). **Fl.**: white. Certainly the daintiest of all our borages. A compact little plant known for its pigmented roots, which stain the objects they touch a rich violet. When dry, the plants often turn cinnamon-brown. Sandy areas of both deserts, up to 4,000 ft.; Tex., to northern Baja Calif., Ore., and Utah. The two illustrations show different forms of this very variable plant.

490. SLENDER FORGET-ME-NOT. *Cryptantha gracilis* (L., slender). **Fl.**: white. Individualized by its densely soft-hairy calyx. Found in the piñon belt of the higher desert ranges of Calif., such as the Providence and Panamint mountains; north to Mono Co.; east to Colo. and Ariz.

491. FLEXUOUS FORGET-ME-NOT. *Cryptantha dumetorum* (L., of thickets. **Fl.**: white. Generally supporting itself by climbing up among woody-branched bushes, thus effectively protecting itself from sheep and other browsing animals. The leaves have numerous hair-bearing pustules. Creosote-bush area of the Mohave and northern Colorado deserts; to western Nevada.

492. NEVADA FORGET-ME-NOT. *Cryptantha nevadensis* (of Nevada). **Fl.**: white. A very distinctive species. The small corolla is inconspicuous, but the long-hairy, slender calyx lobes (8–12 mm. long) immediately draw our attention. The stems are often flexuous and support themselves on shrubs. A plant of wide altitudinal range (500–6,000 ft.) from the deserts of Calif.; to Ariz. and Utah.

493. DARWIN FORGET-ME-NOT. *Cryptantha inaequata* (L., "unequal," since there is inequality in the lengths of the nutlets, one of the four being longer than the rest). **Fl.**: white. A leafy, much-branched annual, with roots yielding a violet dye; largely confined to mountains of the eastern border of Calif.; and southern Nev.

494. BEARDED FORGET-ME-NOT. *Cryptantha barbigera* (L., "bearded," with respect to the hairy herbage). **Fl.**: white. Coarse-stemmed, very bristly, large-leaved annual, 10–20 in. tall. Of frequent occurrence in coarse sands of the low basins of both Calif. deserts as far north as Inyo Co.; to Utah and Ariz.

495. WING-NUT FORGET-ME-NOT. *Cryptantha pterocarya* (Gr., "wing nut," since most of the nutlets have almost knife-like, winged margins). **Fl.**: white. Very noticeable are the fat, lively-green, handsomely marked "fruits." The species and its varieties are wide-ranging from Calif. to Ida., Utah, and Tex.

496. ARCHED-CALYXED FORGET-ME-NOT. *Cryptantha recurvata* (L., curved back. "The species may be distinguished at sight from all the related ones by its recurved fruiting calyx"—Coville). **Fl.**: white. A low, much-branched annual, rarely found, but distributed widely in the desert ranges of Calif.; to Nev. and Colo.

497. WESTERN FORGET-ME-NOT. *Cryptantha circumscissa* (L., "cut around," because the upper half of the fruiting calyx falls away when the nutlets are ripe). **Fl.**: white. Sandy areas of the Mohave D.; Baja Calif., Wash., Colo., Utah, and even to Patagonia.

498. WOODY FORGET - ME - NOT. *Cryptantha racemosa* (L., "clustered," like grapes, with reference to the racemose inflorescence). **Fl.**: white. An intricately branched, stemmy perennial, 1–2 ft. tall, finding lodgment among rocks, particularly on steep slopes. The numerous small leaves, at first deep green, dry early in the season and the plants soon appear the color of straw. The main stems are woody; the herbage is hispid. Desert mountains, up to 4,600 ft.; to northern Calif.

499. GOLDEN FORGET-ME-NOT. *Cryptantha confertiflora* (L., crowded-flowered). **Fl.**: yellow. Coarse, bristly-stemmed perennial, 6–20 in. tall. The stems occur in clusters from a branched root crown. Rocky slopes of the piñon belt from 4,000 to 9,000 ft., Mohave D. to the Sierra, and the White Mts.; east to Utah.

500. NARROW-LEAVED FORGET-ME-NOT. *Cryptantha angustifolia* (L., narrow-leaved). **Fl.**: white. Branching annual, with stems 2–10 in. long, especially abundant and widespread in disturbed ground along roadsides. When the seeds are ripening, numerous black harvester ants may be seen busily gathering them in quantity. Widely distributed from the Calif. deserts; east to Tex. and south to Sonora and Baja Calif.

501. TUFTED FORGET-ME-NOT. *Cryptantha virginensis* ("of the virgin," in reference to Virgin Valley, Utah). **Fl.**: white. Usually biennial. Among rocks in the mountainous areas of the northern Mohave D., 3,800–9,000 ft.; to southwestern Utah.

502. ROUGH-STEMMED FORGET-ME-NOT. *Cryptantha holoptera* (Gr., "completely winged," said of the seed). **Fl.**: white. A borage of very upright habit and rough-hairy herbage; one of the rarest of the genus. It grows from 8 in. to 2 ft. tall in the coarse sands of washes on the lower Colorado D. from the Palm Springs area to the Colorado R.

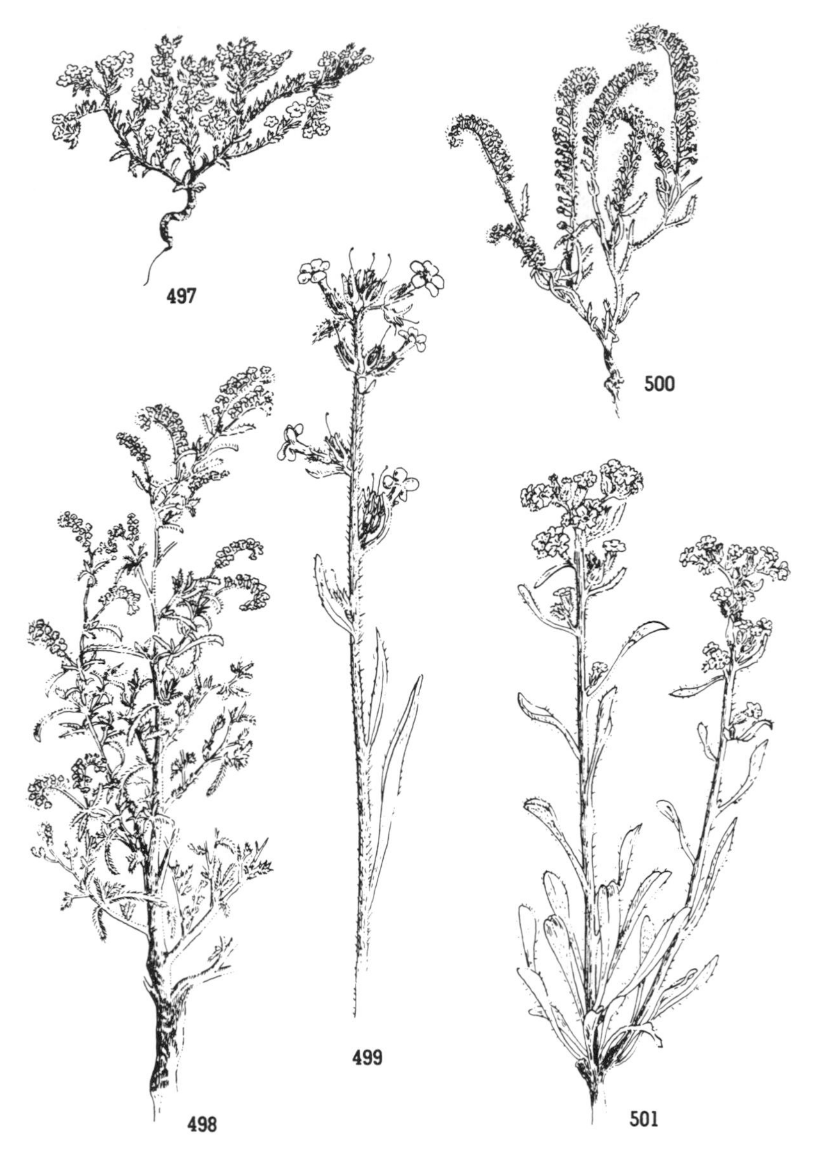

497. *Cryptantha circumscissa*
498. *Cryptantha racemosa*
499. *Cryptantha confertiflora*
500. *Cryptantha angustifolia*
501.* *Cryptantha virginensis*

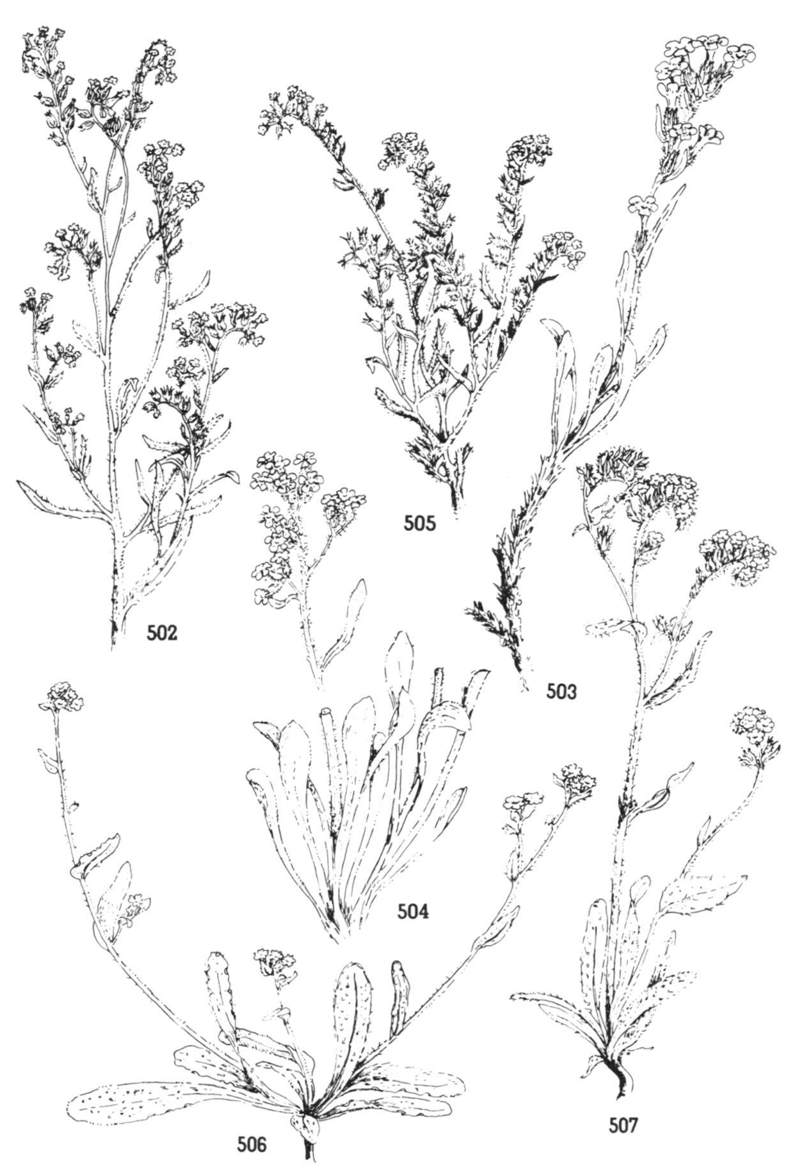

502. *Cryptantha holoptera*
503.* *Cryptantha flavoculata*
504. *Cryptantha tumulosa*
505.* *Cryptantha costata*
506. *Plagiobothrys arizonicus*
507. *Plagiobothrys Jonesii*

503. SULPHUR-THROATED FORGET-ME-NOT. *Cryptantha flavoculata* (L., golden-eyed). **Fl.**: white, with yellow throat. Perennial, with stems 4–12 in. tall, and rounded or obtuse basal leaves. Up to 8,000 ft., on dry, rocky slopes of the Mohave D.; east to Wyo. and Colo.

504. PIÑON FORGET-ME-NOT. *Cryptantha tumulosa* (L., "hilly," referring to the dorsal surface of the nutlets). **Fl.**: white. Perennial, 4–10 in. tall, arising from a branched root crown. Rocky slopes of piñon belt of the Providence, the New York, and the Ivanpah mountains; and Charleston Mts. of Nev.

505. ASHEN FORGET-ME-NOT. *Cryptantha costata* (L., "ribbed," referring to the sepals). **Fl.**: white. A species with ashen stems and leaves, known from the sand hills and gravelly flats of the lower Colorado D. from the Palm Springs region to Yuma. Not common. The plant is low and the branches stiff and leafy. Small, leaf-like bracts occur along the flowering spikes.

506. ARIZONA POPCORN-FLOWER. *Plagiobothrys arizonicus* (Gr., "oblique," or side pit, in reference to the position of the attachment scar on nutlets of the first-described species; of Arizona). **Fl.**: white. Loosely branched annual, immediately recognized in the field by the purple midribs and margins of the leaves. When pulled from the ground, even the root is seen to be impregnated with a purple dye, which is said to react to acids and alkalies just like litmus. Often about rocks or under shrubs on arid slopes and flats of the western Colorado and Mohave deserts; east to Nev. and N.Mex.

507. JONES POPCORN-FLOWER. *Plagiobothrys Jonesii.* (M. E. Jones, 1852–1934, who for fifty years botanized over Western America from Washington to central Mexico. Much of his early collecting was done on foot or horseback. Later he used a bicycle, pushing it loaded with specimens over rocky roads. During his last years a model T Ford was his means of conveyance. He was first to visit many of the least accessible desert basins and ranges, and his collections, now housed in the herbarium at Pomona College, are among the most extensive in the West. Many of his pungently written articles on botanical subjects are contained in a series of papers, *Contributions to Western Botany.*) **Fl.**: white. Erect, up to 1 ft. high. Very bristly. Mountains of eastern and northern Mohave D.; to Ariz.

VERBENACEAE. Verbena Family

508. GOODDING VERBENA. *Verbena Gooddingii* (an old Roman name used by Pliny, derived from Gr., signifying "holy herb"; L. N. Goodding—see **622**). **Fl.**: purplish. A perennial species, generally with several ascending stems. Known in California from dry, limestone canyons and slopes of the Clark and Providence mountains; to Tex.

509. WRIGHT LIPPIA. *Lippia Wrightii* (A. Lippi, eighteenth-century French physician and traveler in Abyssinia; Charles Wright —see **209**). **Fl.**: white. Aromatic shrub, with spreading branches, known in California from the Providence Mountains. The large genus *Lippia* to which it belongs is mostly American, the remaining few species being found in Africa and other warm parts of the globe. *Lippia nodiflora* is a nearly related perennial herb with subcylindric flower heads and purple corollas, known from the Colorado D. near Yuma.

LABIATAE. Mint Family

510. HORSEMINT. *Monarda pectinata* (Nicolas Monardes, botanical writer and physician of Seville, "who published in 1569–71 an account of the West Indies in which among other things occurs the first full account of the Tobacco"; L., comb-like, i.e., with comb-like teeth [on the seeds]). **Fl.**: white. A rather low, slender, perennial herb known on our deserts from the New York Mts.

511. THYME PENNYROYAL. *Hedeoma thymoides* (Gr., name of a sweet-smelling herb; L., thyme-like). **Fl.**: purplish. A perennial, aromatic herb, 4–8 in. high, the stems and leaves ashen-haired. Arid slopes of the Clark and Providence mountains; to Nev. and Tex.

512. THISTLE SAGE. *Salvia carduacea* (L., "salvia," an herb used for healing; L., thistle-like). **Fl.**: lavender. A handsome annual which invades the desert areas from the west. The plant is covered with a dense spider-web- or cotton-like wool, and bears large flowers of exquisite form and color, with lacquer-red anthers. The leaves form a basal rosette. It generally occurs in scattered groups on sands and loose-soil flats in the western parts of the California deserts, seldom ranging above 3,700 ft. altitude. Most common on the Mohave D. The common name, "sage," was given the *Salvias* because of their supposed power of making a person wise or sage.

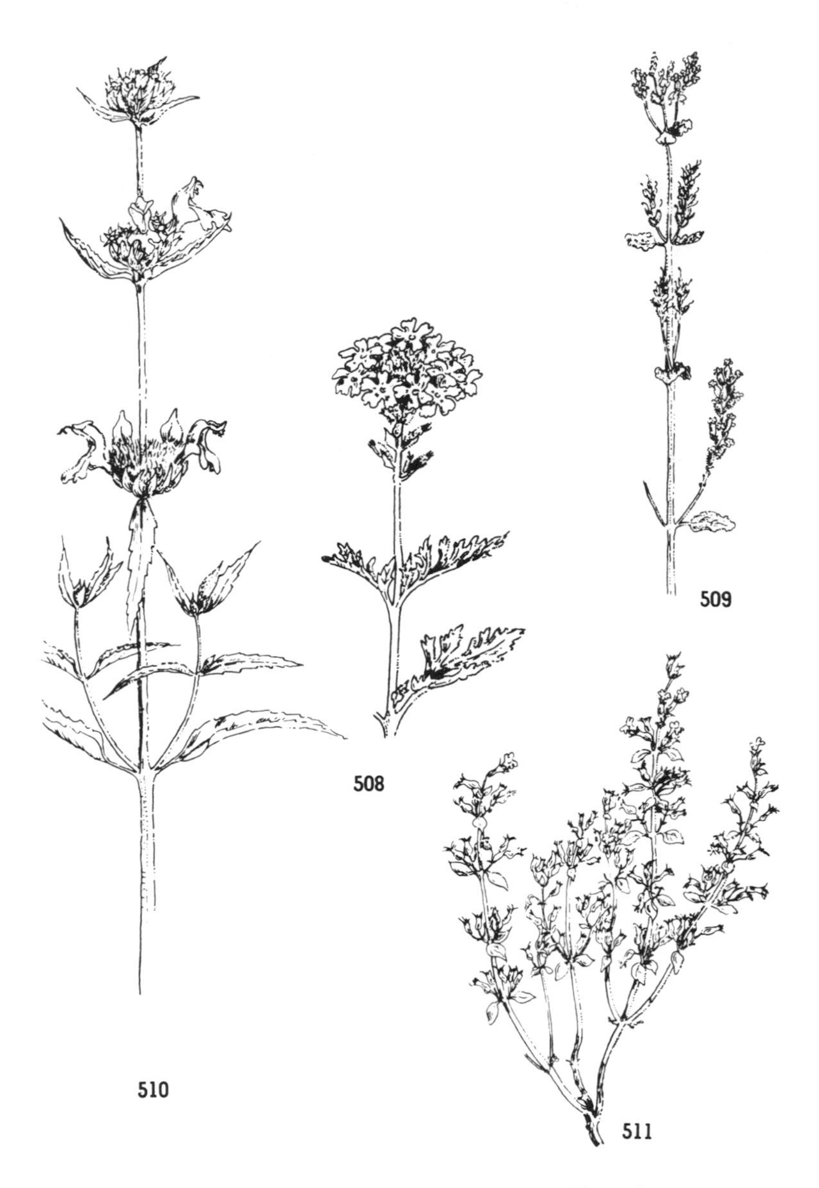

508. *Verbena Gooddingii*
509.* *Lippia Wrightii*
510. *Monarda pectinata*
511. *Hedeoma thymoides*

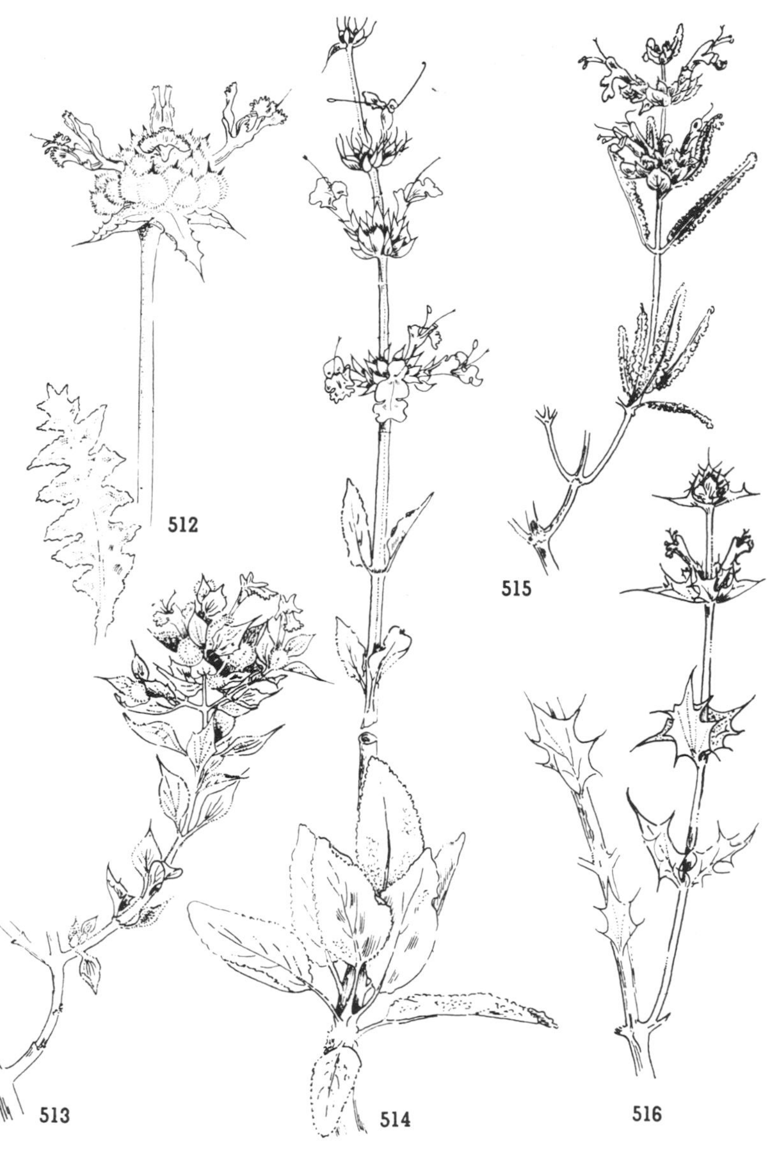

512. *Salvia carduacea*
513. *Salvia funerea*
514. *Salvia Vaseyi*
515. *Salvia eremostachya*
516. *Salvia Greatae*

513. DEATH VALLEY SAGE. *Salvia funerea* (L., pertaining to a funeral, i.e., the Funeral Mountains). **Fl.**: light violet. One of the most remarkable shrubby plants of the Death Valley area. Its spiny leaves and branches are made ashy-white by a dense covering of branching hairs. Though first found in the Funeral Mountains, it is known to occur in narrow canyons and small embranching gullies on benches in several of the neighboring ranges. It grows in the partial shade of rocks. Note how the bracts in this species, as well as in *S. Graetae* and *S. Columbariae,* exceed the flowers in length.

514. WAND SAGE. *Salvia Vaseyi* (George Vasey, 1822–1893, well-known authority on grasses). **Fl.**: white. A white-leaved species known best from canyons and mountainsides to the west of the Salton Sink but also found in the Morongo Pass area. Diagnostic field marks are the long, wand-like, flowering branches and long, sharp-pointed bracts and calyx teeth. It sometimes hybridizes with *S. apiana* and *S. eremostachya.*

A variety of the nearly related coastal white sage, *S. apiana compacta,* with flowers arranged in a condensed panicle resembling a spike, occurs along the desert's western borders.

515. SANTA ROSA SAGE. *Salvia eremostachya* (Gr., "desert Stachys," in allusion to the likeness to plants of the genus *Stachys*). **Fl.**: blue to rose-color. A shrub a yard or less tall. Known only from arid canyons about the base of the Santa Rosa Mts. Desert travelers may see it along the Pines-to-Palms highway. The branchlets are ashen with spreading, glandular hairs.

516. OROCOPIA SAGE. *Salvia Greatae* (Louis Greata, 1857–1911, amateur San Francisco botanist, who with Dr. H. M. Hall—see **275**—made a lengthy trip in the early 1900's in search of California Compositae, traveling with a horse named Molly and a buckboard fitted with water casks and an umbrella). **Fl.**: lavender. Closely related to *S. funerea* is this rare, spiny-leaved shrub of the gravelly washes of the Orocopia Mts. to the north of the Salton Sea. Like the Death Valley sage, its holly-like leaves and branches are ashen-gray. Its ability to withstand long periods of drought, when it loses practically all its leaves, is most remarkable. Salt Creek Wash is the type-locality.

517. NARROW-LEAVED MONARDELLA. *Monardella linoides* (Gr., like *Linum,* flax). **Fl.**: pale lavender. A strong-scented perennial, with silvery stems springing from a woody base. Common among rocks of the piñon-juniper belt on mountains of both deserts.

518. ROSE SAGE. *Salvia pachyphylla* (L., thick-leaved). **Fl.:** blue; bracts purple. A compact shrub, often forming low, broad, round mats 3 or 4 ft. in diameter. The effect given by the rose-colored flowering bracts and the contrasting, dark violet-blue flowers is most striking. Though nowhere common, it is a species of wide range in the piñon belt from northern Baja Calif. north along the mountains of the west side of the Colorado D. to the Joshua Tree Nat. Mon., thence to some of the arid ranges to the east and west of the Death Valley Sink.

519. MOHAVE PENNYROYAL. *Monardella exilis* (diminutive of *Monarda,* a genus it resembles; L., slender, feeble). **Fl.:** white. This is the desert's only annual *Monardella.* It grows from 4 in. to 1 ft. high, in sandy soils of the western Mohave D. When crushed it gives off a most pleasant aroma. The white-margined calyx teeth and white bracts veined with green are good field marks.

520. ROCK PENNYROYAL. *Monardella Robisonii* (William Robison, who with Dr. Carl Epling collected the type at Keye Ranch in the Joshua Tree National Monument in 1934). **Fl.:** pale blue. An aromatic, perennial species often met with among rocks of the Little San Bernardino Mountains, where it grows in rounded clusters. June, July. A plant quite similar to *M. linoides* (see **517**).

521. LOW GERMANDER. *Teucrium depressum* (Teucer, a king of Troy, who is said to have first used plants of this genus medicinally; L., depressed, i.e., growing near the ground). **Fl.:** pale blue. Annual herb, about 1 ft. high, slightly aromatic; belonging to a large genus of nearly one hundred species confined mainly to the Old World. Rare, eastern Colorado D.; to Tex., West Indies, and South America. "Germander" is probably a corruption of Scamander, a Trojan river.

522. DESERT LAVENDER. *Hyptis Emoryi* (Gr., "turned back," referring to the lower lip of the flower; Maj. W. H. Emory, 1812–1887, Director of the Mexican Boundary Survey). **Fl.:** violet-blue. Lavender-scented shrub belonging to a large genus, mostly South American. The bark is ashen-gray; the leaves are covered with a scurfy mat of hair. The agreeable sweet and turpentiny odor of the leaves is especially noticeable after desert showers. Being very sensitive to frost, *Hyptis* is confined to the warmer washes, canyons, and rocky hillsides of the Colorado Desert and related areas of southern

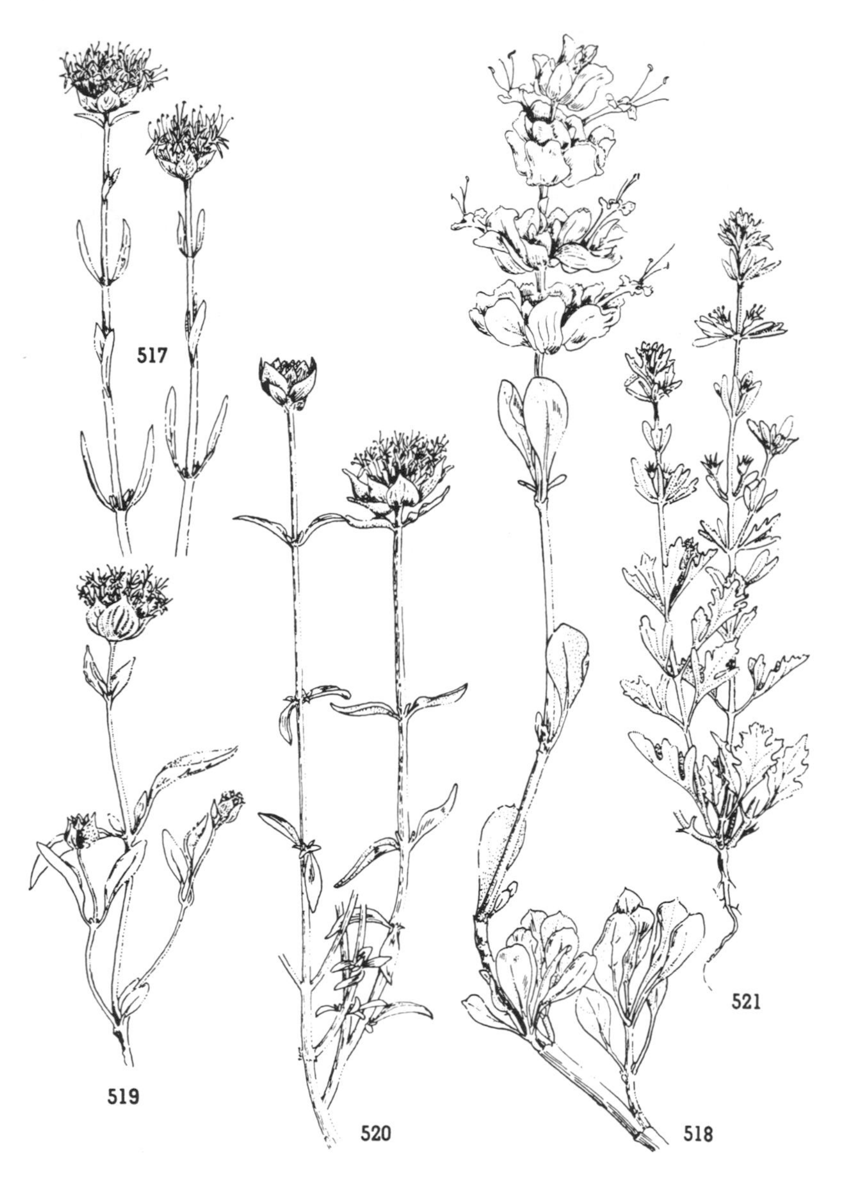

517. *Monardella linoides*
518. *Salvia pachyphylla*
519. *Monardella exilis*
520. *Monardella Robisonii*
521.* *Teucrium depressum*

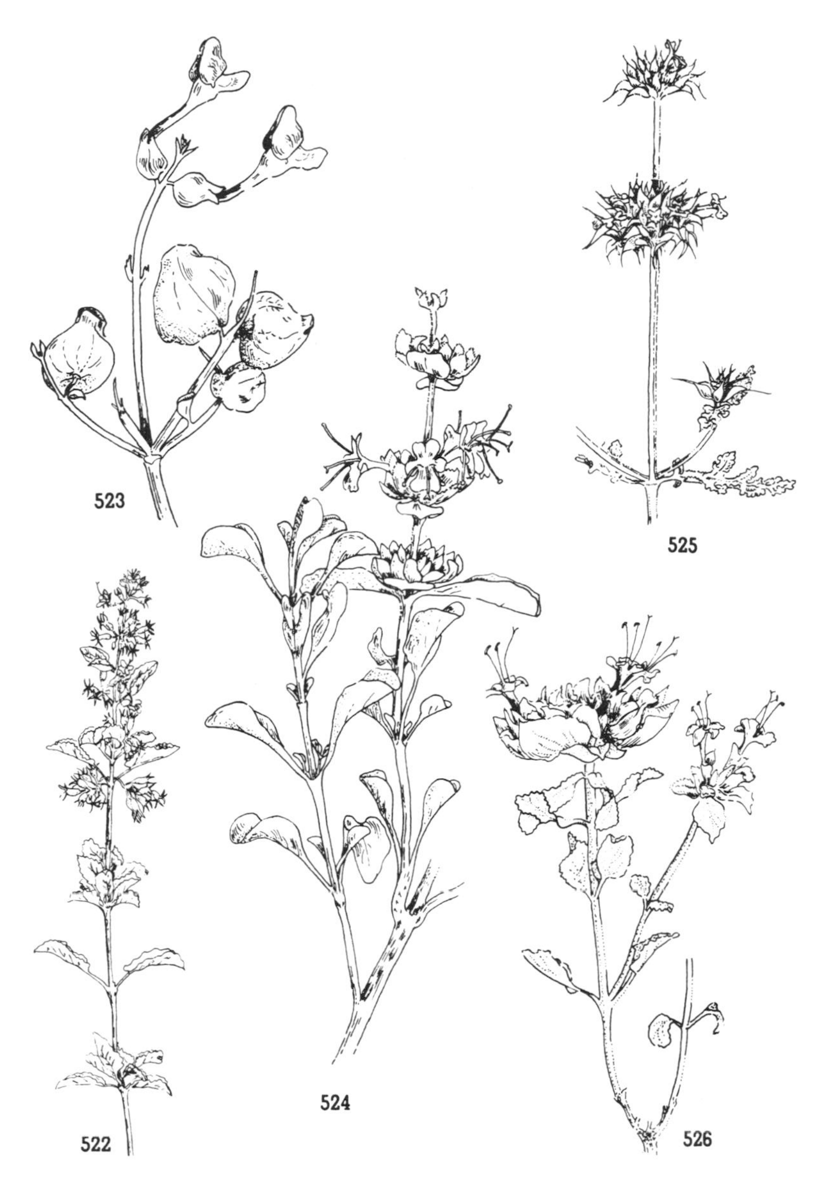

522. *Hyptis Emoryi*
523. *Salazaria mexicana,* ×1
524. *Salvia carnosa pilosa,* ×1
525. *Salvia Columbariae*
526. *Salvia mohavensis*

Arizona. A rust fungus (*Puccinia distorta*), forming black stem swellings, is responsible for many of the dead limbs. Apiarists rank desert lavender as a most valuable bee plant. Both verdins and plumbeous gnatcatchers nest in its branches and freely use the woolly calyx tubes as lining material for their nests.

523. PAPER-BAG BUSH. *Salazaria mexicana* (Don José Salazar, Mexican Commissioner on the Boundary Survey; of Mexico). **Fl.**: upper lip velvety dark purple, lower lip whitish. A very handsome rounded shrub, presenting a singular appearance because of its numerous inflated, papery pods, which in age are often tinged with rose. The little white-tailed desert ground squirrels climb in among the papery fruits and extract the seeds. It is more or less common over the northern Colorado and Mohave deserts; to Utah and Mex. Though most commonly found along washes, it is also met with on rocky hillsides.

524. GREAT BASIN BLUE SAGE. *Salvia carnosa pilosa* (L., fleshy, red; L., "hairy," with respect to the herbage). **Fl.**: blue. A very variable, handsome shrub, with erect branches 1–2 ft. high, widely dispersed in all the states west of the Rockies, and most commonly found in association with the three-toothed sagebrush (*A. tridentata*). Following spring rains it flowers profusely. In California it and all of its five subspecies are confined to the Mohave D. *S. carnosa Gilmani* of the Death Valley area has dark blue flowers.

525. CHIA. *Salvia Columbariae* (L., like [Scabiosa] Columbaria). **Fl.**: blue. An annual, generally of short stature and with leaves mostly basal. Chia is ubiquitous in dryish, open situations over a wide area in middle and southern Calif., southern Nev., southwestern Utah, Ariz., and Baja Calif. It has the greatest altitudinal range (0–7,000 ft.) of any of the California salvias. The seeds once formed a staple article of diet among the Indians.

526. MOHAVE SAGE. *Salvia mohavensis* (L., of Mohave [River], where first collected). **Fl.**: pale blue or white. A low (1–3 ft.), rounded, aromatic shrub, with dark green, reticulately veined leaves and large, whitish bracts; most handsome but seldom seen in flower, for it blooms at a time (late June) when there are few visitors in the desert. It grows in rocky situations in the upper creosote-bush and juniper belts of the Joshua Tree Nat. Mon. east and northward; to Nev. A very long-flowered form has recently been found by the author in the Ivanpah Mts.

SOLANACEAE. Potato Family

527. FRÉMONT THORNBUSH. *Lycium Fremontii* (Lycia, in ancient geography, a district of southwest Asia Minor; Captain J. C. Frémont—see **315**). **Fl.**: pale violet. A shrub, 3–6 ft. tall, having herbage with numerous glandular hairs. Note that the calyx lobes are much shorter than in *L. brevipes*. The fruit is red and juicy. Partial to alkaline soils of the Colorado D.

528. SQUAW-THORN. *Lycium Torreyi* (John Torrey—see **91**). **Fl.**: lavender-purple. Erect shrub, 4–6 ft. high, with bright green leaves and but few, short thorns. Both flowers and fruits occur in clusters. The juicy, many-seeded berries are bright, shining red and though rather insipid were eaten by native peoples. Occasional along borders of sandy washes; also along streams such as the Colorado and Mohave rivers; to Tex. and Mex.

529. PARISH THORNBUSH. *Lycium Parishii* (S. B. Parish—see **603**). **Fl.**: lavender. A very spiny, pubescent shrub, 3–9 ft. tall. The fruits are red. The calyx tube is very short, not more than 3 mm. long. A rare species of the southeastern Colorado D., Vallecitos Mts., Palm Springs.

530. RABBIT-THORN. *Lycium pallidum oligospermum* (L., "pallid," because of the herbage; Gr., few-seeded). **Fl.**: white to lavender, fading to pale yellow. When in full flower the handsomest of our rabbit-thorns. A rough, very thorny, rounded shrub, 1½–3 ft. high, with numerous, stoutish, glaucous, zigzag branchlets. The colorful flowers occur in profusion, appear early in April, and last but a few days. The several-seeded, hardish fruits are greenish-white or greenish-purple. From the region about Barstow north to the Death Valley area.

531. ANDERSON THORNBUSH. *Lycium Andersonii* (Dr. C. L. Anderson—see **191**). **Fl.**: light lavender. Exceedingly branchy shrub, 1–4 ft. high, with short, sharp, needle-like spines and small, somewhat pear-shaped, succulent green leaves. The numerous small, ovoid berries are bright tomato-red. Widely spread over both deserts; southern Nev. and Ariz.

532. NARROW-LEAVED THORNBUSH. *Lycium Andersonii deserticola* (L., desert-dwelling). On the Colorado Desert there is often seen side by side with Anderson thornbush this variety, *deserticola*, characterized by its longer leaves (1–1¼ in.). Quail eat the juicy leaves and berries of both the species and its variety.

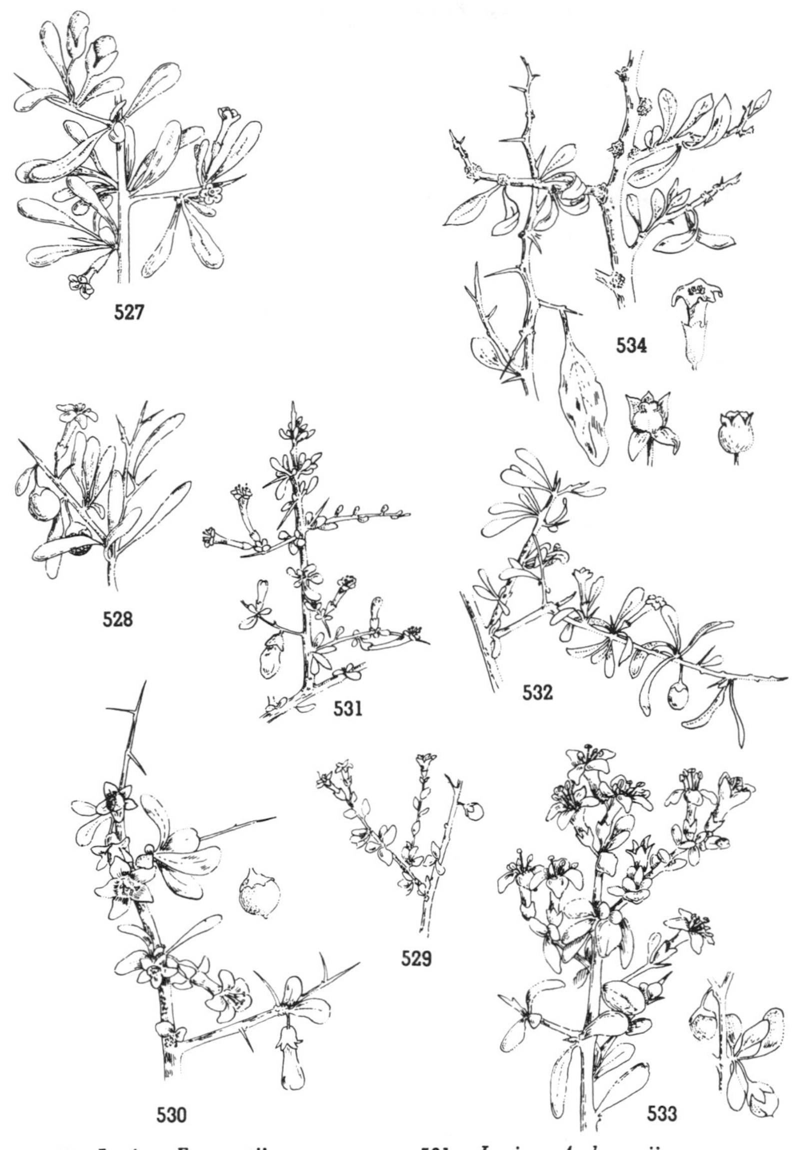

527. *Lycium Fremontii*
528. *Lycium Torreyi*
529. *Lycium Parishii*
530. *Lycium pallidum oligospermum*
531. *Lycium Andersonii*
532. *Lycium Andersonii deserticola*
533. *Lycium brevipes*
534. *Lycium Cooperi*

533. FRUTILLA. *Lycium brevipes.* (L., "short-footed," in allusion to the shortness of the flower stem as compared with that of *L. chilensis.* Really an inapt name, for there is considerable variation in the lengths of the pedicels. Bentham, who named the plant, based his description on a poor fragment.) **Fl.**: lilac. Rigid, spiny shrub, 3–7 ft. tall, individualized by its short-tubed, widespreading flowers, its long calyx lobes, and the fine-hairiness of its herbage. The small, tomato-like fruits are many-seeded. This is a Mexican species, having its northern limits on the Colorado D. Often listed under *L. Richii.*

534. PEACH-THORN. *Lycium Cooperi* (Dr. J. G. Cooper—see **132**). **Fl.**: greenish-white. Coarse shrub, with stout stems 2–4 ft. high, generally occurring in aggregations on dry desert mountain flats and slopes or on the desert floor. The smooth bark, at first reddish-brown, turns black on older stems and is inclined to peel. During the late summer the plants lose most of their leaves and go into a drought rest, causing the blackish, thorn-armed stems to be very conspicuous. The several-seeded, greenish fruits, peculiarly constricted just above the middle, distinguish this thornbush from all others of its range. Very common on both deserts; east to southern Nev.

535. WESTERN JIMSON. *Datura meteloides* (Hindu name, dhatura; L., like *Metel,* the "mad Solanum" of the Greeks). **Fl.**: white, suffused with violet. This Western representative of the Eastern Jimson, so common in the coastal counties of southern Calif., is now well established in the desert area along the main travel routes below 3,000 ft.; to Tex. and Mex. In England it is used as a greenhouse perennial.

536. SMALL DATURA. *Datura discolor* (L., of different colors). **Fl.**: white, with purplish tinge in the throat. Distinguished in the field from the Western Jimson-weed by its smaller flowers and nodding, stout-spined fruits. It is an annual very common in autumn in sand washes and irrigated areas of the Colorado D. near Mecca; to Ariz. and Mex. The blackish larvae of the striped datura beetle, *Lema nigrovittata,* eat both the leaves and stems, causing them to die.

The Cahuillas made a decoction from datura and administered it to boys at the time of their initiation. "As soon as the boys had taken it," says Miss Lucile Hooper, "they would begin to dance, but shortly would become very dizzy. They were then all put in a dark corner. It is asserted that drinking this decoction made the mind clearer . . . . By the next night the bad effects of the narcotic had worn off and the boys usually felt about normal." During the five succeeding nights they were taught the enemy songs, how to dance, and how to use the gourd rattles as an accompaniment.

537. IVY-LEAVED GROUND-CHERRY. *Physalis hederaefolia* (Gr., "a bladder," referring to the inflated calyx; Gr., *Hedera,* name for the English ivy, and, L., "a leaf"). **Fl.**: pale yellow. A perennial species, largely of the eastern Mohave D.; to Tex. and Mex. The short, viscid hairs covering the herbage are unbranched.

538. FENDLER GROUND-CHERRY. *Physalis Fendleri.* (August Fendler, who was induced by Asa Gray and Dr. Engelmann to undertake in 1846 a botanical exploration of the country around Santa Fe. He collected also in Texas and Venezuela.) **Fl.**: yellow, with brown center. A perennial, with long, white, fleshy, radish-like root; 6–12 in. high. The herbage is densely fine-hairy, with stellate or branching hairs. Providence Mts.; to Colo. and Mex.

539. HAIRY GROUND-CHERRY (not illustrated). *Physalis pubescens* (L., hairy). **Fl.**: yellow with dark center. Diffusely branched annual, having herbage covered with long, soft hairs. Introduced from southeastern U.S.; known from Needles, Fort Yuma, and San Diego.

540. THICK-LEAVED GROUND-CHERRY. *Physalis crassifolia* (L., thick-leaved). **Fl.**: pale tawny-yellow. The common perennial species of the desert. The plants, 8–16 in. high, spring from long, white, radish-like roots. Both deserts; to Tex. and Mex.

541. DESERT TOBACCO. *Nicotiana trigonophylla* (Jean Nicot, French ambassador to Portugal, who introduced tobacco into France about 1560; Gr., triangular leaf). Biennial or perennial, dark green herb, 1–2 ft. high; most of the parts ill-smelling and viscid. "In aspect . . . . very different from any of the other species of the genus in N.A." (Setchell). One has only to go to the low, rocky canyons and washes to find the plants in numbers. In autumn they are often about the only green thing to be seen. Both deserts; to Tex. and Mex. It was used for smoking by the Yuma and Havasupai Indians. In order to promote the growth of large plants they cut down mesquite trees, burned them on unbroken soil, and then scattered the tobacco seeds in the dead ashes.

SCROPHULARIACEAE. Figwort Family

542. TWINING SNAPDRAGON. *Antirrhinum filipes* (Gr., "like a snout," in reference to the aspect of the flowers; L., "thread-footed," alluding to the tortuous, filamentous flower stalk). **Fl.**: bright yellow. A most handsome, climbing annual, with long, slender, bright green stems and leaves. It climbs by twisting its filamentous flower stalks around the branchlets of shrubs offering it protection. The early flowers, borne about the base of the plant, are often cleistogamous, i.e., fertilized in the bud without the opening of the flower. Confined to the creosote-bush area of both deserts; eastward to Ariz. and southern Utah.

543. COULTER SNAPDRAGON. *Antirrhinum Coulterianum* (Dr. Thos. Coulter—see **755**). **Fl.**: white to pale yellow. Erect annual, with numerous twining side branchlets, which enable it to climb up among or beside other plants. A cismontane species reaching the arid mountains of the Joshua Tree Nat. Mon.

544. KING SNAPDRAGON (not illustrated). *Antirrhinum Kingii* (Clarence C. King, 1842–1901, geologist connected with the California Geological Survey in the 1860's, in charge of the U.S. Geological Exploration of the Fortieth Parallel, 1870–1880, and author of *Mountaineering in the Sierra Nevada.* During the winter of 1865–66, as scientific aid to General McDowell, he engaged in an exploration of the desert region of southern California). **Fl.**: white, with purple veins. An erect annual, 4–16 in. high, met with in washes and clay flats from the more arid portions of the Inyo and Panamint mountains of Calif.; to Nev., Ida., and Ore.

545. BLUE MAURANDYA. *Maurandya antirrhiniflora* (Dr. Maurandy, teacher of botany at Carthagena; L., *Antirrhinum*-like flowers). **Fl.**: light blue. Maurandya is a small genus of perennial herbs mainly confined to Mexico, all with showy flowers. This slender vine is known in California only from the limestones of the Providence Mts. of the eastern Mohave D. Eastward it is found in southern Ariz. to western Tex. and Mex.

546. ROCK MAURANDYA. *Maurandya petrophila* (Gr., rock-loving). **Fl.**: pale yellow. Spiny-leaved, low, tufted perennial, known from limestone crevices of Titus Canyon, Death Valley Nat. Mon.

547. EATON FIRECRACKER. *Penstemon Eatonii.* (Using the original spelling, without the first *t*, Gr., "fifth stamen," in allusion to the remarkable development of the sterile filament, which bears no

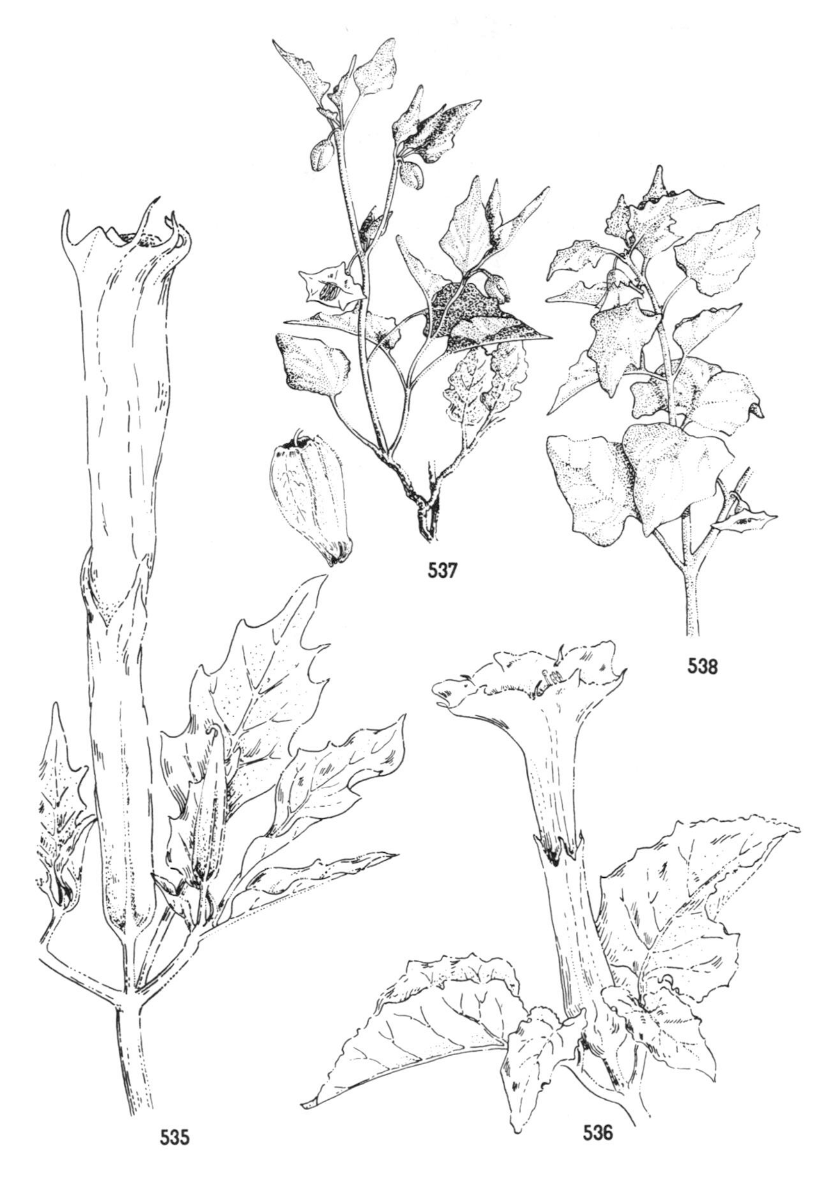

535. *Datura meteloides*
536. *Datura discolor*
537. *Physalis hederaefolia*
538. *Physalis Fendleri*

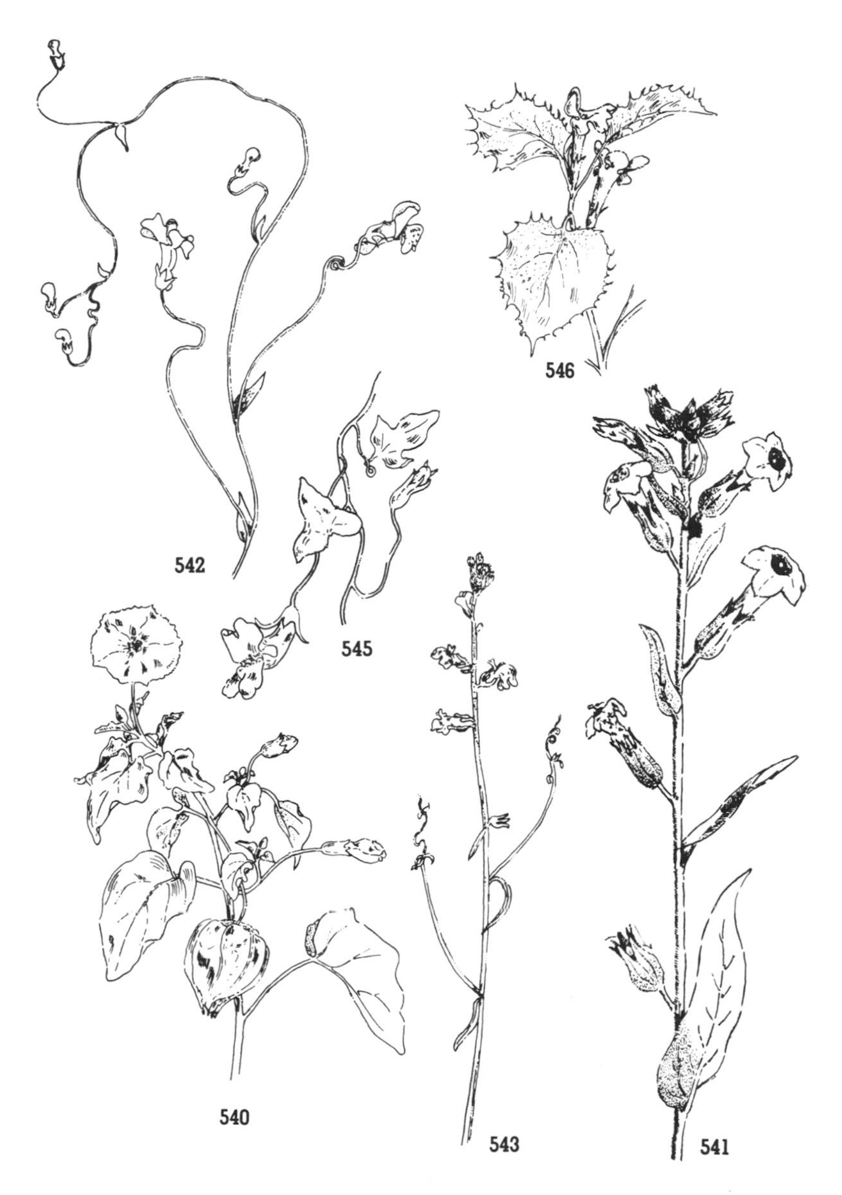

540. *Physalis crassifolia*
541. *Nicotiana trigonophylla*
542. *Antirrhinum filipes*
543. *Antirrhinum Coulterianum*
545. *Maurandya antirrhiniflora*
546.* *Maurandya petrophila*

anther but is often adorned with a brush of yellow hairs; hence the common name, "beard-tongue," for members of this genus. Prof. D. C. Eaton, 1834–1895, botanist, of Yale University, and student of ferns.) **Fl.**: scarlet. A showy, perennial herb, with several coarse stems and deep green herbage. On gravelly slopes among rocks in Joshua Tree Nat. Mon.; Old Dad, Providence, and Panamint ranges; east to Nev., Utah, and Ariz.

548. WESTGARD PENSTEMON. *Penstemon scapoides* (L., "scape-like," the flowering stems being essentially leafless). **Fl.**: blue, A plant 12–18 in. high, found in rocky gorges of the juniper-piñon belt. Very local in the Inyo Mts. and adjacent canyons.

549. UTAH FIRECRACKER. *Penstemon utahensis* (of Utah, since the type collection was taken in that state). **Fl.**: scarlet-red. A short-statured beard-tongue, with several stems 8–12 in. tall. Mountains of the eastern Mohave D.; to southern Nev., Ariz., and Utah.

550. THURBER PENSTEMON. *Penstemon Thurberi* (George Thurber—see **319**). **Fl.**: pink to rose. Stems shrubby, 10–16 in. high. The plants generally occur in colonies on loose, sandy soils. Joshua Tree Nat. Mon.; San Felipe, Calif.; to N.Mex. and Baja Calif.

551. MONO BEARD-TONGUE. *Penstemon monoensis* (of Mono [County]). **Fl.**: wine-red. Blooming in May. Piñon-juniper belt of the arid ranges bordering Owens Valley. Not common.

552. BUSH PENSTEMON. *Penstemon microphyllus* (Gr., small-leaved). **Fl.**: yellow. A handsome shrub, 3–6 ft. high, found in rocky situations in the mountains (1,500–5,000 ft.) along the western edge of the Colorado D., in the Joshua Tree Nat. Mon., and ranges of the eastern Mohave D., eastward to Ariz. It is one of the principal food plants of the very variable checker-spot or chalcedon butterfly (*Euphydryas chalcedona*), which in southern California is "the most common butterfly that waves its wings in the air."

553. INYO BEARD-TONGUE. *Penstemon floridus Austinii.* (L., abounding with flowers; S. W. Austin, husband of Mary Austin, who wrote *Land of Little Rain.* The type was collected on a Fourth of July holiday in 1899 while he was in charge of the United States Land Office in Bishop.) **Fl.**: pinkish-purple. Piñon-juniper belt of the Inyo, Panamint, and Grapevine ranges, chiefly north of Death Valley, and in adjacent Nev. From the region of Independence and Westgard Pass northward one finds typical *P. floridus,* in which the corolla is much inflated and is slipper-shaped.

554. LIME PENSTEMON. *Penstemon calcareus* (L., "pertaining to lime," in reference to its habitat). **Fl.**: deep pink. A short-statured, small-flowered species, growing 2–10 in. high, in crevices of lime rock, in the Providence Mts. and mountains at the north end of the Death Valley Sink. The sterile filament is strongly bearded with coarse, golden yellow hairs.

555. MOHAVE BEARD-TONGUE. *Penstemon pseudospectabilis* (L., "false spectabilis," in reference to its similarity to a common species of coastal southern California). **Fl.**: rose-purple. A tall herb (2–4 ft.), with several erect stems and blue-glaucous herbage; found usually in sandy washes in the Sheephole, Chuckawalla, and Turtle mountains; Ariz.

556. DEATH VALLEY PENSTEMON. *Penstemon fruticiformis* (L., in the form of a shrub). **Fl.**: whitish, with lavender-blue lobes and purplish pencilings extending into the throat. A whitish-leaved shrub, with numerous tough, erect stems 1–1½ ft. high. Frequent in canyons of the Panamint and Argus ranges, to the west of Death Valley.

557. SCENTED PENSTEMON. *Penstemon Palmeri* (Dr. Edward Palmer—see **474**). **Fl.**: flesh-color, often tinged with rose. In limestone soils of the mountains of the eastern Mohave Desert this tall (2–5 ft.), extraordinarily sweet-scented beard-tongue comes to perfection in mid-May. If the plants occur in numbers, as often happens in gravelly washes, the sight is indeed a fine one. The sweet nectar is much sought by bees, who are guided to the nectaries by the prominent purple pencilings within the corolla.

558. CLEVELAND PENSTEMON. *Penstemon Clevelandii connatus* (Daniel Cleveland—see **754**; L., "grown together," in reference to the upper pairs of leaves). **Fl.**: purplish-red. Several stems, 16–30 in. high. Canyons bordering the western edge of the Colorado D. (3,000–4,500 ft.). The typical form of the species, in which the upper leaves are distinct from each other, grows along the desert borders of San Diego Co. and south to Baja Calif.

559. MOHAVE BEARD-TONGUE. *Penstemon Clevelandii mohavensis* (of the Mohave). **Fl.**: purplish-red. The upper leaves are distinct, the sterile filament is only about ¼ in. long, and the corolla is somewhat constricted at the mouth. Joshua Tree Nat. Mon. and the Sheephole Mts.

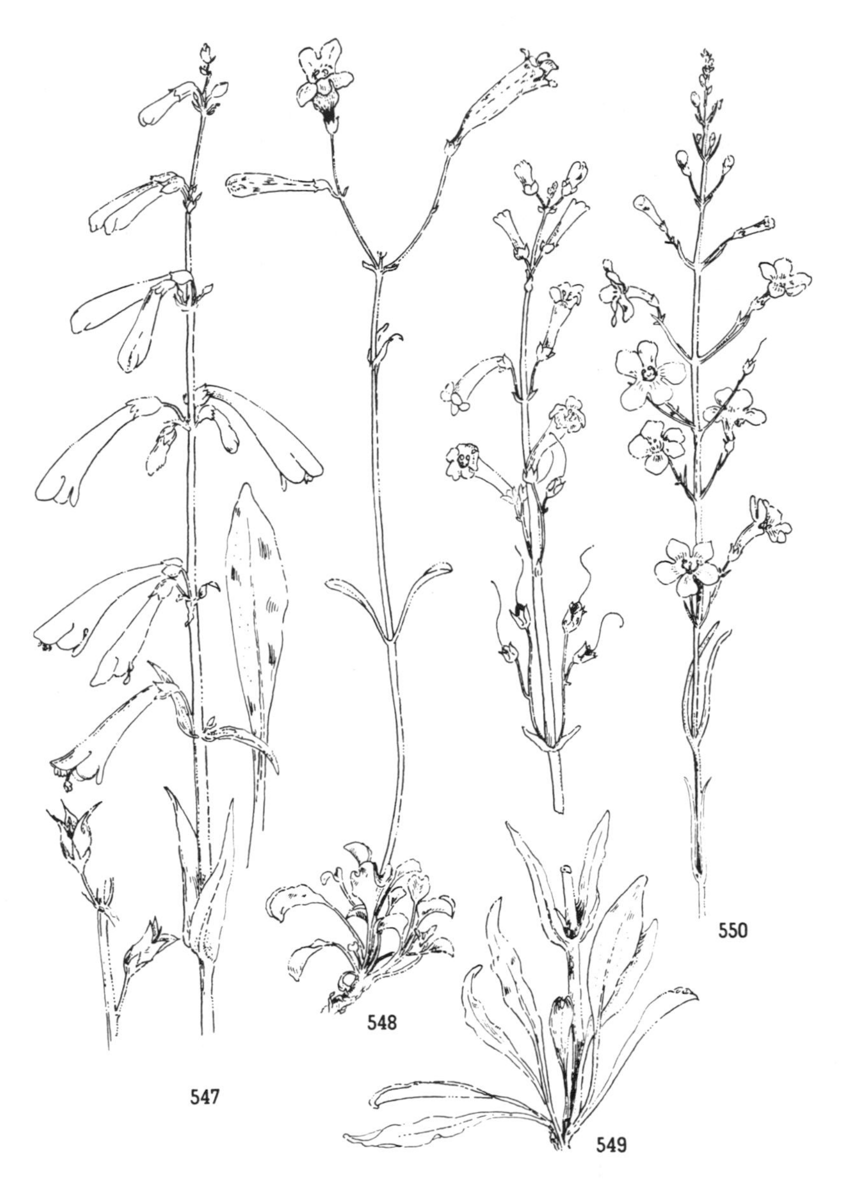

547. *Penstemon Eatonii*
548. *Penstemon scapoides*
549. *Penstemon utahensis*
550. *Penstemon Thurberi*

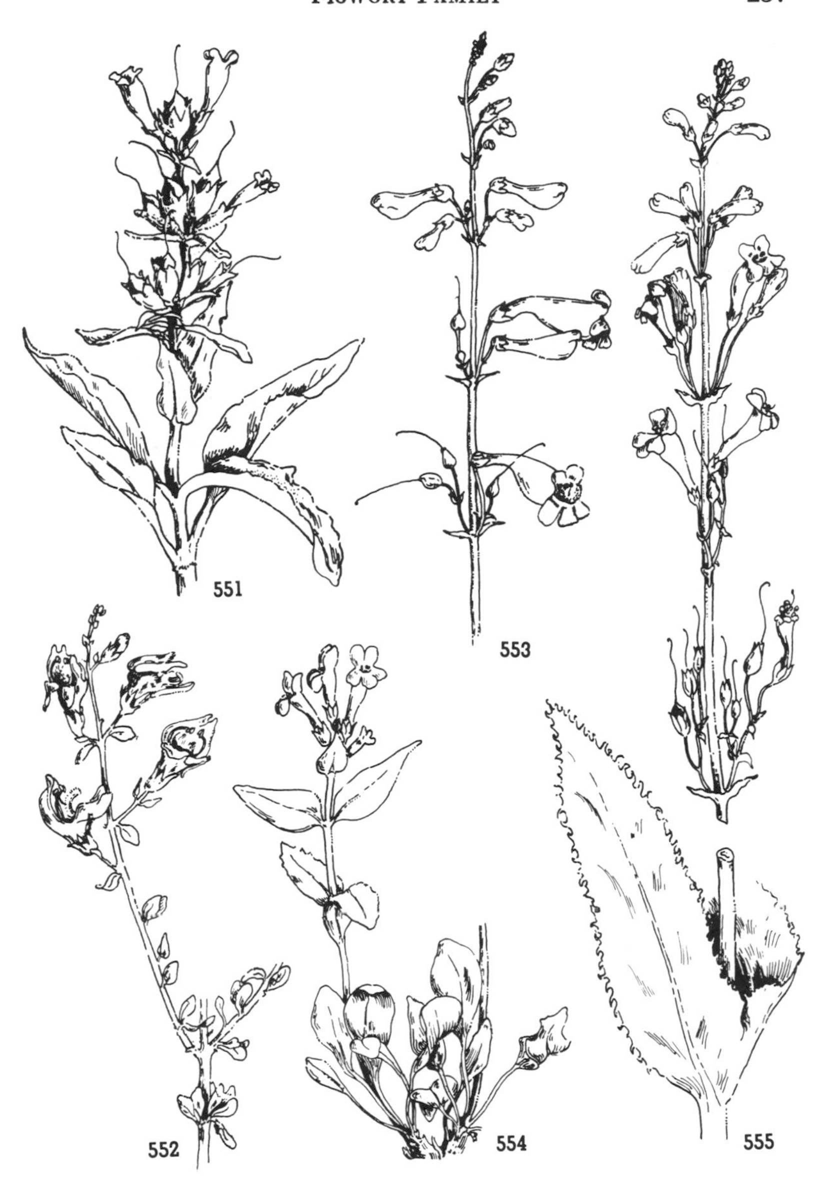

551. *Penstemon monoensis*
552. *Penstemon microphyllus*
553. *Penstemon floridus Austinii*
554. *Penstemon calcareus*
555. *Penstemon pseudospectabilis*

560. STEPHENS PENSTEMON (not illustrated). *Penstemon Stephensii* (Frank Stephens, 1849–1937, well-known mammalogist and collector, of San Diego). **Fl.**: flesh-color to pink-lavender. A rare relative of the Mohave beard-tongue, known only from the Providence Mts. Notable for its narrowly tubular corolla and very short (1/8 in.) calyx.

561. OWENS VALLEY BEARD-TONGUE. *Penstemon confusus patens* (L., "confused," in reference to its previous wrong identification; L., "open," in reference to the inflorescence). **Fl.**: light or rose-lavender. Entirely herbaceous, short-lived perennial with few, blue-glaucous stems, up to a foot or so in height. Canyons on both sides of Owens Valley from Lone Pine northward to southern Mono Co.

562. WESTERN DESERT PENSTEMON. *Penstemon incertus* (L., "uncertain," with reference to affinities). **Fl.**: blue-purple. Similar in habit to the preceding but differing in many ways regarding the flower; sometimes a dense bush 3 ft. across, but otherwise quite herbaceous to a narrow base. Occasional around the western borders of the Mohave D., particularly between Antelope Valley and Inyo Co.

563. THOMPSON PENSTEMON. *Penstemon Thompsoniae* (Mrs. Thompson, little-known collector, of Utah). **Fl.**: blue. A gray mat only an inch or two high and six inches or more across. Known in California only from near Coliseum Mine, Clark Mt., eastern Mohave D.; more frequent in southern Nev. and Utah, and northwest Ariz. Just over the state line in the ranges of Clark Co., Nev., at somewhat higher elevations, one finds *P. Thompsoniae Jaegeri*, with more open inflorescence and fewer but woodier stems. It is to be looked for in the Mohave ranges.

564. WHITE-MARGINED PENSTEMON. *Penstemon albomarginatus* (L., white-bordered). **Fl.**: light pink, tinged with purple. As indicated by the specific name, both leaves and calyx lobes have whitish, translucent margins. The pale green, crowded stems are seldom over 8 in. long. The short corolla is densely yellow-bearded within. A rare species of the central Mohave D.; eastward to southern Nev. and northwestern Ariz.; occurring in drifting sand.

565. SCARLET BUGLER. *Penstemon centranthifolius* (L., "leaf like Centranthus," the familiar centranth or red valerian of our gardens). **Fl.**: scarlet. One to several stems (1–3 ft.). Fairly common along the western montane borders of our deserts.

566. SHOWY PENSTEMON. *Penstemon speciosus* (L., showy, beautiful). **Fl.**: bright blue or blue-purple. A herbaceous species, with several ascending stems 4–18 in. high. The sterile filament is usually bearded near the apex. Slopes (5,000–7,000 ft.) of mountains on the northern borders of the Mohave D.; north to Ore. and Wash.

567. MOHAVE OWL CLOVER. *Orthocarpus purpurascens ornatus* (Gr., upright fruit; L., becoming purple; L., ornamented). **Fl.**: deep velvet-red, the lower lip tipped with rich yellow. The handsome flowers and purple bracts make this one of the most showy of the owl clovers. Confined to the western half of the Mohave D., where it grows at the same low altitude as the creosote bush. It is a low plant, 4–6 in. high. It flowers in late April and May, when the season reaches its climax for the annuals.

568. DESERT PAINTBRUSH. *Castilleja angustifolia* (D. Castilleja, Spanish botanist; L., narrow-leaved). Spike red. A flash of brilliant red among shrubs generally calls our attention to this handsome paintbrush, which comes into flower very early in the spring. It is often met with in rocky desert (2,000–7,000 ft.), from the Little San Bernardino Mts. north and east to the yellow pine and spruce forests of western Canada and Colo. (A wide range indeed!)

569. WOOLLY PAINTBRUSH. *Castilleja foliolosa* (L., "leaflet-bearing," in reference to almost distinct divisions of the leaves). Spike vermilion. A brushy, perennial species, 8–20 in. high, with herbage white-woolly throughout. Chaparral slopes from San Diego Co. to northern Calif., and at the desert's edge. The bright color display of most paintbrushes, as in this case, is due to the brilliant hues of the conspicuous bracts and calyces, which frequently quite conceal the flowers.

570. YELLOW PAINTBRUSH. *Castilleja plagiotoma* (Gr., "obliquely cut," in reference to the broad, stubby lobes of the calyx). Spike greenish-yellow. This weak-stemmed, rather woolly perennial has a habit of protecting itself from grazing animals by growing up through the branches of low shrubs. Western Mohave D., 3,000–7,000 ft.

571. LONG-LEAVED PAINTBRUSH. *Castilleja linariaefolia* (leaves like *Linaria*, common toadflax). Spike scarlet, owing partly to the long-exserted flowers. A very slender, tall-stemmed perennial, found occasionally in dry soils but mostly about springs and seeps of the higher desert areas; eastward to Wyo. and N.Mex. Adopted in 1917 as Wyoming's state flower, to which Professor Aven Nelson,

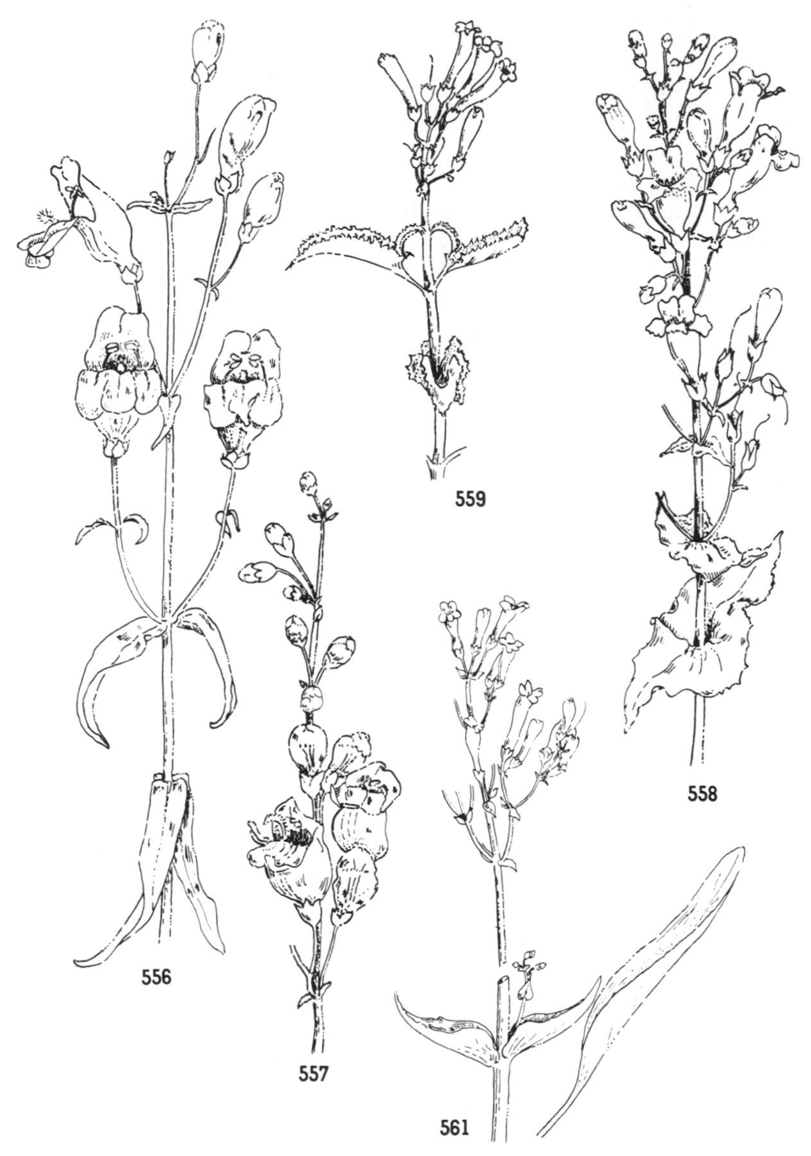

556. *Penstemon fruticiformis*
557. *Penstemon Palmeri*
558. *Penstemon Clevelandii connatus*
559. *Penstemon Clevelandii mohavensis*
561. *Penstemon confusus patens*

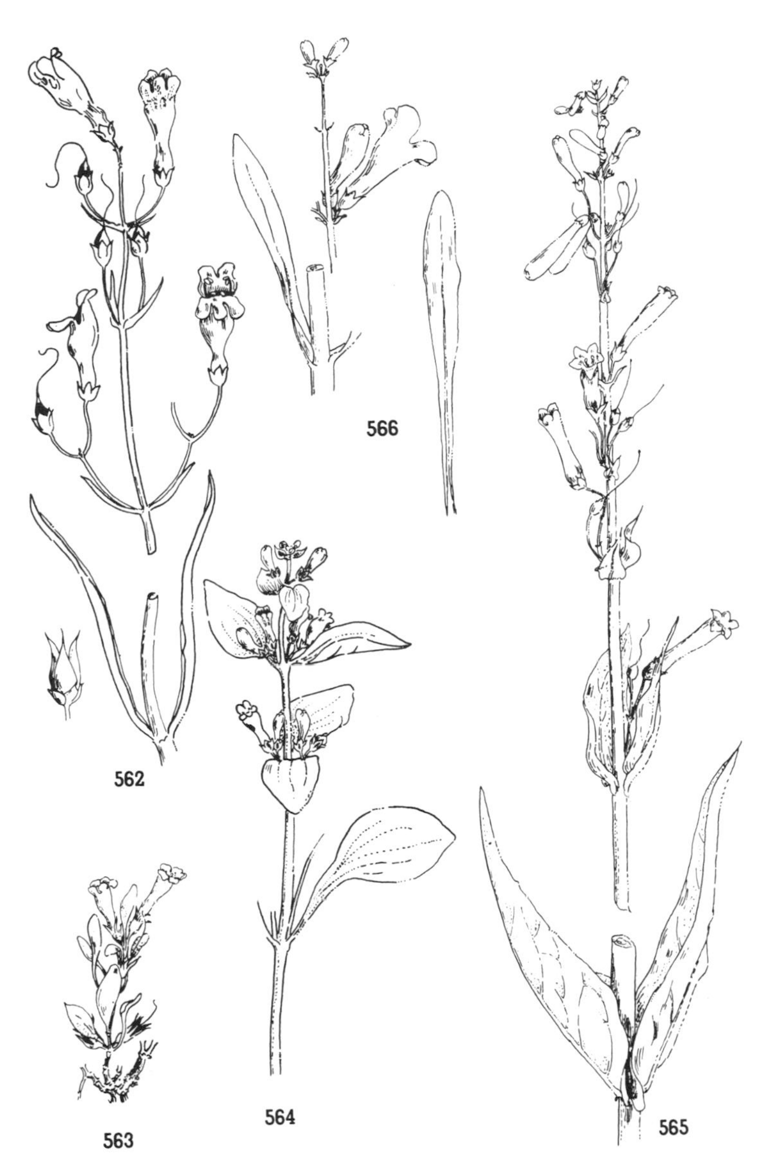

562. *Penstemon incertus*
563.* *Penstemon Thompsoniae*
564.* *Penstemon albomarginatus*
565. *Penstemon centranthifolius*
566. *Penstemon speciosus*

most famous Wyoming botanist, objects on the basis of its being a semiparasitic plant.

572. ROCK-MIDGET. *Mimulus rupicola* (L., "a little mime," i.e., like a jester, mimic, or actor in a play, alluding to the gaping mast-like corolla; L., rock-dweller). **Fl.**: rose to almost white, with golden throat and large maroon spot at base of the corolla lobes. Tiny, thin-leaved, delicate annual. "The conspicuous flowers," says Dr. Coville, "the habitat of the plant, in the shaded rock crevices, and its small size all combine to give the species charm. The plant sometimes begins to flower when it is less than an inch in height and has only basal leaves." As the flowers age, the flower stalk bends in a sigmoid curve, at the same time twisting about 180°, the result being that the capsules are somewhat sickle-shaped. Limestone canyons, 900–2,500 ft. in mountains about the Death Valley Sink.

573. BIGELOW MIMULUS. *Mimulus Bigelovii* (Dr. J. M. Bigelow—see **338**). **Fl.**: reddish-purple. Usually a branched annual, not over 10 in. high, with reddish stem and thin leaves. Canyons and washes, up to 6,500 ft., on both deserts; Nev. and Ariz.

The variety *cuspidatus* may be readily recognized by the decidedly sharp-pointed, sticky leaves. Barstow to Death Valley and western Nev.

574. RED-STEMMED MIMULUS. *Mimulus rubellus* (L., red). **Fl.**: red to yellow. An erect annual, 1½–6 in. high, with reddish herbage. The calyx teeth are of equal length, and the corolla is only slightly irregular. The yellow-flowered form is known from juniper areas of mountains of the eastern Mohave D.

575. PARISH MIMULUS (not illustrated). *Mimulus Parishii* (S. B. Parish—see **603**). **Fl.**: pinkish to white. A simple- or basal-branching, stout-stemmed annual, 5–16 in. high, with sticky or glandular herbage. The broad-based, sessile leaves are gray-green. Along streams below 7,000 ft., southwestern borders of the Mohave D. and along Whitewater Creek on the Colorado D.

576. BUSH MONKEY FLOWER. *Mimulus longiflorus* (L., long-flowered). **Fl.**: light yellow to light salmon. A low, shrubby, cismontane species found among granitic rocks of the higher mountains of the Joshua Tree Nat. Mon.

577. SPOTTED MIMULUS. *Mimulus montioides* (L., like *Montia*, a genus including "miner's lettuce"). **Fl.**: bright yellow, commonly purple-dotted. A small annual, not over 4 in. high. Ordinarily a high

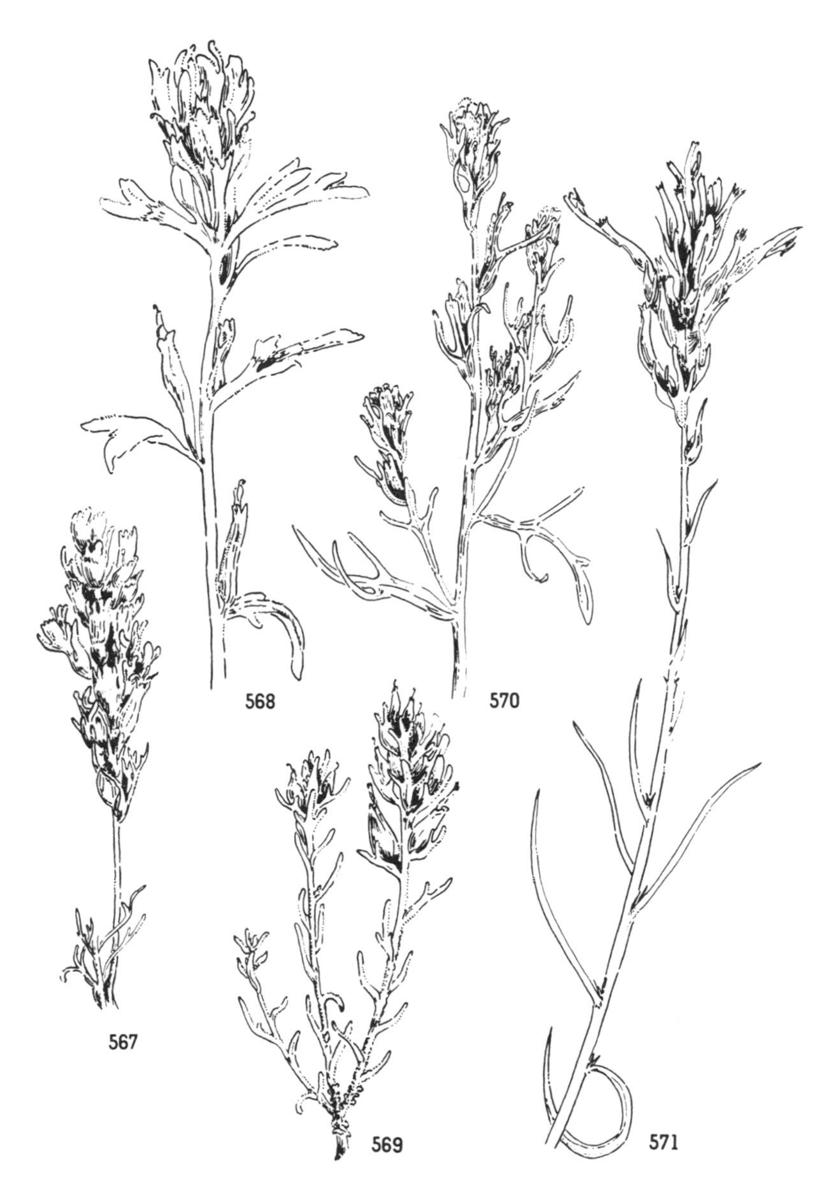

567. *Orthocarpus purpurascens ornatus*
568. *Castilleja angustifolia*
569. *Castilleja foliolosa*
570. *Castilleja plagiotoma*
571. *Castilleja linariaefolia*

montane species but known from the Death Valley area by collections made in Keane Canyon at 3,500-ft. altitude.

578. MOHAVE MIMULUS. *Mimulus mohavensis* (of the Mohave [River], where first collected). **Fl.**: corolla reddish-purple, with white-bordered limb and crimson throat. A simple, much-branched annual, not more than 5 in. high, dense with very short-stemmed flowers. 2,000–3,000 ft., in and about the Ord and Calico mountains.

579. FRÉMONT MIMULUS. *Mimulus Fremontii* (John C. Frémont—see **315**). **Fl.**: rose-red, with darker throat. Annual, not over 10 in. high, ordinarily smooth but sometimes with long, shaggy hairs. A cismontane species which has invaded the Mohave D. near Hesperia.

580. FALSE MIMULUS. *Mimetanthe pilosus* (Gr., "imitator blossom," i.e., resembling *Mimulus;* L., hairy). **Fl.**: bright yellow. Softly hairy annual; often mistaken for a *Mimulus*, but examination will show that the calyx is not plainly 5-angled as in the monkey flowers. Brown spots are generally found on the lower lobe of the corolla. Moist banks of ephemeral seeps about desert springs; cismontane southern Calif.; to Ore., Nev., and Ariz.

581. PURPLE BIRD'S-BEAK. *Cordylanthus eremicus* (Gr., club flower; Gr., of solitudes). **Fl.**: purplish. Not only the flower but also the entire plant has a purple tinge, which makes a fine contrast with the picric-yellow root. The name "bird's-beak" has reference to the flower, which resembles the broad beak of a young bird. Locally common in the piñon forests of the Inyo and Panamint mountains; to Nev.

582. NEVIN BIRD S-BEAK. *Cordylanthus Nevinii* (Rev. Joseph C. Nevin, 1835–1913, of Los Angeles, brilliant linguist and botanical collector. With W. S. Lyon, first to collect on Catalina Island). **Fl.**: purple and yellowish-white. A tall, loosely branched annual, blooming in late summer. The leaves and stems are a peculiar yellowish-green. Dry slopes, 5,000–8,000 ft., of the mountains of the southern Mohave and northwestern Colorado deserts.

583. GHOST-FLOWER. *Mohavea confertiflora* (Mohave River, where first collected by Frémont; L., crowded-flowered). **Fl.**: faint cream; lips ornamented with fine lines of purple dots. Erect annual, 3–16 in. high, often branched and very viscid (sticky). The odd form of the large, silky corolla makes it unique among desert flowers.

Common both in sands and alluvium of washes and on rocky hillsides of the eastern Mohave and Colorado deserts.

584. LESSER MOHAVEA. *Mohavea breviflora* (L., short-flowered). **Fl.**: lemon-yellow. An early spring annual of the middle and northern Mohave D.; most plentiful in washes draining into Death Valley Sink. The stem is 2–6 in. high, simple, or branched from the base. In the flower there are but two fertile stamens, the other three being abortive. Kelso and mid-Mohave to the Death Valley area; to western Nev.

ACANTHACEAE. Acanthus Family

585. CHUPAROSA. *Beloperone californica* (Gr., "dart or arrow band," in allusion to the oblique band connecting the unequal cells of the dart-shaped anther; of California). **Fl.**: corolla dull red, "nopal red." Low, rounded shrub, the single California representative of a tropical American genus. Found at low altitudes along borders of sandy washes and among rocks of the western and northern borders of the Colorado Desert. During the dry season the leaves turn yellow and fall, leaving the branches almost naked. The Papago Indians eat the blossoms. Linnets and Gambel sparrows bite off and eat the nectar-filled flower bases. Black-chinned hummingbirds probe the blossoms for both insects and nectar.

BIGNONIACEAE. Bignonia Family

586. DESERT WILLOW, DESERT CATALPA. *Chilopsis linearis* (Gr., "lip-like," alluding to the lower lip of the flower; L., "linear," in reference to the narrow, willow-like leaves). **Fl.**: pink. Not a true willow, but rather a catalpa. A small tree or large shrub, with crooked and leaning, black-barked trunks. It ordinarily grows from 6 to 15 ft. high. The handsome, violet-scented flowers follow periods of rain and soon give way to long, narrow seed pods, which remain dangling on the stems long afterward. The durable wood is prized for fence posts by desert folk. In Mexico a tea made from the dried flowers is considered to be of medicinal value. Chilopsis is the food plant of the white-winged moth (*Eucaterva variaria*), whose larvae build the tough, gauzy, buff-colored pupa cases, about an inch long, so often found on the branchlets during late spring. Desert willow is a wash-inhabiting species of southern Nev. and the Mohave and Colorado deserts of Calif; to Tex. and Mex. This tree is for the most part absent from the Death Valley area.

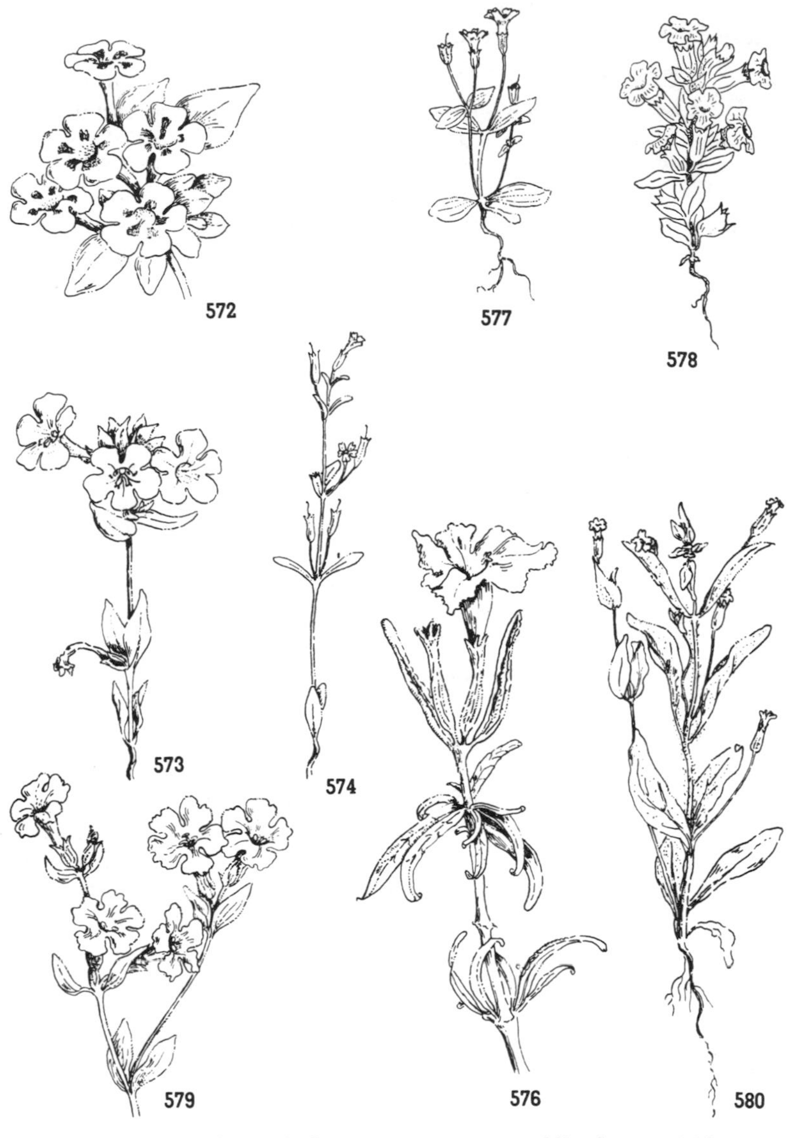

572.* *Mimulus rupicola*
573. *Mimulus Bigelovii*
574. *Mimulus rubellus*
576. *Mimulus longiflorus*
577. *Mimulus montioides*
578. *Mimulus mohavensis*
579. *Mimulus Fremontii*
580. *Mimetanthe pilosa*

581. *Cordylanthus eremicus*
582. *Cordylanthus Nevinii*
583. *Mohavea confertiflora*
584. *Mohavea breviflora*
585. *Beloperone californica*

MARTYNIACEAE. Martynia Family

587. DESERT UNICORN PLANT, ELEPHANT-TUSKS. *Martynia altheaefolia* (John Martyn, 1699–1768, for thirty years professor of botany of Cambridge University; L., "leaf like that of *Althaea*," an Old World plant genus which includes the marsh mallow). **Fl.**: vivid yellow. This is a rare, small-leaved perennial, of the scenic Vallecito and Chocolate mountains of the Colorado D.; to Ariz. and Mex. The fruit is a woody pod, ending in two curved, prong-like appendages that are capable of engaging the fetlocks of burros and sheep that happen to step on it—thus most ingeniously disseminating the seeds. Indian basket makers made black patterns from the tough covering of the fruits. The large-fruited *Martynia louisianica*, native to the Mid-West, is an escape found near Palm Springs.

The small-flowered species, *M. parviflora*, a common Arizona plant along the Gila River, was introduced into Death Valley eighty years ago by a brother of Hungry Bill, a Shoshone Indian, who visited Fort Mohave and found the Indians there making black patterns in their baskets from fibers of the fruits. He procured seeds and planted them in Johnson Canyon; the plants still flourish there.

OROBANCHACEAE. Broomrape Family

588. PIÑON STRANGLEROOT. *Orobanche fasciculata* (Gr., strangle vetch; L., in bundles). **Fl.**: yellowish, with darker veins. A root parasite, 2–6 in. high. Occasional in the piñon forests of desert mountains as well as in the mountains nearer the coast.

589. BURRO-WEED STRANGLER. *Orobanche Ludoviciana Cooperi* (Ludovici = Louis, i.e., of Louisiana Territory; J. G. Cooper —see **132**). **Fl.**: purplish, the lower lip sometimes yellowish. Like *Ammobroma* and *Pholisma*, this is a root parasite. The host plant is usually burro-bush. A variety, *latiloba*, parasitic on *Hymenoclea salsola*, is found principally on sandy areas of the Colorado D.; also N.Mex. and Sonora.

PLANTAGINACEAE. Plantain Family

590. PURSH PLANTAIN. *Plantago Purshii* (L., name of Plantain; Frederick Pursh—see **185**). **Fl.**: whitish. A short, tufted annual, of both Calif. deserts. The leaves and spikes are white-woolly; the linear bracts are generally without scarious (thin, dry) margins.

591. SPINY PLANTAIN. *Plantago spinulosa oblonga* (L., "furnished with diminutive spines"; L., "oblong," the calyx divisions except the outermost being broadly oblong). **Fl.**: whitish. Erect, light green annual, with short, cylindrical, compact spikes less than 1½ in. long. The lower bracts are aristate, i.e., each tipped by a bristle or spine. The spines are most noticeable in well-matured plants. Rare, Colorado D.; to Ariz, and N.Mex.

592. WOOLLY PLANTAIN (not illustrated). *Plantago insularis fastigiata* (L., insular; L., with erect, parallel branching). **Fl.**: whitish. This small, white-woolly plantain, having bracts with green midribs, is very common in spring on desert clay flats and bordering mountains. The smooth little seeds when ripe are reddish-yellow. Both deserts; to Utah and Ariz.

RUBIACEAE. Madder Family

593. NARROW-LEAVED BEDSTRAW. *Galium angustifolium diffusum* (Gr., "milk," which some species were used to curdle; L., "narrow-leaved"; L., "diffuse," with reference to the branching). **Fl.**: greenish-white. Erect, shrubby species, with long, lax branches, having dioecious flowers solitary on slender pedicels. Dry, rocky places along the western edge of the Colorado D. and in the Little San Bernardino Mts.

594. DESERT BEDSTRAW. *Galium stellatum eremicum* (L., "like the rays of a star," originally with respect to the stellate hairs investing the herbage; but this is a misnomer, for the hairs are actually short, stiff, simple ones. It would have been better if the name had applied to the star-like arrangement of the leaves. Gr., "lonely," in sense of desert-dwelling). **Fl.**: corolla greenish-white. A shrubby perennial species, 6 in. to 1½ ft. high, found among rocks of dry slopes below 4,500 ft. of the Colorado and Mohave deserts; to Ariz. and Nev. The old English name, bedstraw, was given to plants of this genus because they were sometimes used in bed ticks in place of straw.

595. MUNZ BEDSTRAW. *Galium Munzii* (P. A. Munz—see **342**). **Fl.**: brownish. Many-stemmed from a partially woody base; the herbage with short, stiff hairs. Dry, rocky slopes of Mohave D., north to Lone Pine. The variety *carneum* (L., flesh-like, red), with reddish flowers, is known from the Panamint Mts.

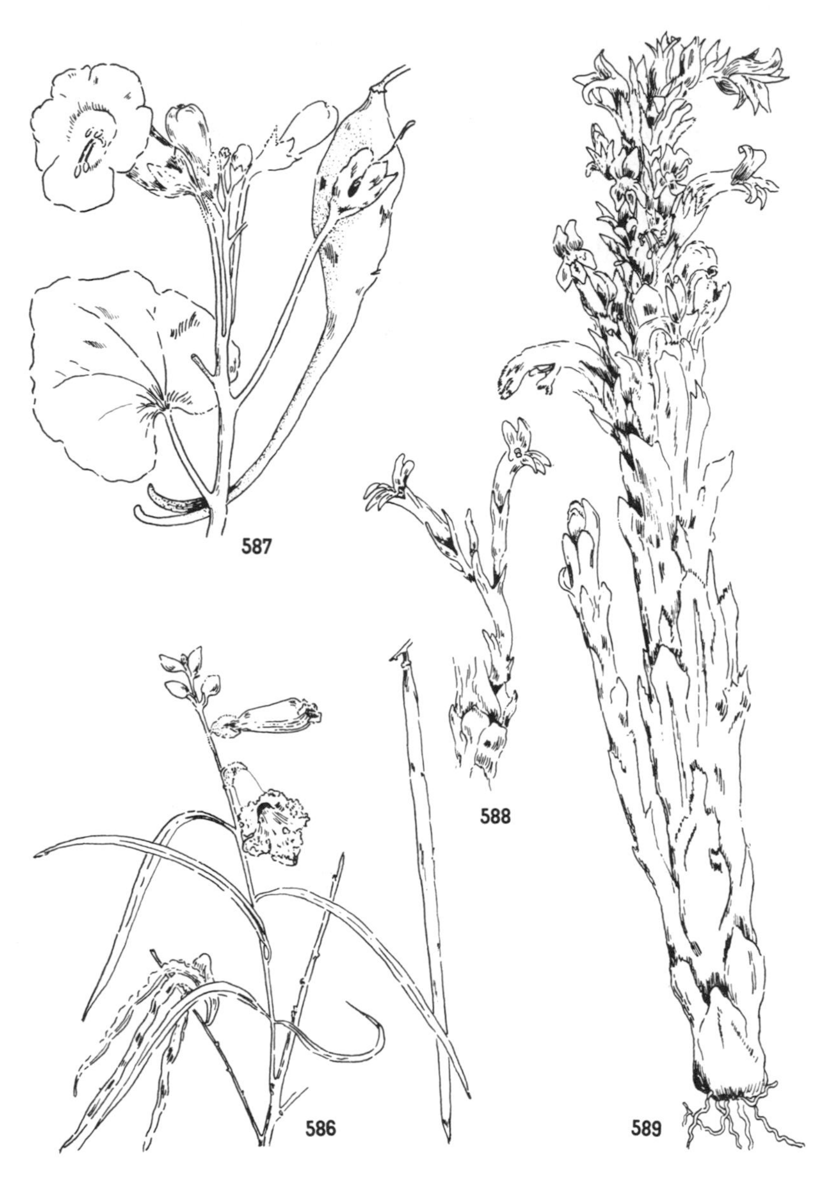

586. *Chilopsis linearis*
587.* *Martynia altheaefolia*
588. *Orobanche fasciculata*
589. *Orobanche Ludoviciana Cooperi*

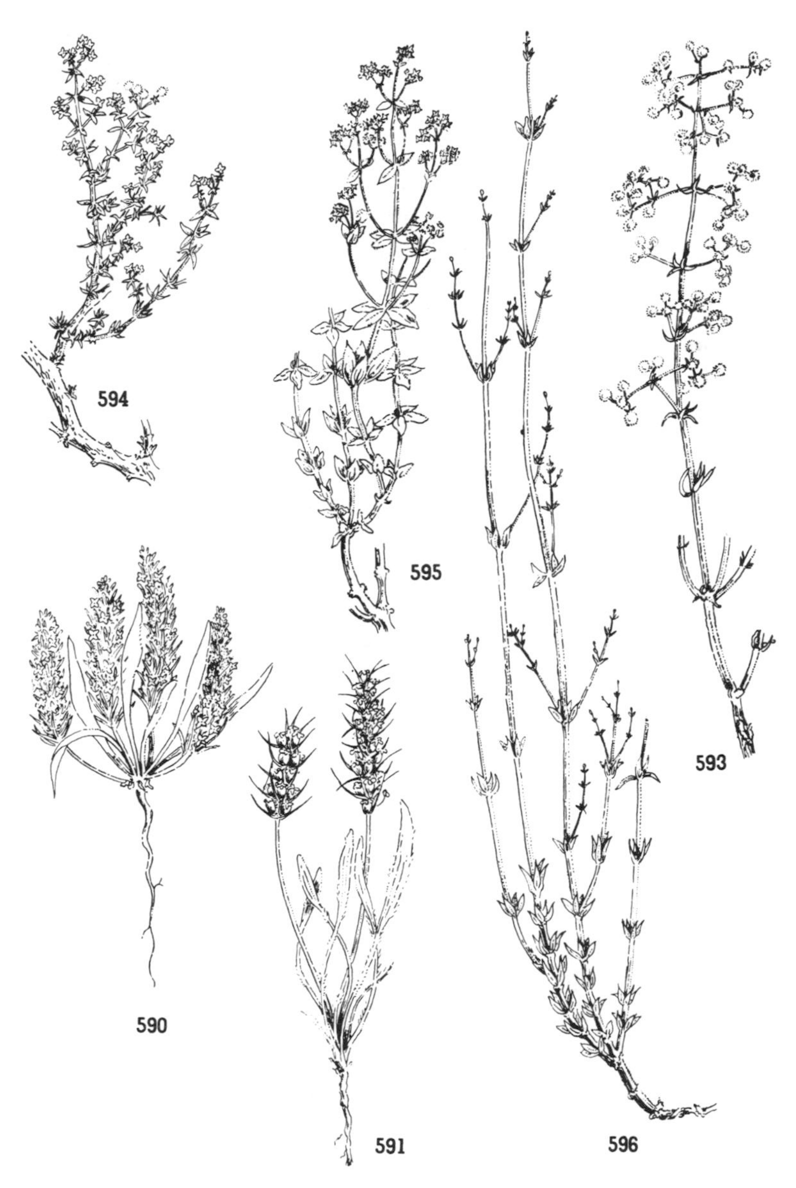

590. *Plantago Purshii*
591. *Plantago spinulosa oblonga*
593. *Galium angustifolium diffusum*
594. *Galium stellatum eremicum*
595. *Galium Munzii*
596. *Galium Rothrockii*

596. SLENDER-BRANCHED BEDSTRAW. *Galium Rothrockii* (Dr. Joseph Trimble Rothrock, Professor of Botany, University of Pennsylvania, and later Pennsylvania State Forest Commissioner. Surgeon on the Wheeler Exploring Expedition, 1873–1875). **Fl.**: red. Perennial from a woody rootstalk, 1–2 ft. tall. The drawing is of a plant in bud. Found in the mountains of the eastern Mohave D.; thence eastward to Ariz and N.Mex., to Mexico.

597. GREAT BASIN BEDSTRAW (not illustrated). *Galium proliferum* (L., "developing descendants," in reference to the peculiar production of the flowers, i.e., instead of occurring singly they form a chain-like series). **Fl.**: greenish-yellow. Ascending-stemmed annual, known from the mountains of the eastern Mohave D., through southern Nev. to Utah and Tex. Identifying characters are the few (2–4) leaves in a whorl and the almost sessile flowers.

598. RECLINING BEDSTRAW (not illustrated). *Galium aparine* (Gr., bedstraw). **Fl.**: white. A weak-stemmed annual, with leaves 6–8 in a whorl. The fruit is armed with hooked bristles. Providence Mts.; to Atlantic Coast. Native of Europe. In Europe used medicinally and as food for geese.

CUCURBITACEAE. Gourd Family

599. PALMATE-LEAVED GOURD. *Cucurbita palmata* (L., name of the gourd; L., "palmate," the leaves being palmately lobed). **Fl.**: yellow. A most handsome species, with trailing stems 4–6 ft. long, spreading radially on the sands. The ripe yellow gourds, 2½–4 in. in diameter, are called "coyote melons" by local Indians because as they say "they are good enough for coyotes."

The robust, brilliantly colored moth, *Melittia gloriosa*, which mimics a wasp, is on the wing during June and July and its females deposit their eggs singly on the stems of the dead herbage of this gourd. "The eggs," says Commander C. M. Dammers, "hatch in a few days and the young larvae follow down the stems and burrow into the large tuberous root, there to spend two years feeding before reaching maturity. The larva then leaves the root and pupates below the surface of the soil close by. The pupa is armed with a remarkable spade-like projection of the prothorax which it uses to dig its way up to the surface when ready to emerge. The empty brown pupal cases armed with peculiar rings of retrorse spines, may be found sticking out of the soil close to the plants. This moth is known to lay its eggs on *C. foetidissima* in Arizona, but in California is only known on *C. palmata*." Sand and clay soils of the Colorado and eastern Mohave deserts.

600. FINGER-LEAVED GOURD. *Cucurbita digitata* (L., fingered [-leaved]). **Fl.**: yellow. Best known from the dry, sandy plains of the low deserts of Ariz. and N.Mex. Locally known from the Colorado D. near Whitewater.

601. CALABAZILLA. *Cucurbita foetidissima* (L., "most fetid," on account of its rank odor when crushed). **Fl.**: golden yellow. This large-leaved gourd, so common in coastal southern California, is now widely found in the desert area, especially along highways. After winter frosts kill the vines, the yellow gourds, 2–3 in. in diameter, are very conspicuous. Called "Chili Cojote" by the early Californians; they held the crushed root in high esteem as a cleansing agent in washing clothes, but were insistent that the clothes should be well rinsed afterward, since particles clinging to the fabrics were found to be very irritating to the skin.

602. BRANDEGEA. *Brandegea Bigelovii* (T. S. Brandegee; Dr. John M. Bigelow—see **338**). **Fl.**: white. Perennial herb, with large, thick roots and thin leaves, upon whose upper surfaces are numerous disk-like pustules. Found trailing among branches of trees and shrubs bordering sandy washes and canyon bottoms of the Colorado D. It blooms during March–April, and later following summer cloudbursts. The leaves are very variable in form and size. The leafy vines often almost smother small shrubs over which they climb.

Mr. Brandegee, in whose honor the genus was named, was known principally for his botanical collections made on the west coast of Mexico and Baja California. He and Mrs. Brandegee spent their last years in active work at the University of California Herbarium. An index to the enthusiasm of the Brandegees for botanical studies in the field is shown by the fact that when in 1889 Mr. Brandegee married Mrs. Katherine (Layne) Curran in San Diego, the two "chose to take their wedding trip *on foot* to San Francisco, botanizing along the way."

LOBELIACEAE. Lobelia Family

603. PARISHELLA. *Parishella californica.* (W. F. and S. B. Parish, pioneer botanical collectors of southern California. They lived on a ranch in San Bernardino and spent such time as they could spare from their work in wide travels in desert and mountains. For these trips they had fitted up a wagon with low-hung body and broad tires adapted for carrying their barrels of water, barley, hay, and camp outfit. Samuel Parish was the more active botanist and maintained deep interest in his science until his death in Berkeley in 1928.

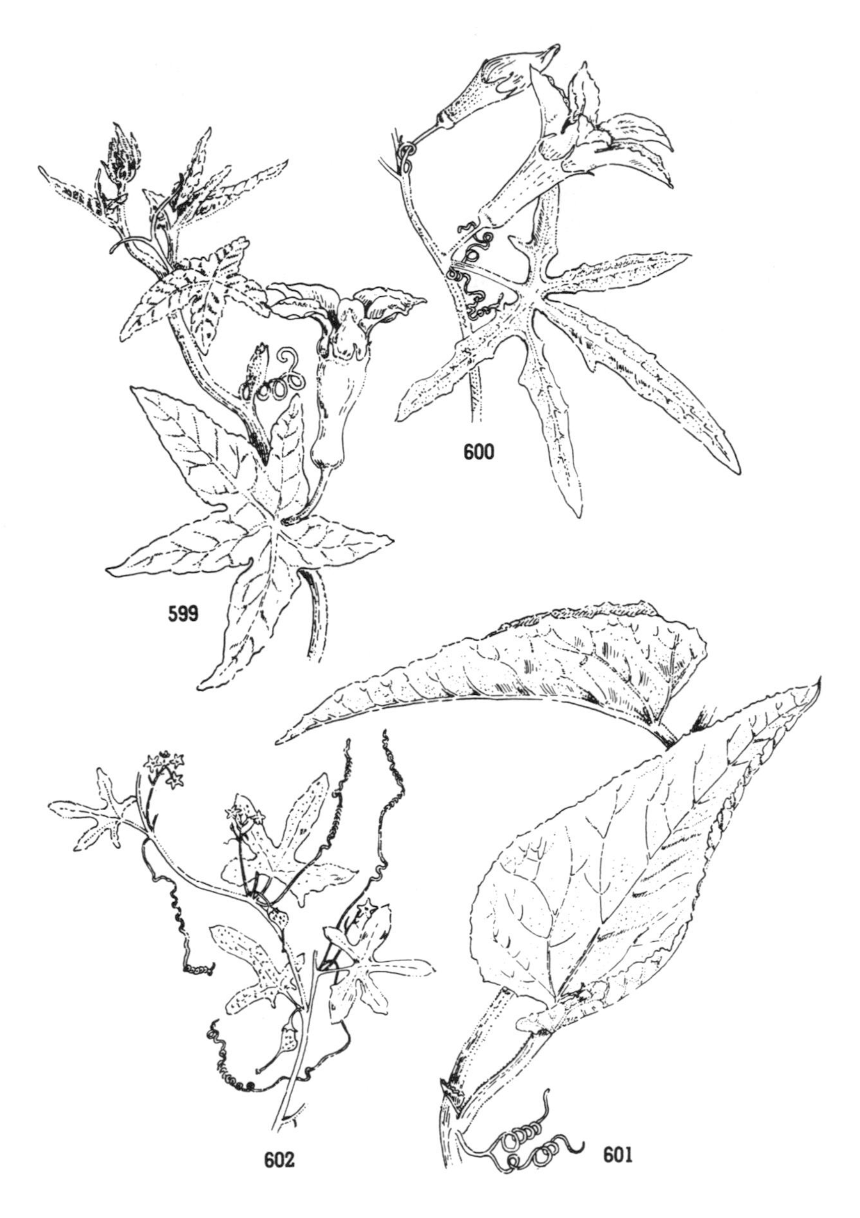

599. *Cucurbita palmata*
600. *Cucurbita digitata*
601. *Cucurbita foetidissima*
602. *Brandegea Bigelovii*

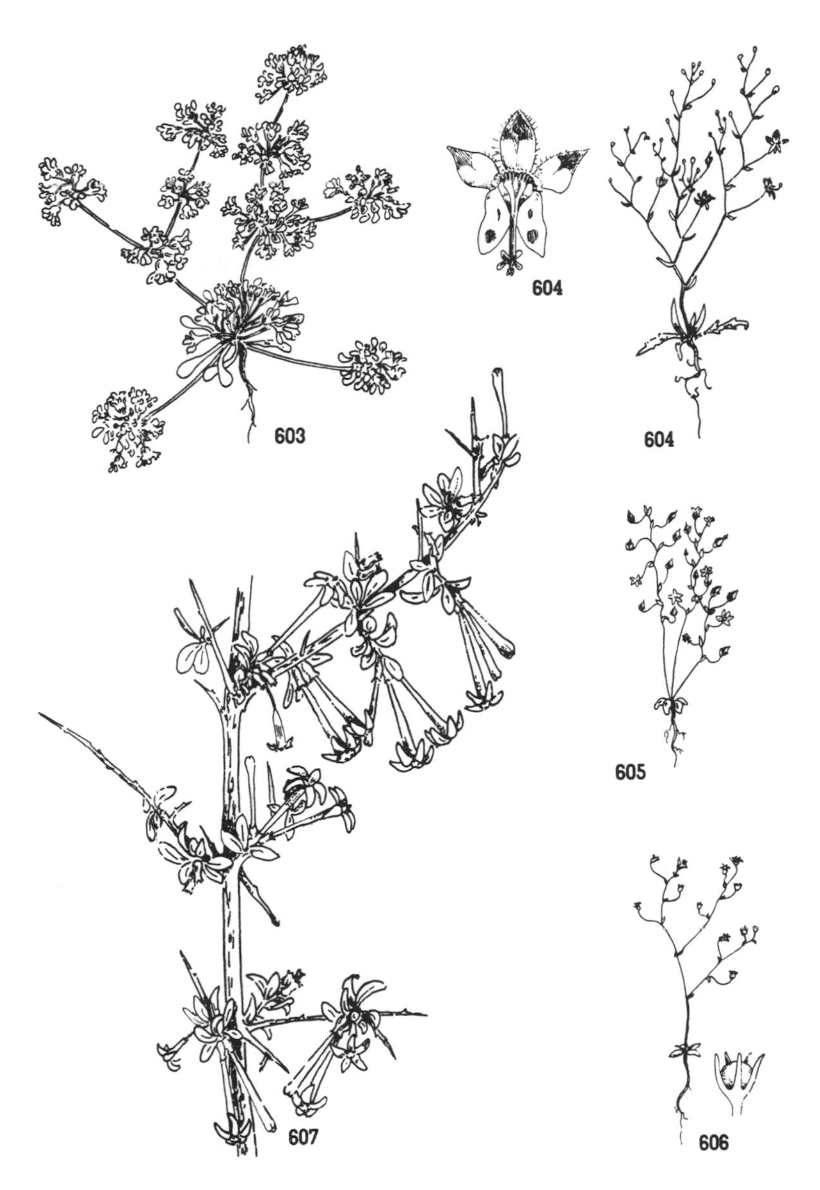

603.* *Parishella californica*
604. *Nemacladus rigidus rubescens*
605. *Nemacladus longiflorus*
606. *Nemacladus ramosissimus gracilis*
607. *Symphoricarpos longiflorus*, ×1

He corresponded widely and was intimately acquainted with many of the leading botanists of his day. Of California.) **Fl.**: white. Small, spreading, winter annual, with leaves and flowers in a radical tuft; found on sandy or clay flats of the mid-Mohave D. Of local distribution and rarely collected.

604. RIGID-STEMMED THREAD PLANT. *Nemacladus rigidus rubescens* (Gr., "thread stem," from the singular, prostrate thread-like stem; L., "stiff"; L., "reddening"). **Fl.**: corolla white, touched with red, brown, and yellow. A tiny plant, so slender-stemmed that it is scarcely noticed by the average traveler. Several varieties are recognized. Common in sandy soils of both deserts; eastward to Utah and N.Mex. The arrangement and coloring of the parts of the small flower make it appear, when examined under a lens, as beautiful as an orchid. Especially appealing are the slender, glass-like rods arranged like tines of a fork on the processes of the two anterior filaments.

605. LONG-FLOWERED THREAD PLANT. *Nemacladus longiflorus* (L., long-flowered). **Fl.**: white, with a reddish stripe on the back side of each petal. The mature, fusiform seed vessels appear bright as if covered with a lacquer. A cismontane plant, which occurs along the western edge of the Mohave D., as in the Joshua Tree Nat. Mon.

606. SMALL-FLOWERED THREAD PLANT. *Nemacladus ramosissimus gracilis* (L., very much branched; L., slender). **Fl.**: white tinged with pink in the bud. The capsules are almost globular and are shorter than the calyx. Plains and slopes of the creosote-bush and piñon belt of southern Nev., Ariz., and southern Calif.

CAPRIFOLIACEAE. Honeysuckle Family

607. LONG-FLOWERED SNOWBERRY. *Symphoricarpos longiflorus* (Gr., "fruit borne together," from the clustered berries; L., long-flowered). **Fl.**: pink. This handsome, exceedingly sweet-scented shrub, 1–4 ft. high, is common in the piñon-juniper belt of the Inyo and other mountains of the Death Valley region; east to Nev. and Utah.

COMPOSITAE. Sunflower Family

608. ARROW-LEAF. *Hofmeisteria pluriseta* (W. Hofmeister, German botanist, who made many fundamental discoveries in botanical science, particularly with reference to the life history of the mosses, ferns, and conifers. Von Sachs declared his results "magnificent beyond all that have been achieved before or since in the domain of descriptive botany." L., very many-bristled). **Fl.**: whitish. Intricately branched, weak-stemmed fragrant shrub, 1–2½ ft. high, generally growing among rocks or in rock crevices of canyon walls. When growing in shaded places the leaf petioles are often twice as long and more vividly green. Both deserts; to Ariz.

609. BROWN TURBANS. *Malperia tenuis* (anagram of Palmeri; L., "thin, fine," alluding to the stem or perhaps to the filiform style). **Fl.**: brownish. A very rare annual with erect, branching, reddish stems 8–12 in. high. The scarious-margined flower bracts are very distinctive. On arid, stony benches in canyons of the extreme southwestern Colorado D.; to Baja Calif. Often growing with *Chaenactis carphoclina* (see **715**) and, when that plant is in its early budding stage, often confused with it.

610. SWEET BRICKELLIA. *Brickellia Watsonii* (Dr. J. Brickell, early botanist of Savannah, Ga.; Sereno Watson—see **459**). **Fl.**: whitish. Intricately branched shrub, 8–12 in. high, with light green, fragrant leaves; blooms in August. Rocky ledges and canyon walls of the higher mountains of the eastern Mohave D.; to Nev. and Utah.

611. GUM-LEAVED BRICKELLIA. *Brickellia multiflora* (L., many-flowered). **Fl.**: whitish. An erect shrub, 1–2 yds. tall, with 3-nerved, somewhat sickle-shaped, gummy leaves (the upper leaves small and linear). Found in sandy washes of the eastern Sierra Nevada; east to western Nev.

612. SHRUBBY BRICKELLIA. *Brickellia frutescens* (L., shrub-like). **Fl.**: whitish. Low, rigid shrub, 8–14 in. high, found among rocks at 2,000–3,000 ft. altitude along the western borders of the Colorado D.; to Nev., Baja Calif., and Sonora.

613. KNAPP BRICKELLIA. *Brickellia Knappiana* (M. A. Knapp, plant collector on the Mohave River in 1888). **Fl.**: whitish. Slender, willow-like perennial, 2–8 ft. high, with white-barked branches and somewhat viscid leaves. A rare plant of the Panamint range at about 3,000 ft. altitude.

614. WOOLLY BRICKELLIA. *Brickellia incana* (L., "gray" or "hoary," with white or gray pubescence). **Fl.**: maroon. A handsome, rounded bush, with silver-green leaves. The numerous flowers are soon followed by very beautiful and conspicuous hemispherical seed heads, with silvery bristles, which glisten brightly in the sunshine. A plant wholly confined to gravelly washes of the middle and eastern Mohave D. and the Colorado D. east of the Salton Sea. A

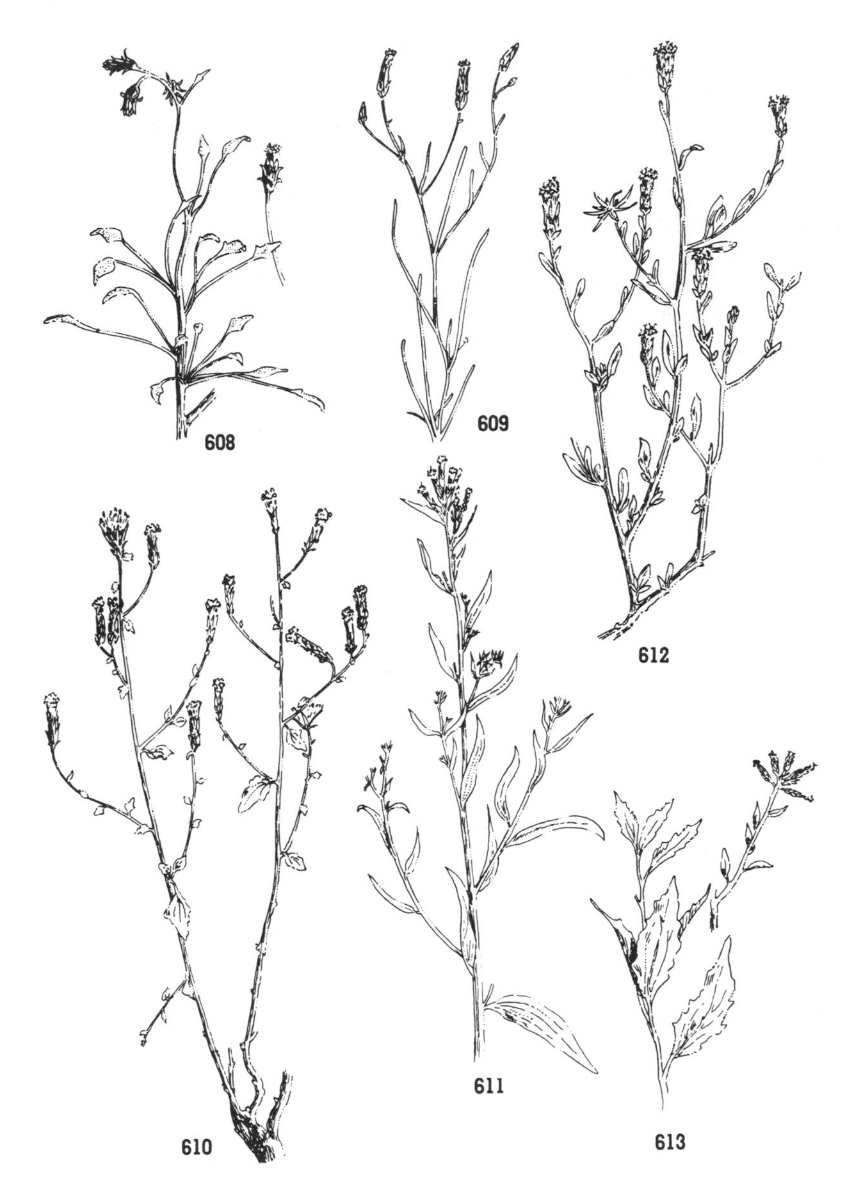

608. *Hofmeisteria pluriseta*
609. *Malperia tenuis*
610.* *Brickellia Watsonii*
611. *Brickellia multiflora*
612. *Brickellia frutescens*
613.* *Brickellia Knappiana*

small black beetle, *Lygaeus lateralis*, is seen in great numbers on the flowers, especially in July.

615. SPEAR-LEAVED BRICKELLIA. *Brickellia arguta* (L., "sharp, pungent," probably with reference to the pointed involucral bracts). **Fl.**: yellowish. A small shrub, with zigzag stems and rigid, bright green leaves. The dying leaves turn a clear tan in midsummer. Mostly confined to rocky situations of the northern Mohave and western Colorado deserts. The variety *odontolepis* (Gr., tooth-scaled), with conspicuously toothed involucral bracts, is a rare plant of western Colorado D. (as at Piñon Wells).

616. DESERT BRICKELLIA. *Brickellia desertorum* (L., of the deserts). **Fl.**: yellowish. An intricately branched, brittle-stemmed bush, common among rocks and in rock crevices over most of our deserts. It blooms in midsummer. The leaves are very variable in form, as shown in the illustrations.

617. PIÑON BRICKELLIA. *Brickellia oblongifolia linifolia* (L., oblong-leaved; L., linear-leaved). **Fl.**: whitish. A small bush, 8–15 in. high, with numerous stems springing from a woody base. The fine striae of the bracts of the flower head produce a most beautiful effect. Spottily distributed in stony situations in the higher mountains of both deserts; to Colo. and N.Mex.

618. YELLOW-GREEN MATCHWEED. *Gutierrezia lucida* (named in honor of some member of the noble Spanish family, Gutierrez; L., "bright, shining," referring to the herbage). **Fl.**: yellow. In autumn this many-stemmed, yellowish-green perennial is very common. It often occurs in almost pure stands, especially on neglected cleared areas about deserted homesteads. The resinous stems burn furiously, giving off a dense black smoke. Bees visit it in numbers. Dry hills and mesas, 3,000–6,000 ft., Mohave D. to Inyo Co., and Little San Bernardino Mts. The somewhat similar *G. Sarothrae* (Gr., a broom) is much less common, being confined mostly to the desert slopes of the San Bernardino Mts. Its erect, resin-dotted leaves and more numerous ray and disk florets serve to distinguish it from the yellow-green matchweed.

619. EYTELIA. *Amphipappus Fremontii* (Gr., "double-pappused," the pappus being double; J. C. Frémont—see **315**). A handsome little bush, with whitish stems and yellow-green foliage; occasional in dry, rocky places. Providence Mts. to the Coso Mts.; Nev. and Utah. Named eytelia in honor of Carl Eytel, 1873–1927, desert artist and traveler, and good friend of many botanists. He long resided at Palm Springs.

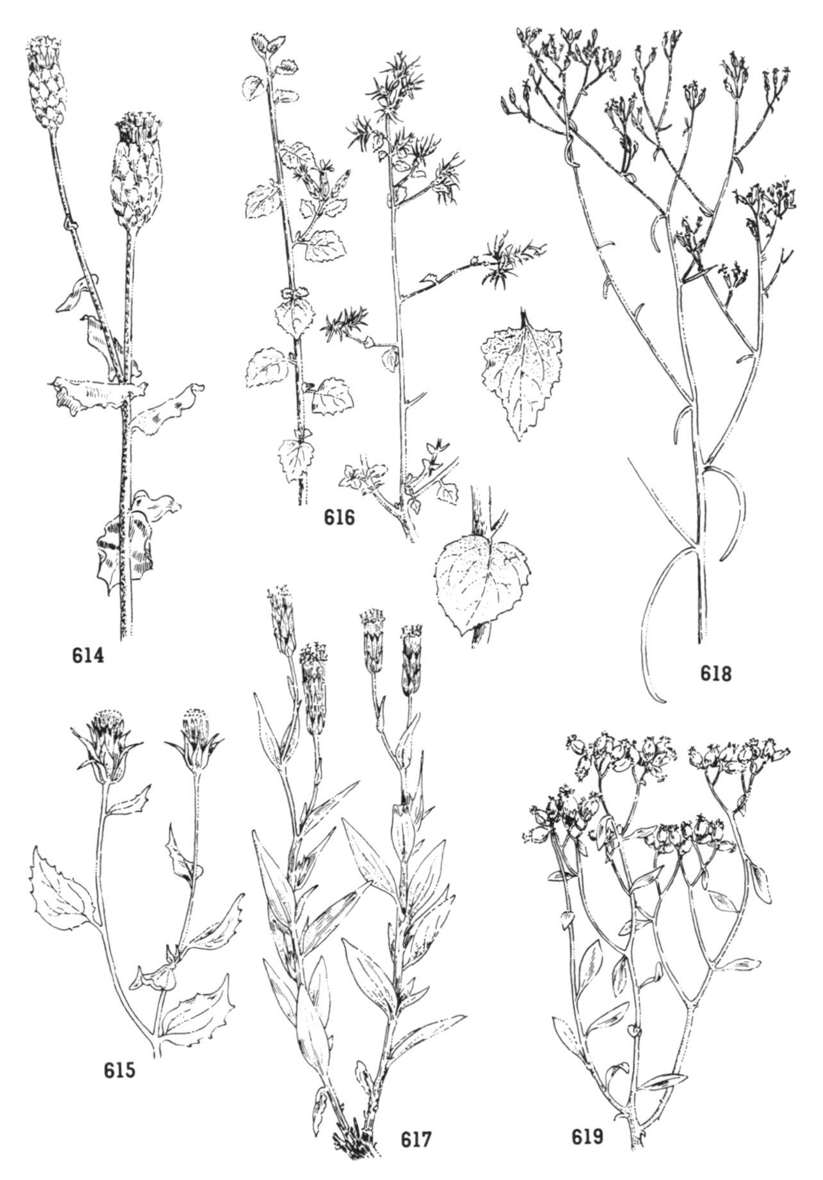

614. *Brickellia incana*
615. *Brickellia arguta*
616. *Brickellia desertorum*
617. *Brickellia oblongifolia linifolia*
618. *Gutierrezia lucida*
619. *Amphipappus Fremontii*

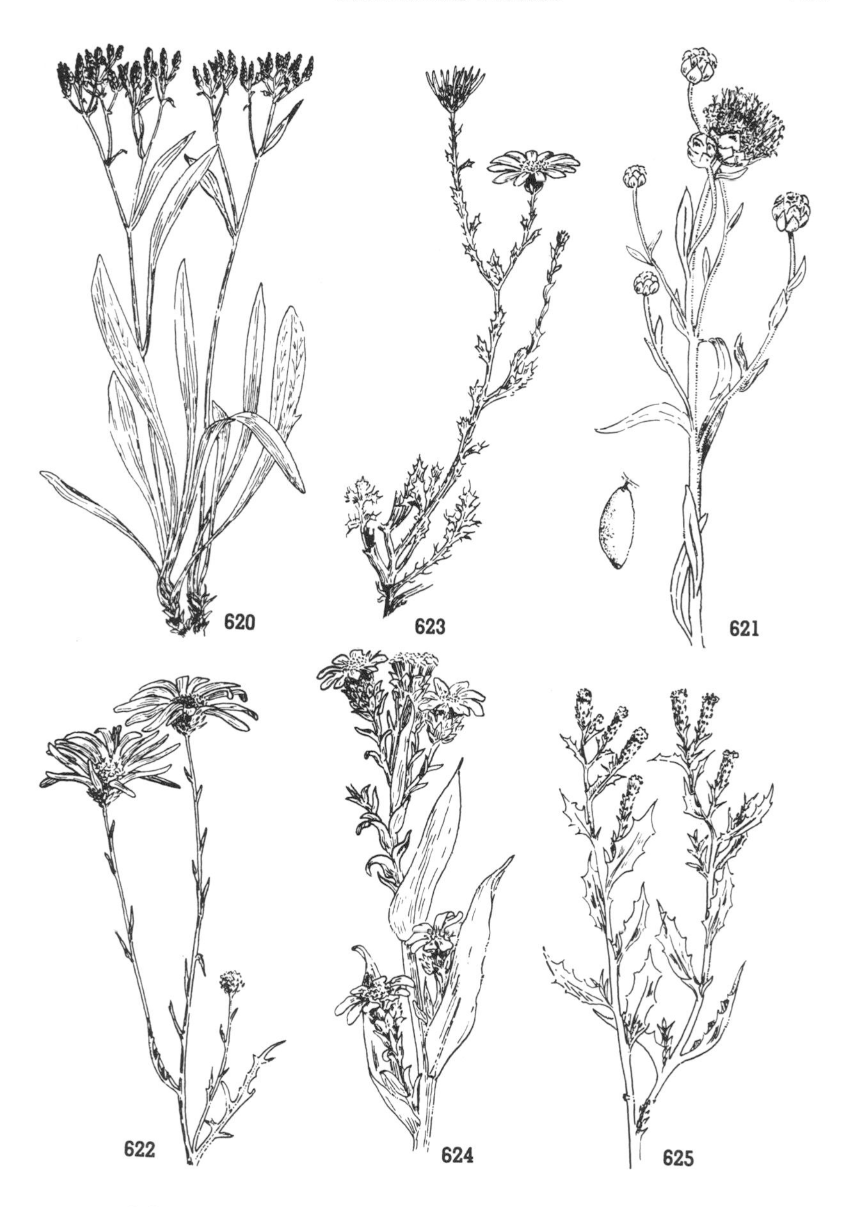

620.* *Solidago Petradoria*
621. *Acamptopappus sphaerocephalus*
622. *Haplopappus spinulosus Gooddingii*
623. *Haplopappus gracilis*
624. *Haplopappus paniculatus*
625.* *Haplopappus brickellioides*

620. ROCK GOLDENROD. *Solidago Petradoria* (L., "to make whole," because of reputed wound-healing properties; Gr., "rock," and *Doria*, early name for the goldenrod). **Fl.**: yellow. Very short-stemmed, I should say, for a goldenrod, being but 4–11 in. tall. The tufts of basal leaves remain through several seasons even after drying. Dry limestone slopes of the Providence and Clark mountains, from 3,500 to 6,000 ft.; to Wyo. and Tex.

621. GOLDENHEAD. *Acamptopappus sphaerocephalus* (Gr., unbending or stiff pappus; Gr., sphere-headed). **Fl.**: pale yellow. Small shrub, 8–16 in. high, with slender, striate stems. Flats and benches of western Colorado and southern Mohave deserts; to Utah and Ariz. On the south and west Mohave D. it is replaced by the variety *hirtellus* (L., hairy), with rough, hairy stems. In the ranges east of Death Valley and adjacent Nev. is found a goldenhead with bright yellow, radiate heads, greener involucres, and broader leaves. It is known as Shockley goldenhead, *A. Shockleyi.* During late April and in May almost every goldenhead bush has numbers of the pure white, inch-long, silken cases of the goldenhead bagworm, *Colcophora acamptopappi* (see illustration), hanging from it. The small, brown, fat larva inside each is very lively and squirms vigorously about whenever its case is disturbed. Take a case from its mooring and hold it up to your ear, and you will hear a sound like the rattle of a paper bag.

622. SPINY GOLDENBUSH. *Haplopappus spinulosus Gooddingii* (Gr., *haplo*, "simple," in reference to the simple pappus ring, + L., "pappus"; L., "minutely thorny"; Leslie M. Goodding, Associate Botanist, Soil Conservation Service, United States Department of Agriculture, former collector for the University of Wyoming). **Fl.**: pale yellow. Perennial, from a branching, woody root crown, 8–16 in. high. Dry slopes west of Needles; to Nev. and Ariz.

623. SLENDER GOLDENBUSH. *Haplopappus gracilis* (L., slender, scanty). **Fl.**: yellow. Annual herb, 2–10 in. high, with radiate heads; the involucral bracts and leaf lobes are tipped with bristles. Higher ranges of the eastern Mohave D.; to Colo.

624. SMOOTH-STEMMED GOLDENBUSH. *Haplopappus paniculatus* (L., panicled). **Fl.**: yellow. Smooth, herbaceous-stemmed perennial, with thickish leaves. Antelope Valley to Death Valley; Ore. and Nev.

625. HOLLY-LEAVED GOLDENBUSH. *Haplopappus brickellioides* (like *Brickellia*). **Fl.**: yellow. Shrubby-stemmed perennial, of the creosote-bush belt of the Death Valley region; to western Nev., as at Ash Meadows.

626. DESERT ROCK GOLDENBUSH. *Haplopappus cuneatus spathulatus* (L., wedge-shaped; L., spatulate). **Fl.**: yellow. A low, spreading shrub, with resinous, glossy, green leaves, commonly occupying space between rocks on slopes of granitic mountains of the western Colorado and Mohave deserts; to Nev. It comes into flower in October.

627. ALKALI GOLDENBUSH. *Haplopappus acradenius* var. ? (Gr., "pointed-glanded," each of the involucral bracts having a large gland at its tip). **Fl.**: yellow. Mohave D.; to Nev. On the Colorado D. this is replaced by the variety *eremophilus,* with denticulate leaves. In the typical species the leaves are entire, i.e., not toothed.

628. ALKALI GOLDENBUSH. *Haplopappus acradenius* var. ? Another plant exhibiting geographic variations. Known from alkaline areas of the eastern Mohave D.

629. COOPER GOLDENBUSH. *Haplopappus Cooperi.* (J. G. Cooper—see **132**). **Fl.**: bright yellow. Low, flat-topped shrub, 9 in. to a foot high, often occurring in almost pure stands in the flat, wind-swept basins and higher mesas of the Mohave D. The numerous small leaves crowding the upper stems and 1- (or 2-) rayed flower heads are aids in identification. Rare in interior cismontane southern Calif.; to Nev.

630. LINEAR-LEAVED GOLDENBUSH. *Haplopappus linearifolius interior* (L., linear-leaved; L., "of the interior," i.e., "inland"). **Fl.**: yellow. A round, bushy shrub, 1½–2 ft. high, with stout, woody branches and resinous leaves. Conspicuous in spring because of the numerous showy flower heads. (See photo, p. 264.) Both deserts; to Colo. and Baja Calif. The type was collected on the Darwin Mesa of Inyo Co.

631. HISPID GOLDEN ASTER. *Chrysopsis hispida* (Gr., golden aspect; L., rough, bristly). **Fl.**: yellow. A slender-stemmed, branching, perennial herb, with glandular-hairy stems and leaves. Rocky places, from 3,000 to 4,000 ft. altitude at Morongo Pass and Little San Bernardino Mts.

632. BLACK-BANDED RABBIT BRUSH. *Chrysothamnus paniculatus* (Gr., gold bush; L., panicled). **Fl.**: yellow. This species is considered the most primitive of the genus. It and the closely related *C. teretifolius* are placed by botanists in a special division of rabbit

630. A Mohave flower field showing LINEAR-LEAVED GOLDENBUSH in full flower

Photo by Avery Edwin Field

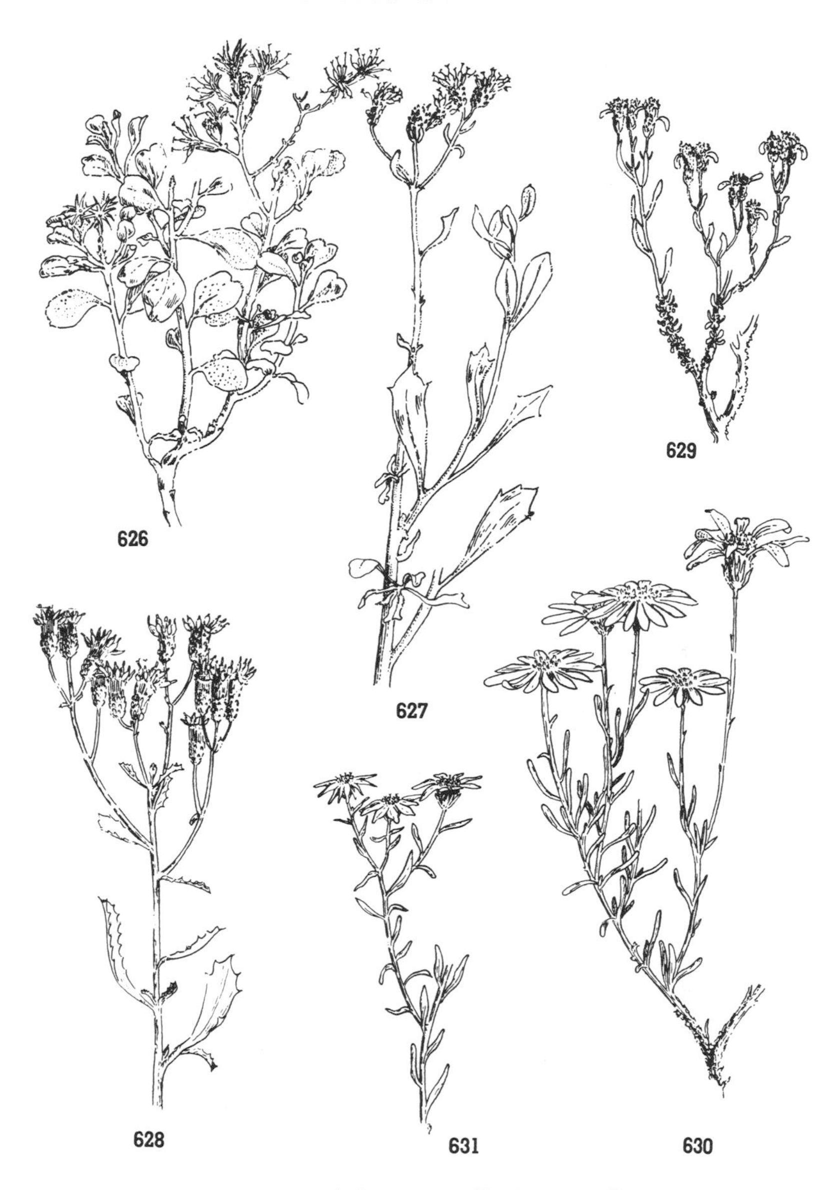

626. *Haplopappus cuneatus spathulatus*
627. *Haplopappus acradenius*
628. *Haplopappus acradenius*
629. *Haplopappus Cooperi*
630. *Haplopappus linearifolius interior*
631. *Chrysopsis hispida*

brushes owing to the fact that in them the herbage is resinous-dotted, the "dots plainly showing as definite depressions." It is a rather common, large-sized, rounded shrub, with prominent, straight, tan-colored stems, and is confined mostly to dry-surfaced streamways showing some moisture beneath. Its range includes southwestern Utah, northern Ariz., and the Colorado and Mohave deserts. Its most common plant associates are the scale-broom and the cheese-bush. Blooms May–October. A peculiar, little-known, smut fungus so commonly causes blackish bands on the young twigs that the common name, black-banded rabbit brush, is given to it. The bands are each about ½ in. broad and often so plentiful that they are noticeable even from some distance.

633. RABBIT BRUSH. *Chrysothamnus depressus* (L., depressed, i.e., growing near the ground). **Fl.**: yellow. Low bush, 4 in. to 1 ft. high, forming dense clumps. Dry, rocky slopes of the upper parts of the Clark and Providence mountains; to Utah and Colo.

634. MOHAVE RUBBERBRUSH. *Chrysothamnus nauseosus mohavensis* (L., nauseating; of the Mohave [River], where first collected). The many varieties of *Chrysothamnus nauseosus* are called rubberbrushes because of the high rubber content of the herbage. Though commercial extraction is not yet practical, it is estimated that a yield of 300,000,000 pounds of good grade of rubber from this plant is available if needed. Rubberbrushes are all handsome plants in autumn when covered with yellow bloom, especially when, as often happens, they occur in large, dense aggregations. The stems and leaves are light gray-green and very bitter to the taste. When burned they give off a dense, black, sooty smoke. The almost leafless variety *mohavensis* here illustrated is a common shrub of the Joshua Tree Nat. Mon. and the western borders of the Mohave D. Two other varieties, *viridulus* and *ceruminosus*, are known from the western Mohave D.

635. TERETE-LEAVED RUBBERBRUSH. *Chrysothamnus teretifolius* (L., having a rounded-off or well-turned leaf). **Fl.**: yellow. Flattish or rounded shrub, 1–3 ft. high, very leafy and with gray stems. The Paiute Indians chewed the wood and bark of this plant to get a crude chewing gum containing rubber. It was this custom that induced the first studies on the rabbit brush as a possible source of rubber during the World War. Occasional on dry, stony hillsides and canyon bottoms of both deserts, from 3,000 to 5,000 ft. Most common in the White and Inyo ranges of California.

636. STICKY-LEAVED RABBIT BRUSH. *Chrysothamnus viscidiflorus* (L., sticky-flowered). **Fl.**: yellow. A round-topped, bushy plant, blooming in late summer and autumn. The shrubs furnish good browse for sheep. The animals eat the flowering shoots with special relish. Joshua Tree Nat. Mon., Santa Rosa and San Jacinto mountains; to Wash.; Rocky Mts.

637. MECCA ASTER. *Aster cognatus* (Gr., "a star," from the radiated appearance of the flower; L., "related," i.e., related to *Aster Orcuttii*). **Fl.**: lavender with yellow center. A bushy aster, 1–2 ft. high, with whitish stems and deep green leaves covered with fine, glandular hairs. The involucral bracts are very long, narrow, and glandular. This is the handsome, spreading aster growing in sandstone and clay crevices at the bases of the vertical walls of Painted, Hidden Springs, and Box canyons near Mecca on the Colorado D.

In gypsum soils of canyons on the southwest side of the Salton Sink, especially those west of Imperial Valley, grows the quite similar but even more handsome Orcutt aster, *A. Orcuttii*, distinguished from the Mecca aster principally by its hairless herbage. Both species bloom soon after winter rains or when summer cloudbursts afford moisture. The beautifully patterned, satin-brown, night-flying moth, *Thyerion ligeae*, rests during the daytime in the center of the flower of the Mohave aster. Orcutt's aster harbors a somewhat similar moth but of satin-white color and with a handsome fringe of silky hairs.

Aster Orcuttii was named after C. R. Orcutt, 1864–1929, zealous, lifelong natural history collector, of San Diego, California. He did a lot of valuable botanical work in the southern Colorado Desert, especially in the canyons of its southwestern border. Beginning in 1884, he printed and edited for some ten years *The West American Scientist*, at that time the only medium that existed in western America for the publication of natural history notes and short articles. One of the later numbers was curiously printed on the back side of wallpaper and advertised everything Orcutt was interested in, from sea shells and minerals to a resort hotel which he was promoting.

638. WHITE ASTER. *Aster Leucelene* (Gr., *Leucelene*, the former genus name). **Fl.**: white. A low plant, with crowded stems from a woody base. The leaves are glandular and covered with stiffish hairs. The preferred habitat is dry limestone slopes of the high mountains of the eastern Mohave D.; to Kan., Tex., and Mex.

639. MOHAVE ASTER, DESERT ASTER. *Aster abatus* (L., beaten down, lowly). **Fl.**: violet to lavender or almost white, with

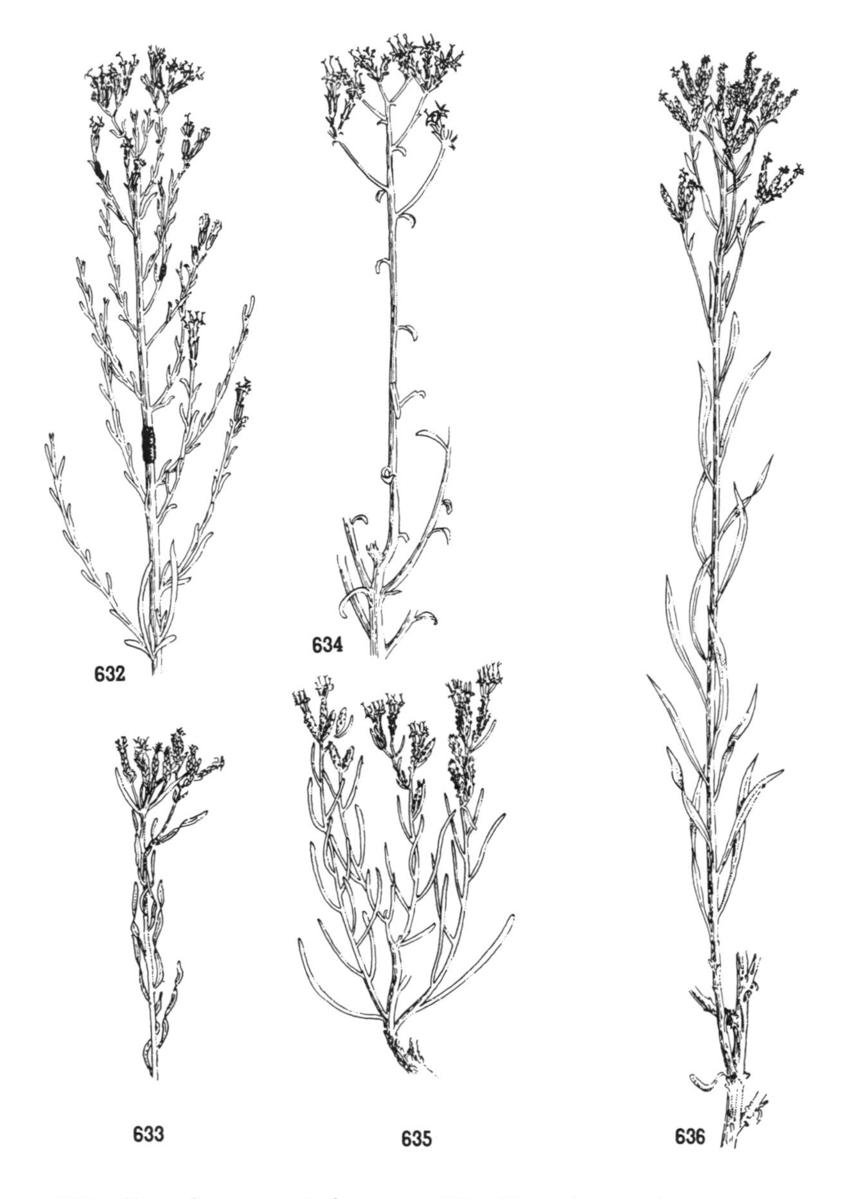

632. *Chrysothamnus paniculatus* 633. *Chrysothamnus depressus*
634. *Chrysothamnus nauseosus mohavensis*
635. *Chrysothamnus teretifolius* 636. *Chrysothamnus viscidiflorus*

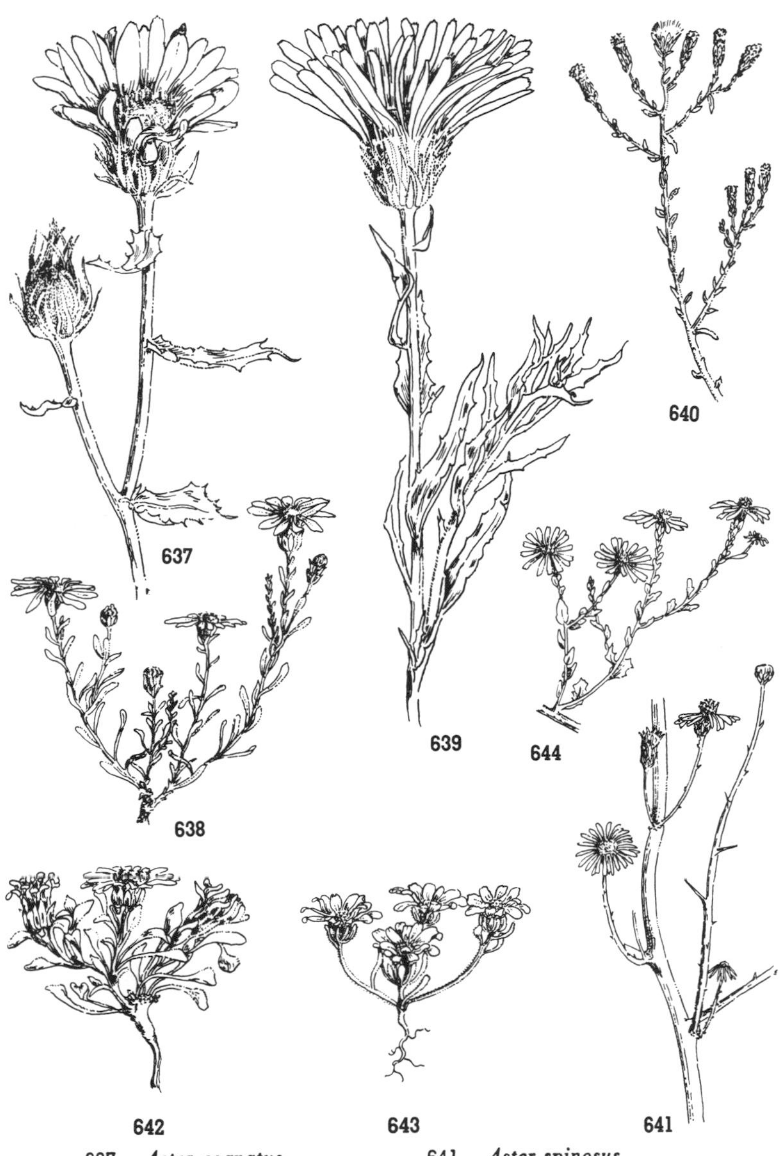

637. *Aster cognatus*
638. *Aster Leucelene*
639. *Aster abatus*
640. *Aster carnosus*
641. *Aster spinosus*
642. *Townsendia scapigera*
643. *Monoptilon bellioides*
644. *Psilactis Coulteri*

yellow center. A most ornamental species and often very plentiful. Sometimes as many as twenty flower heads are on a single plant. The usual habitat is rocky hills and slopes, often exceedingly dry ones. During summer the almost barren stems and the few somewhat twisted leaves turn brown or straw-colored, so that the plants are quite inconspicuous. Mohave and northern Colorado deserts, below 6,000 ft.

640. ALKALI ASTER. *Aster carnosus* (L., fleshy, reddish). **Fl.**: light pink to purplish. Rigid, much-branched shrub, 1½–3 ft. high, with pale, slender branches and rayless flowers. Generally found in alkaline meadows. Western Mohave D.; to Ariz. and Nev.

641. MEXICAN DEVIL-WEED. *Aster spinosus* (L., thorny). **Fl.**: white, drying brown. A weed of frequent occurrence along roadways, field borders, and canals of the irrigated portions of the Imperial Valley. Its large size (3–9 ft. high), spiny nature, and easy propagation make it a pest wherever it grows. The almost leafless stems and scraggly flowers which soon turn brown do not commend it as a thing of beauty. But the butterflies come to its rescue. During the sunshiny days myriads of alfalfa-yellow and cabbage-white butterflies continually hover over it and feed on the flowers, so that from a distance the plants appear as if loaded with large, colorful blossoms. Colorado D.; to Utah, Tex., and Central America.

642. GROUND-DAISY. *Townsendia scapigera* (David Townsend, botanical associate of Dr. Wm. Darlington, 1782–1863, botanist of Pennsylvania who wrote that "delightfully gossipy volume entitled *Flora Cestrica*"; L., scape-bearing). **Fl.**: rays purple to almost white; disk yellow. A plant of the piñon-juniper woodlands of the arid southern end of the Inyo Mts.; to Nev. and Wyo. The plants are generally wedged between small stones and lie flat, close to the surface of the ground. The flowers open between nine and ten in the morning and close between five and six in the evening.

643. MOHAVE DESERT-STAR. *Monoptilon bellioides* (Gr., "one bristle," in reference to the solitary bristle of the pappus of one species; of the form of plants of the Composite genus *Bellis*). **Fl.**: white to pinkish, with yellow center. A gay little annual, which clings close to the desert sands, often forming mats 4–6 in. across. The pappus consists of numerous unequal bristles. Common on the low sandy areas of both deserts; to Utah and Ariz. The very similar species, *M. bellidiforme*, is confined in California to the Mohave D. Among its distinguishing features are the minute crown of scales and the single, long, plumose awn making up the pappus.

644. SILVER LAKE DAISY. *Psilactis Coulteri* (Gr., naked ray; T. Coulter—see **755**). **Fl.**: rays white to light lavender; center of head yellow. A charming, annual "round-bushed daisy," plentiful after wet seasons in the loose sands about the borders of Silver Lake and similar environments from Barstow east to Ariz. and south to Sonora. Blooms from May to October.

645. TIDY FLEABANE. *Erigeron concinnus* (Gr., "early old man," i.e., appearing old early in life, some of the species being downy-hoary when young; L., neat, elegant). **Fl.**: violet or rose to nearly white. Perennial, from the limestone mountains of the eastern Mohave D.; to Canada and Rocky Mts. The variety *aphanactis*, with yellow, rayless flowers, is known from dry, stony slopes of the Argus and Inyo mountains; to Utah. A single flower is shown above to the right.

646. PYGMY FLEABANE. *Erigeron uncialis* (L., *uncia*, a twelfth part of anything; here taken as an inch, in reference to the shortness of the scape). **Fl.**: whitish to pink. Slender-stemmed perennial, from a stout, branched root crown. Found in limestone crevices of high mountains of the eastern Mohave D.; to Nev. The name flea-bane, applied to many of the *Erigerons*, is derived from the English "flea" and "bane," it once being supposed that fleas were driven away by the odors of certain species.

647. NEVADA FLEABANE. *Erigeron nevadensis* (of Nevada). **Fl.**: light lavender, with yellow center. A pale-herbaged perennial, with several to many stems 4–10 in. high, arising from a branched root crown. Among piñons in the Inyo range, north to Modoc Co. Often growing with it is its dwarf variety *pygmaeus*, with very narrow leaves, an inch or more long, and head considerably smaller.

648. PARISH FLEABANE. *Erigeron Parishii* (S. B. Parish—see **603**). **Fl.**: rose to lavender, with yellow center. One of the many plants found by Mr. S. B. Parish many years ago on his botanizing trips to Cushenbury Canyon on the desert slope of the San Bernardino Mts. Not a particularly handsome plant, and, so far as known, restricted to the locality where first found. It blooms in May and June.

649. COVILLE ERIGERON. *Erigeron foliosus Covillei* (L., leafy; F. W. Coville—see **179**). **Fl.**: purplish-violet. A perennial, with several stems 1–1½ ft. long. The herbage is grayish because of the presence of coarse, stiff hairs. Known from the Little San Bernardino Mts. to Owens Valley.

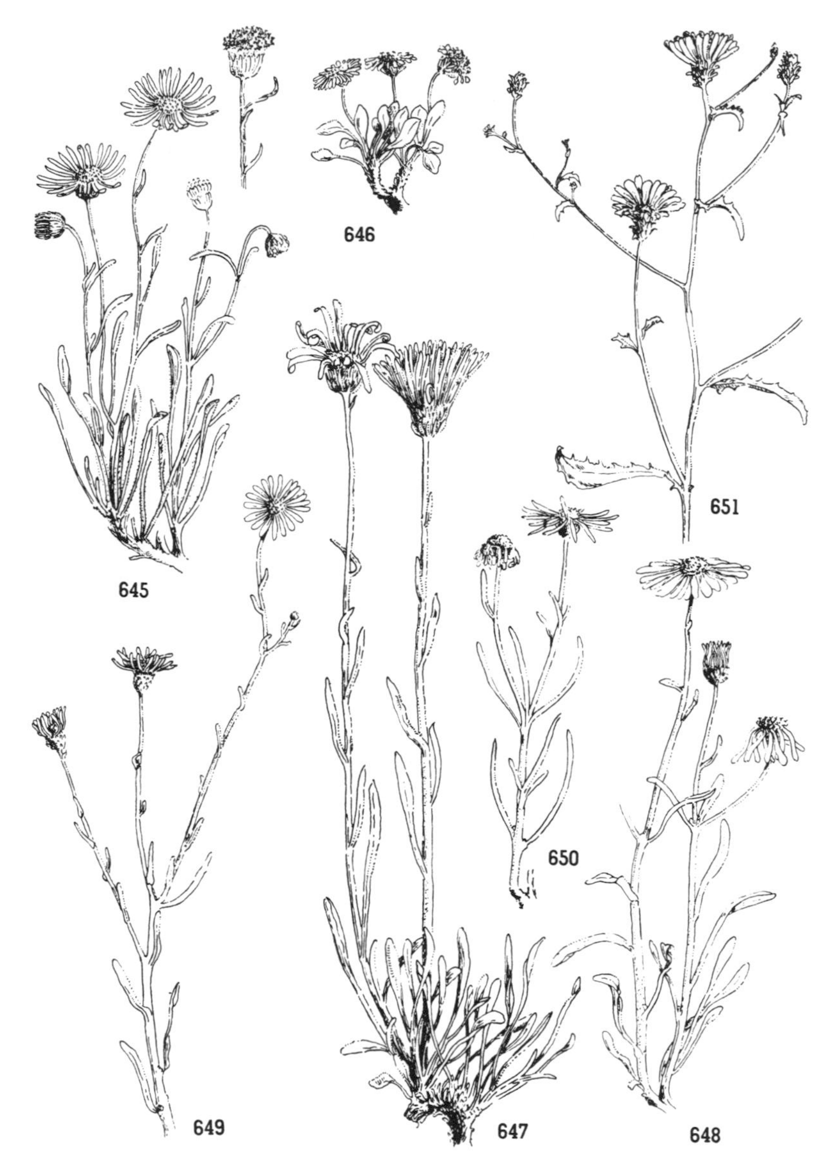

645. *Erigeron concinnus*
646. *Erigeron uncialis*
647. *Erigeron nevadensis*
648. ***Erigeron Parishii***
649. *Erigeron foliosus* ***Covillei***
650. *Erigeron filifolius*
651. ***Aster canescens***

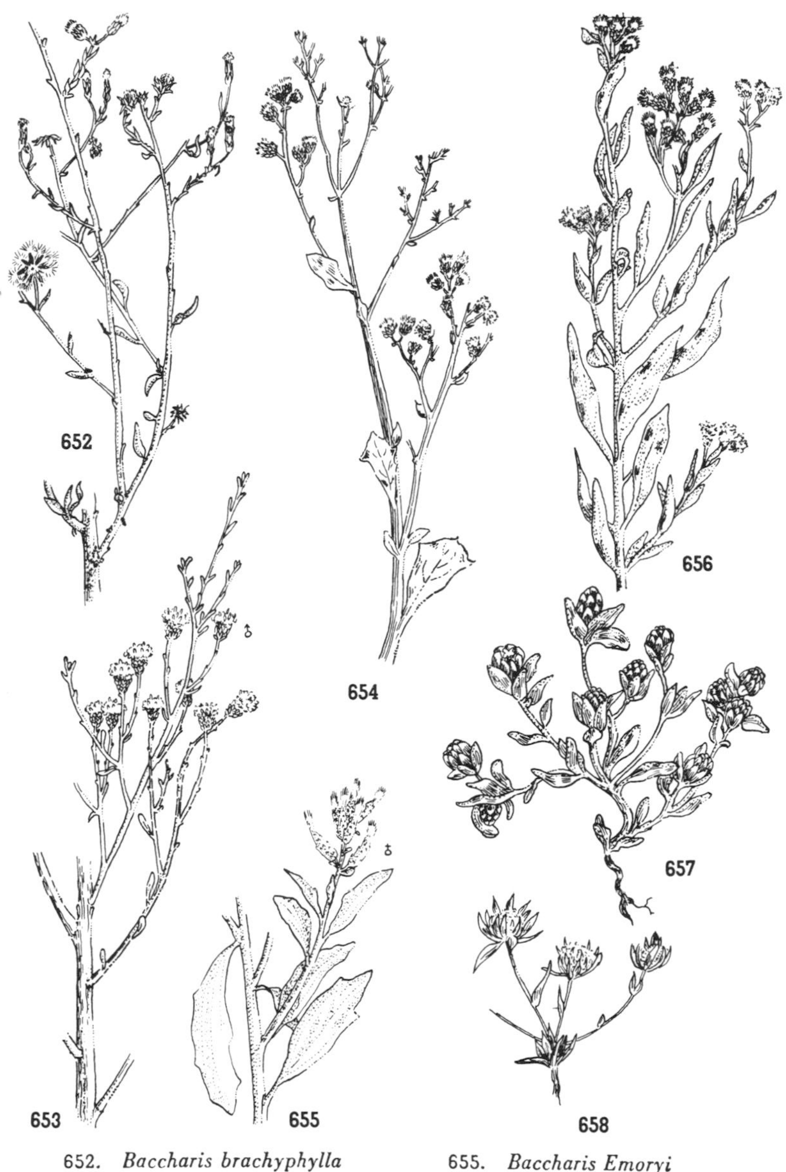

652. *Baccharis brachyphylla*
653. *Baccharis sarothroides*
654. *Baccharis sergiloides*
655. *Baccharis Emoryi*
656. *Pluchea sericea*
657.* *Stylocline gnaphalioides*
658.* *Stylocline micropoides*

650. THREAD-LEAVED FLEABANE. *Erigeron filifolius* (L., thread-leaved). **Fl.**: rays violet to white; center yellow. A many-stemmed, silvery-leaved perennial, growing in little clumps 4–6 in. across and as high. Close to *E. Parishii;* but the pappus is single instead of double and the leaves are narrower. Mountains of the western and eastern Mohave D. among piñons.

651. PIÑON ASTER. *Aster canescens* (L., covered with fine hairs). **Fl.**: rays, if present, violet to purple. Leafy-stemmed biennial or perennial, 6–30 in. high, often several-branched. The outermost green-tipped bracts are recurved in age. Piñon-juniper belt of mountains of the northern and eastern Mohave D.; to Ariz. *A. tephrodes* (Gr., ash-colored) is much like the piñon aster but taller, more hairy, and with larger heads. Yuma, Imperial Valley; to Nev., N.Mex.

652. SHORT-LEAVED BACCHARIS. *Baccharis brachyphylla* (Gr., name for the god, Bacchus; short-leaved). Perennial, woody-based herb, 2–3 ft. high. Southern San Diego Co. and Morongo Pass; to Ariz.

653. BROOM BACCHARIS, HIERBA DEL PASMO. *Baccharis sarothroides* (Gr., broom-like). **Fl.**: whitish or yellowish. Erect, broom-like, glutinous shrub, 3–6 ft. high, growing mostly in sandy washes. The cream-colored heads are very striking in contrast with the vivid green branchlets and blackish stems. Among certain Indians the twigs are chewed as a remedy for toothache. Salt Creek Wash, also near Gulliday Well of the Colorado D.; east to Ariz., Sonora, Sinaloa, Baja Calif.

654. DESERT BACCHARIS. *Baccharis sergiloides* (like *Sergilus,* the old name for *Baccharis*). **Fl.**: whitish or yellowish. The many green, almost leafless stems (2–3 ft. high) occur in a rounded clump. The numerous heads are usually brown. A water-indicating plant found in washes and canyons of both deserts; to Ariz., Nev., and Sonora.

655. EMORY BACCHARIS. *Baccharis Emoryi* (W. H. Emory—see **522**). **Fl.**: whitish or yellowish. A loosely branched, glutinous shrub with striate stems. The female flower heads are soon conspicuous because of the numerous glaring-white pappus hairs. A great drinker and, therefore, a plant which flourishes best alongside cattails and other water-loving plants. Found at Dos Palmas Spring on the Colorado D.; cismontane southern Calif.; Utah and Ariz.

656. ARROW-WEED. *Pluchea sericea* (N. A. Pluche, Parisian naturalist of the eighteenth century and author of *Spectacle de la Nature,* 1732; L., silky). **Fl.**: pale roseate purple. Slender, willow-like plant, forming low thickets about springs or river bottoms, where it occurs in pure, dense stands. Its straight stems were used by Indians for arrow shafts, also for construction of baskets, cages, and storage bins. Among the Pimas a tea made from the stem tips was used as an eyewash. The green herbage gives off a very agreeable odor. Both deserts, particularly along the Colorado R.; to Tex.

657. EVERLASTING NEST-STRAW. *Stylocline gnaphalioides* (Gr., "column bed," in reference to the form of the receptacle, which is long and slender, with little notches in which the achenes are seated, i.e., a bed for the columnar seeds; L., "like *Gnaphalium,*" the everlasting). A low, white-woolly annual, 1½–7 in. high. A number of desert birds use the soft stems as nest material. Western Colorado D.; to central Calif. and Channel Islands, below 4,000 ft.

658. DESERT NEST-STRAW. *Stylocline micropoides* (Gr., of the form of a small foot). This small, woolly, gray-green plant is often found growing beside large rocks. It too is a favorite nest material for several desert birds. Both deserts; to N.Mex. and northern Colo.

659. DESERT DICORIA. *Dicoria canescens* (Gr., "two bugs," referring to the aspect of the seeds; L., graying). Spreading annual herb, 1–2½ ft. high, with harsh hairs especially on the stems. Active growth begins either in late spring or after summer rains, and the plants come to maturity in late autumn. When wetted by rains or bruised, the herbage gives off a most agreeable odor. Especially abundant on roadside sands of the Colorado D. Local on the Mohave D.; east to Utah. *D. Clarkae,* with larger involucral bracts, is known from the sand hills west of Yuma.

660. SINGLE-FRUITED DICORIA. *Dicoria Brandegei* (T. S. Brandegee—see **602**). A low bush, distinguished by its lanceolate leaves and solitary achenes (seeds). Limited in Calif. to sandy washes of the eastern Colorado D.; to Colo., N.Mex.

661. DWARF FILAGO. *Filago depressa* (L., "thread," referring to the fine, cottony hairs; L., "depressed," i.e., growing near the ground). White-woolly annual, 1–3 in. high, common on sands of both deserts at low altitudes. The Le Conte thrasher almost invariably lines its nest with it. When rains wet the old nests and sand blows into them, this lining material forms into a solid, felty mass which may remain for years in the bushes where the birds nested.

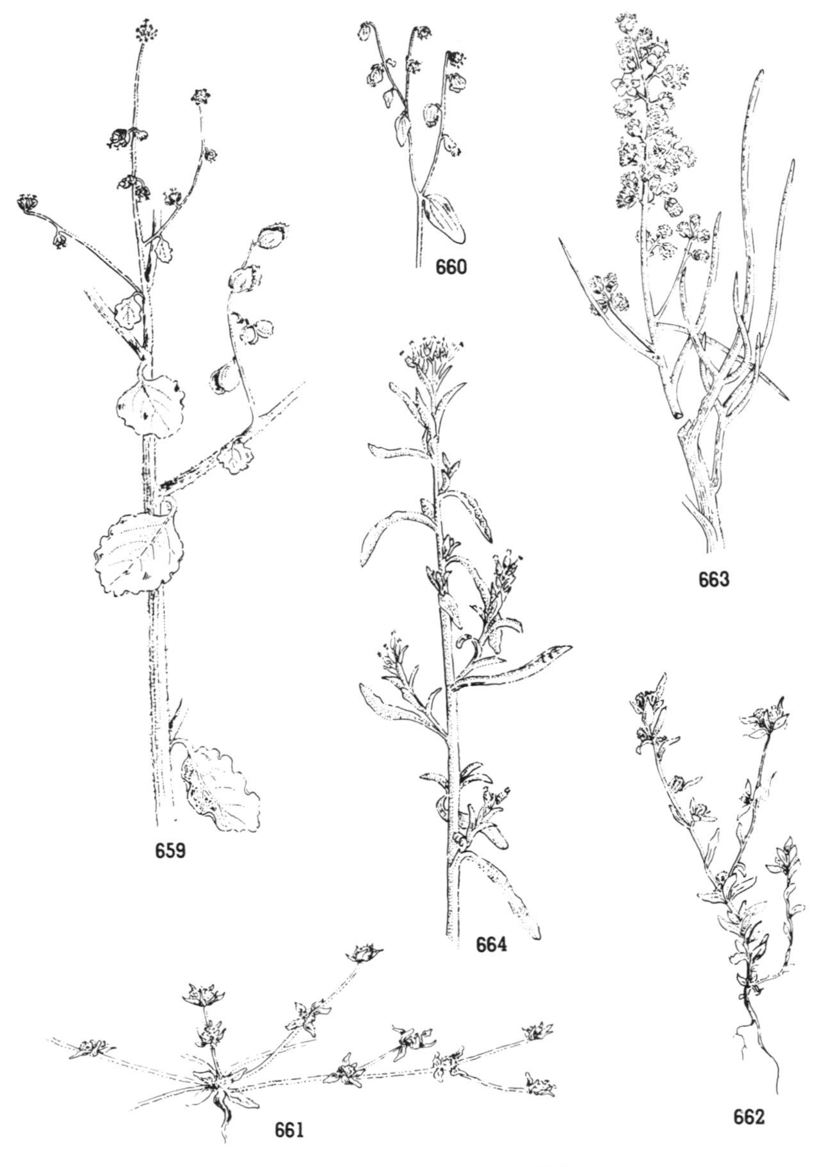

659. *Dicoria canescens*
660.* *Dicoria Brandegei*
661. *Filago depressa*
662. *Filago arizonica*
663. *Oxytenia acerosa*
664. *Gnaphalium chilense*

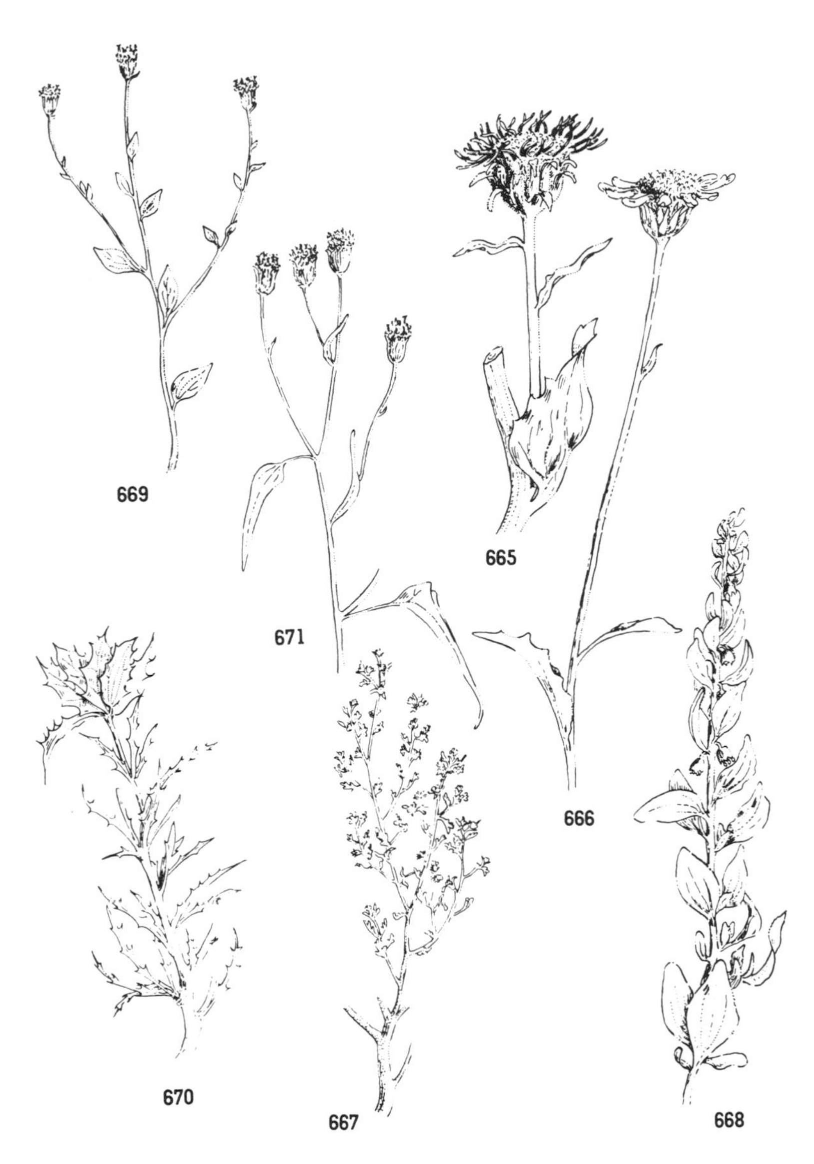

665. *Hulsea heterochroma*
666.* *Hulsea vestita callicarpha*
667. *Iva nevadensis*
668. *Iva axillaris*
669.* *Laphamia fastigiata*
670.* *Hecastocleis Shockleyi*
671.* *Pericoma caudata*

662. ARIZONA FILAGO. *Filago arizonica* (of Arizona). White-woolly annual, 1½–6 in. high, with linear leaves, the upper surpassing the flower head. Both deserts; west to Channel Islands.

663. OXYTENIA. *Oxytenia acerosa* (Gr., "pointed," in reference to the leaves; L., "full of chaff"). **Fl.**: pale buffish-yellow. Erect, shrubby perennial, often with leafless, rush-like stems, 1–2 yds. high. The flowers are unisexual. Closely related to the genus *Iva;* but the heads are erect, not nodding, as in that genus. A Colo. and Ariz. species, now known from alkaline flats of Death Valley.

664. SMALL-FLOWERED CUDWEED. *Gnaphalium chilense* (Gr., "a lock of wool," since the plants are often white-woolly; Chilean). **Fl.**: greenish-yellow. Woolly herb, with several erect stems. The flower heads, at first narrow, open up broadly in age. This is a cismontane species found along the western edges of the deserts.

665. GREAT HULSEA. *Hulsea heterochroma* (Dr. G. W. Hulse, of the U.S. Army, zealous collector and discoverer of the first species *H. nana,* in the "mountains back of San Diego"; Gr., of different colors). **Fl.**: rays purple, disk yellow. Robust, sticky-herbaged annual, 16–30 in. high, with rank odor. The vivid green leaves appear as if varnished. A montane species of southern and central Calif. which is known on the desert from the Panamint Mts.

666. EL CAPAROSSA. *Hulsea vestita callicarpha* (L., clothing or adorning; L., beautiful bract). **Fl.**: rays yellow, sometimes with purplish base. Loosely branched annual or biennial, 8–28 in. high, with leaves mainly basal. Collected on the Mohave D. in mountains of the Death Valley area and Little San Bernardino Mts.; also mountains of southern Calif. nearer the coast.

667. NEVADA POVERTY WEED. *Iva nevadensis* (after *Ajuga iva,* one of the mints whose odor some of the species resemble; of Nevada). A Great Basin annual which comes into California along the Nevada border. The heads are scattered, and the fertile flowers are with evident corolla. First collected in Calif. in Death Valley by the author in 1938.

668. WESTERN POVERTY WEED. *Iva axillaris* (L., "axillary," in reference to the position of the flower heads). A low, many-stemmed alkali- and salt-loving plant, with a somewhat creeping root-stalk. It is widely distributed over the western United States and is found occasionally about springs and seeps on the desert. The bracts of the involucre are united to form a distinct cup.

669. EXALTED LAPHAMIA. *Laphamia fastigiata* ("I. A. Lapham, of Milwaukee, Wisconsin, author of a catalogue of plants of that state and zealous explorer of its botany"; L., high, exalted). **Fl.**: light yellow. A perennial herb, with trifid or entire leaves. The pappus of the achene (seed) consists of a single awn. A rare species known from the Sheep Mts. of Nev. and arid ranges about Death Valley. *L. megacephala* (L., large-headed), with attractive, bright yellow flower heads and round-ovate leaves, occurs as a small, compact bush in rocky areas in the piñon belt of the Inyo and White mountains.

670. PRICKLE-LEAF. *Hecastocleis Shockleyi* (Gr., "each shut up," each flower being enclosed in its own involucre; W. H. Shockley—see **215**). **Fl.**: dull white. Low, smooth-stemmed shrub, with rigid branches and leaves of two kinds: the stem leaves, which are small linear-lanceolate; and the small floral leaves, which are clustered. A species of far-western Nev. reaching Calif. in the mountains near Death Valley.

671. TAILED PERICOME. *Pericoma caudata* (Gr., *peri*, "around," and *come*, "hair," in reference to a tuft of hispid hairs around the edge of the achenes; L., having a tail stuck on). **Fl.**: yellow. Rather tall, widely branching, strongly scented, herbaceous annual, flowering in late summer and autumn. A Colo. and Ariz. species entering the mountains of the Mohave D. along the eastern borders of Calif.

672. WOOLLY-FRUITED BURBUSH. *Franseria eriocentra* (Ant. Franser, eighteenth-century physician and botanist of Madrid; Gr., "woolly-pointed," with reference to the woolly, one-flowered involucre). Spreading, much-branched, rigid bush, 1–3 ft. high, with white-woolly fruits. Confined to washes of the mountains of the eastern and northern Mohave D.; to southern Nev. and northern Ariz.

(Ambrosia)

673. BURROBUSH. *Franseria dumosa* (L., bushy). Described in 1845 from specimens taken by Captain Frémont along the Mohave River. Low, rounded, white-barked shrub, very common in the broad desert basins, where its ashy-white foliage shows up conspicuously against the green of the creosote bush, its most common associate. Only during the spring season are the plants green. The flowers are of separate sexes. Burro bush is our second most widespread and dominant xeric shrub. In spite of its bitterness it is among the preferred foods of donkeys and sheep. Erroneously it is generally called burro *weed;* I propose the name burro bush, for it is not in any sense an obnoxious plant.

672. *Franseria eriocentra*
673. *Franseria dumosa*
674. *Franseria ilicifolia*
675. *Franseria acanthicarpa*
676. *Franseria tenuifolia*
677. *Ambrosia psilostachya californica*
678. *Hymenoclea Salsola*

679. *Viguiera multiflora nevadensis* 680. *Viguiera reticulata*
681. *Viguiera deltoidea Parishii*

674. HOLLY-LEAVED BURBUSH. *Franseria ilicifolia* (L., holly-leaved). The most handsome of our *Franserias* and plentiful in canyons and washes along the northern base of the Chuckawalla Mts. Found also in Baja Calif. The low-domed shrubs, with green, holly-like leaves, are often 5 ft. across. The branches are commonly clothed with numerous, tan, dry leaves of the previous season.

675. ANNUAL BURWEED. *Franseria acanthicarpa* (Gr., thorn fruit). A weed, plentiful to occasional, along roadsides and railways in disturbed ground, distributed doubtless by human agencies. Both deserts; to Wash., Colo., and Tex., below 6,800 ft.

676. WEAK-LEAVED BURWEED. *Franseria tenuifolia* (L., "weak-leaved," in reference to the drooping leaf position). *Perennial* herb, 1–3 ft. high. The small burs have hook-tipped spines. Occasional, Colorado D.; to Colo., Mex.

677. WESTERN RAGWEED. *Ambrosia psilostachya californica* (Anc. Gr. and L. name for a number of plants, also "food of the gods"; Gr., smooth-spiked; L., of California). A widespread weed locally found about cultivated fields and waste places on the desert. Cismontane southern Calif.; to Nev. and northern Calif.

678. CHEESE-BUSH. *Hymenoclea Salsola* (Gr., membrane [i.e., wing] enclosed; *Salsola,* the Russian thistle, which it was thought somewhat to resemble). Bright green, rank-smelling shrub, with numerous light tan stems and unisexual flowers; generally about 2–3 ft. high. The involucres of the pistillate flowers are surrounded by 5–12 silvery to reddish membranous scales, which make a fine show of color if the plants are heavily fruited. Widespread and common in sandy washes of both deserts; to Baja Calif. Called cheese-bush because of the cheesy odor of the crushed herbage. The illustration on the right shows the naked, leafy stem.

679. NEVADA VIGUIERA. *Viguiera multiflora nevadensis* (Dr. A. Viguier, French bookseller and botanist, of Montpellier; L., many-flowered; of Nevada). **Fl.**: yellow. Perennial, with several brownish stems, 1–1½ ft. high. Common in open spaces among junipers in the desert mountains of Inyo Co. and on the desert floor in southern Nev. This species is occasionally seen in cultivation.

680. LEATHER-LEAVED VIGUIERA. *Viguiera reticulata* (L., "forming a network," the leaves showing a dense network of prominent veins beneath). **Fl.**: yellow. Erect, bushy plant, 2–4 ft. across and as high, with white-barked stems and rather harsh-surfaced, somewhat silvery-green leaves. Abundant in the dry, rocky gorges of canyons and slopes about basins bordering Death Valley. Travelers going to Trona will see it in full flower during May in Poison Canyon. It is also common between Ryan Wash and Death Valley Junction. The leaves are sometimes 4 in. long.

681. PARISH VIGUIERA. *Viguiera deltoidea Parishii* (L., deltoid; S. B. Parish—see **603**). **Fl.**: yellow. Low, rounded shrub, with deep green leaves and numerous flowers on long stems. Found along edges of washes and among rocks of the Colo. and Mohave deserts; eastward to Nev. and Ariz., and Baja Calif.

682. HAIRY-LEAVED SUNFLOWER. *Helianthus canus* (Gr., sunflower; L., ash-colored). **Fl.**: ray flowers light yellow. Annual, 1–2 ft. high, the herbage densely covered with stiff, white hairs, especially the leaves. Not common; confined to open, sandy places of the Salton Sink; to Tex. and N.Mex.

683. SILVER-LEAVED SUNFLOWER. *Helianthus niveus.* (L., "snowy," because of the whitened foliage). **Fl.**: rays yellow, disk flowers brown. An annual, 7–12 in. high. A rare plant of the Algodones sand dunes of the Colorado D.; to Baja Calif.

684. LARGE-FLOWERED SUNRAY. *Enceliopsis Covillei* (Gr., of the appearance of *Encelia;* F. W. Coville—see **179**). **Fl.**: lemon-yellow. The most impressive composite of the desert area. It is a perennial, with many scapes 1–2½ ft. high, bearing flowers up to 6 in. broad. The succulent basal leaves are beautifully veined and have a silvery, smooth, felt-like surface; the involucre is a lovely, frosted green. Known from canyons of the west side of the Panamint Mts. *Enceliopsis argophylla,* known from Nev., has smaller flowers.

685. NAKED-STEMMED SUNRAY. *Enceliopsis nudicaulis* (L., naked-stemmed). **Fl.**: yellow. A smaller plant than *E. Covillei,* entering the Death Valley Nat. Mon. from the east; Nev., Ida., and Ariz.

686. ACTON ENCELIA. *Encelia actoni* (Christopher Encel, writer on oak galls; Acton, the type locality, a village of the western Mohave Desert). **Fl.**: cadmium-yellow. This shrubby species has handsome gray-green leaves. It is erroneously considered by some as only a variety of *Encelia frutescens,* which has rayless flower heads. It ranges from the Joshua Tree Nat. Mon. to the middle and western Mohave D. and as far north as Inyo Co.; occurs in eastern cismontane southern Calif., also in Ariz. In its southernmost distribution the plants have silvery, puckered leaves, while over most of the Mohave

682.* *Helianthus canus*
683.* *Helianthus niveus*
684.* *Enceliopsis Covillei*
685.* *Enceliopsis nudicaulis*

686. *Encelia actoni*
687. *Encelia farinosa*
688. *Encelia frutescens*
689. *Geraea canescens*
690. *Bebbia juncea aspera*

Desert the leaf surface is plain. By boiling the leaves and flowers, some of the California Indian tribes procured a wash for the relief of rheumatic pains.

687. BRITTLE-BUSH, INCIENSO. *Encelia farinosa* (L., "mealy," with reference to the leaf surface). **Fl.**: lemon-chrome. A low, dome-shaped bush, common to rocky slopes and benches of the Colorado Desert and occasional on the Mohave Desert. When new, the leaves are light greenish-gray, but later turn almost white. If severe drought ensues, almost all of the leaves are dropped, leaving the fat, purplish or brownish-gray stems to act as water storers until better times are at hand. The crystals of resin which exude from the somewhat woody stems were burned as incense by the early padres. The Indians chewed the resin, also smeared it warm on the body to relieve pains. When melted it was used as a varnish. Lower foothills of the San Bernardino Valley to the Colorado and southern Mohave deserts; to Ariz. and Sonora. In the vicinity of the Iron Mts. of eastern San Bernardino Co. many of the plants have maroon disk flowers.

688. RAYLESS ENCELIA. *Encelia frutescens* (L., shrub-like). **Fl.**: yellow. Rounded, shrubby bush, generally about 2 ft. high, with white stems and green leaves. It is widespread on the driest rocky hills and mesas of both deserts; east to Ariz.

689. DESERT SUNFLOWER, HAIRY-HEADED SUNFLOWER. *Geraea canescens* (Gr., "old," because of the white-villous achenes; L., "graying," with reference to herbage). **Fl.**: yellow. Annual herb, 6 in. to 2 ft. tall, maturing early in the season and often forming wild gardens of luxuriant bloom on sandy basins and along roadsides of both deserts; to Ariz., Sonora, and Utah. This magnificent, sweet-scented species gives us the finest show of massed yellow on the desert. It is one of the dependable sources of food of small rodents, and its seeds are stored in quantity especially by the pocket mice. Several wild bees feed on its fragrant flowers. The nearly related, stout, sticky-stemmed relative, *G. viscida,* found on the desert edge at Jacumba, is about the bitterest plant with which I am acquainted. Hands freely touching it will, in spite of repeated soap washings, retain the bitter principle several days.

690. SWEETBUSH, CHUCKAWALLA'S DELIGHT. *Bebbia juncea aspera* (Michael Schuck Bebb, 1833–1895, distinguished American specialist on willows, and for some time resident of San Bernardino. His services to botany were commemorated by Professor Edw. L. Greene in naming this genus. Said Professor S. C. Sargent at the time of Mr. Bebb's death: "Every important collection of willows made in this country passed through his hands for determination. .... [to him] more than to anyone else in this generation we owe our knowledge of American willows." L., rush-like; L., rough). **Fl.**: yellow. A very common species, reaching perfection along the broad, sandy washes of the Colorado D.; here the rounded, multistemmed bushes reach a diameter of 4–5 ft. When for a short period in spring the quite leafless, whitish stems are almost hidden by a mass of fragrant flower heads, the sweetbush is a handsome plant indeed. During much of the year *Bebbia* remains in a state of drought rest. Blooming proceeds whenever rains furnish water to the roots. It is visited often by bees. Chuckawallas feed greedily on the flowers. Both deserts, but less common on the Mohave D.

691. BIGELOW COREOPSIS. *Coreopsis Bigelovii* (Gr., "bug appearance," in allusion to the achenes of certain species; J. M. Bigelow —see **338**). **Fl.**: yellow. Both deserts, on dry flats, to Tulare Co. In this common annual the stems are naked and the leaves wholly basal. It is distinguished in the field from Douglas coreopsis, *C. Douglasii*, by its larger flowers. The latter is a coastal species reaching only the edge of the deserts; its achenes are generally without a pappus.

692. LEAFY-STEMMED COREOPSIS. *Coreopsis calliopsidea* (Gr., "beautiful aspect," or the name may have been intended to mean "calliopsis-like.") **Fl.**: yellow. A species of the eastern Mohave D., north to San Luis Obispo Co. The leafy stems are ½–1, rarely 2, feet high. In this species the achenes of the disk florets differ from those of the ray flowers, whereas in *C. Douglasii* the achenes are all alike.

693. WHITE TIDY-TIPS. *Layia glandulosa* (G. Tradescant Lay, naturalist under Captain Beechey on the voyage of the "Blossom" and the "Sulphur," 1825–1828; L., provided with glands). **Fl.**: white or pinkish. The leaves and brownish stems are covered with short, stiff hairs. Common in loose, sandy soils, below 3,500 ft., Mohave and northern Colorado deserts; to B.C., N.Mex.

694. AUTUMN VINEGAR-WEED. *Lessingia germanorum ramulosissima* (the distinguished German family, Lessing; L., of the Germans; L., very much branched). **Fl.**: yellow. A short-statured, rounded annual, with marked sour-resinous odor, coming into flower from late June to September. Plentiful in sunny situations among junipers along the Cajon Pass–Victorville road and in similar situations of the western Mohave D., below 7,000 ft.

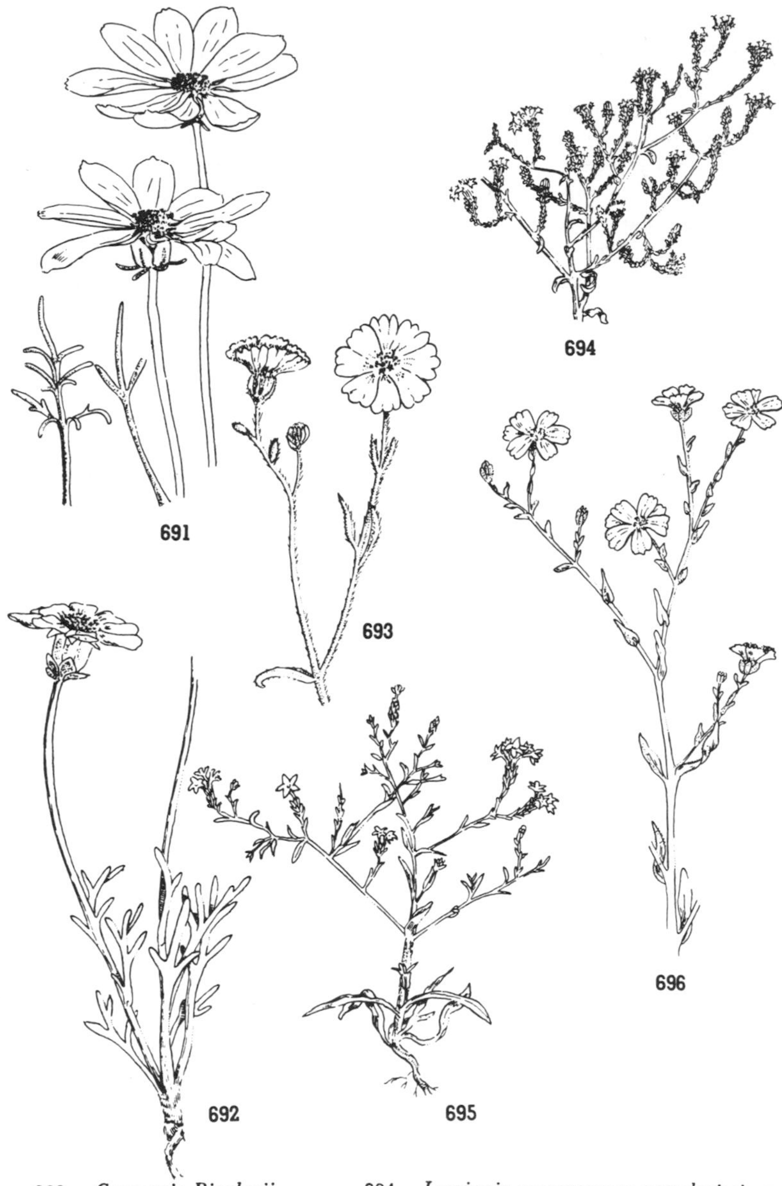

691. *Coreopsis Bigelovii*
692. *Coreopsis calliopsidea*
693. *Layia glandulosa*
694. *Lessingia germanorum ramulosissima*
695. *Lessingia germanorum Peirsonii*
696. *Hemizonia Kelloggii*

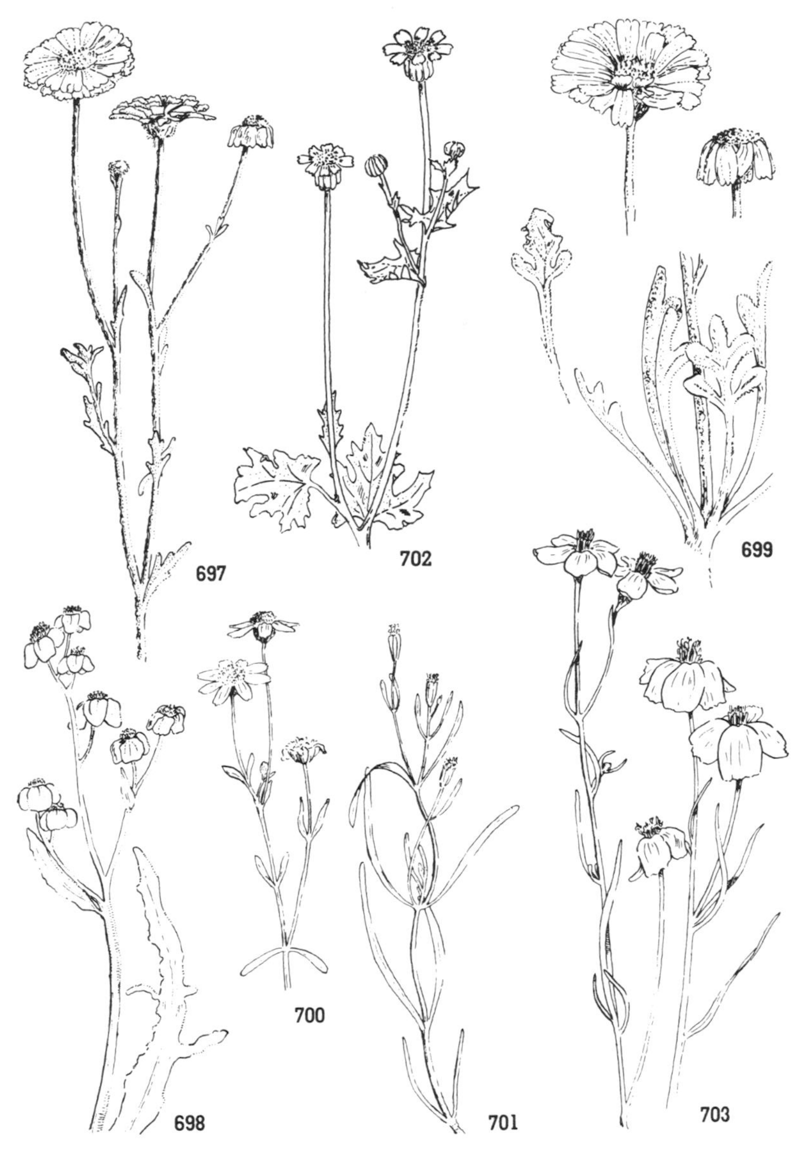

697. *Baileya pleniradiata*
698. *Baileya pauciradiata*
699. *Baileya multiradiata*
700. *Baeria platycarpha*
701. *Baeria microglossa*
702. *Perityle Emoryi*
703. *Psilostrophe Cooperi*

695. PEIRSON VINEGAR-WEED. *Lessingia germanorum Peirsonii* (Frank Peirson, of Pasadena, intelligent student of California botany). **Fl.**: yellow. A low, pungent-odored annual, appearing abundantly in May in flat areas and along roadsides in lower Owens Valley and mountains immediately west.

696. KELLOGG TARWEED. *Hemizonia Kelloggii* (Gr., "half-zone," the achenes being but half-enclosed; Dr. Albert Kellogg, 1813–1886, central California botanist, and one of the seven founders of the California Academy of Sciences, 1853). **Fl.**: yellow. Erect, widely branched, resinous annual. Los Angeles Co. south to San Diego Co., edges of both deserts, San Joaquin Valley.

697. WOOLLY MARIGOLD. *Baileya pleniradiata* (Jacob Whitman Bailey, early American microscopist; L., full-rayed). **Fl.**: yellow. Handsome annual, with leafy stems 4 in. to 2 ft. tall. The numerous flowers are borne on branching peduncles. Common to both deserts, in sandy flats and in disturbed soil along roadsides; to Utah and Sonora.

698. LAX-FLOWER. *Baileya pauciradiata* (L., few-rayed). **Fl.**: lemon-yellow. An erect, perennial herb, $\frac{3}{4}$–$1\frac{1}{2}$ ft. tall, covered with long, loose, woolly, white hairs, which give the plant a general pale green appearance. The rays, at first narrow and lying horizontally, become broad-ovoid and lax and turn down as the flowers age. They vary in number from 5 to 8. Widespread and common in sandy areas of both deserts; Ariz.

699. WILD MARIGOLD. *Baileya multiradiata* (L., many-rayed). **Fl.**: yellow. Much like the woolly marigold, but with woody-based stems and leaves mostly basal. The flower stems are long, and each bears but a single head. Well drained, rocky or gravelly slopes about the Providence, New York, and Clark mountains of the eastern Mohave D.; to Utah, Tex.

700. ALKALI GOLD-FIELDS. *Baeria platycarpha* (Karl Ernst von Baer, eminent Russian zoölogist, of the University of Dorpat, who in 1827 discovered the ovum of mammals, and who first described the three primary germ layers of the animal embryo; Gr., broad and [something] dry, such as a bract or palea). **Fl.**: yellow. Stems purplish; the bracts strongly keeled. Forming vast yellow fields on coarse soils of the western Mohave D.; to middle Calif., where it dominates alkaline flats.

701. SMALL-RAYED BAERIA. *Baeria microglossa* (Gr., small-tongued, because the ray flowers are so small as to be quite inconspicuous). **Fl.**: yellow. A very tender, light green, spring annual growing in thickset stands in damp earth in the shade of rocks. Both deserts, to middle Calif.

702. EMORY ROCK DAISY. *Perityle Emoryi* (Gr., "around callus," many species having callus-margined achenes; Maj. W. H. Emory—see **522**). **Fl.**: yellow center, with white, or rarely yellow, rays. A long-lived, ill-scented, freely branching annual, ½–2 ft. tall, with exceedingly brittle, succulent stems; most common in washes and canyons of the Colorado D. Because of its vivid green foliage and bright, contrasting flowers it is easily noticed against rock walls or about gray-green shrubs beneath which it may seek shelter. Both deserts, to Mex., Channel Islands; north to Ventura Co. on the coast.

703. PAPER-FLOWER. *Psilostrophe Cooperi* (Gr., "to turn bare"; Dr. J. G. Cooper—see **132**). **Fl.**: deep yellow. Rounded shrub, 1–1½ ft. high, with conspicuous flowers which fade to straw-color and turn papery in age. The flowers are very persistent, often remaining on the stems intact for weeks after blooming. Plentiful on the southwestern side of the Chuckawalla Mts. and slopes about mountains of the eastern Mohave D.

704. SPANISH NEEDLE. *Palafoxia linearis* (José Palafox, Spanish general, 1780–1847; L., linear). **Fl.**: deep pink. Abundant annual, 1–2 ft. high, with dark green, exceedingly bitter herbage; plentiful along roadsides in spring and again in autumn, provided there have been summer rains. Both deserts on sandy flats; to Mex. The variety *gigantea*, found on the sand dunes near Yuma, grows 1–2 yds. high!

705. PRINGLE ERIOPHYLLUM. *Eriophyllum Pringlei.* (Gr., woolly leaf; Cyrus G. Pringle, 1838–1911, Vermont botanist, who collected extensively in California and other parts of the Southwest. Over a period of 25 years Pringle made thirty-nine botanizing trips to Mexico, where he collected half a million plants! Many of the trips he made on foot, "carrying with him only some cheese and tortillas, staying away a week or more at a time." "Dr. Asa Gray engaged on his 'Synoptical Flora of North America' assigned to Pringle the investigation of the flora of Mexico, charging him as they sat with a map before them, to ascertain especially the southern limit of distribution of species found in the United States and also to ascertain such related species as might be indigenous to the adjacent regions of Mexico.") **Fl.**: yellow. A queer little plant of the sands; often so low to the ground and with so few leaves that only the small yellow flowers are noticed. Western and middle Mohave D. to the Greenhorn range; to Ariz.

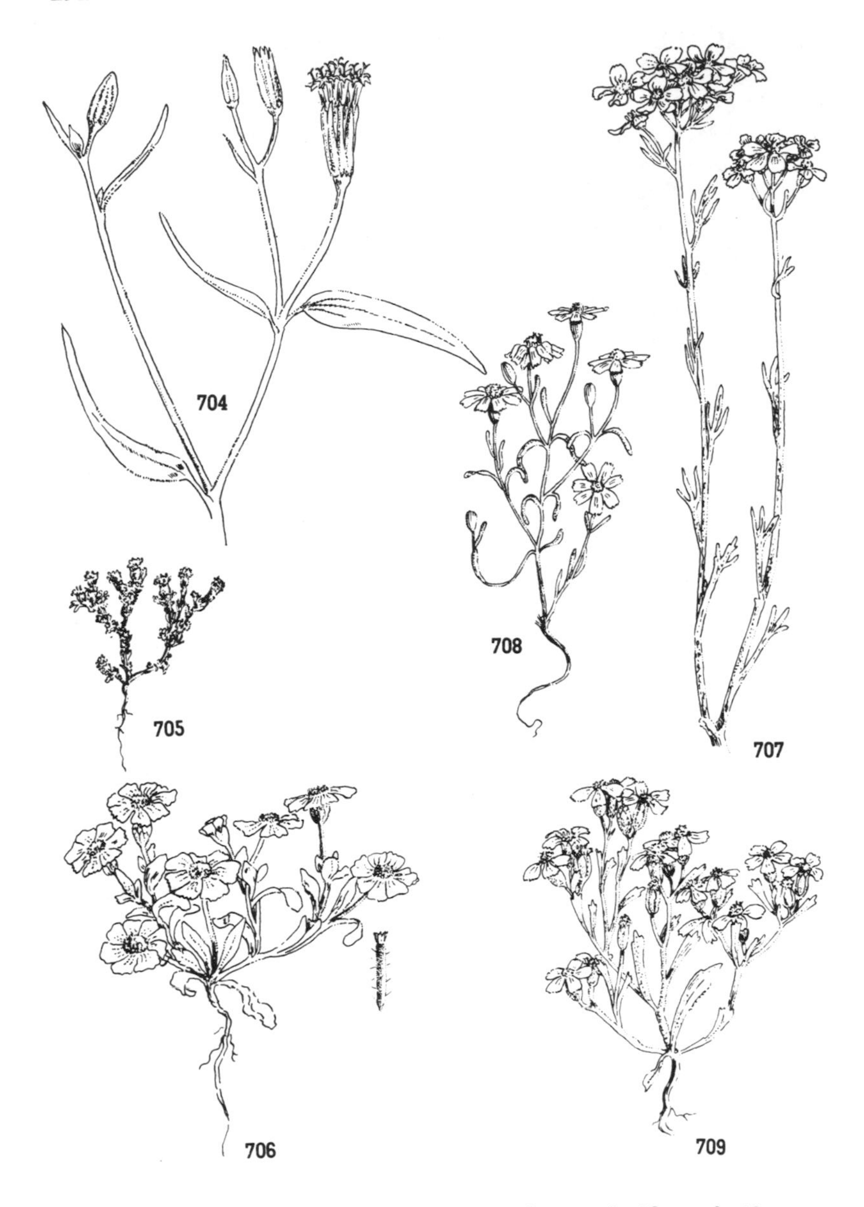

704. *Palafoxia linearis*
705. *Eriophyllum Pringlei*
706. *Eriophyllum Wallacei*
707. *Eriophyllum confertiflorum laxiflorum*
708.* *Syntrichopappus Lemmonii*
709.* *Syntrichopappus Fremontii*

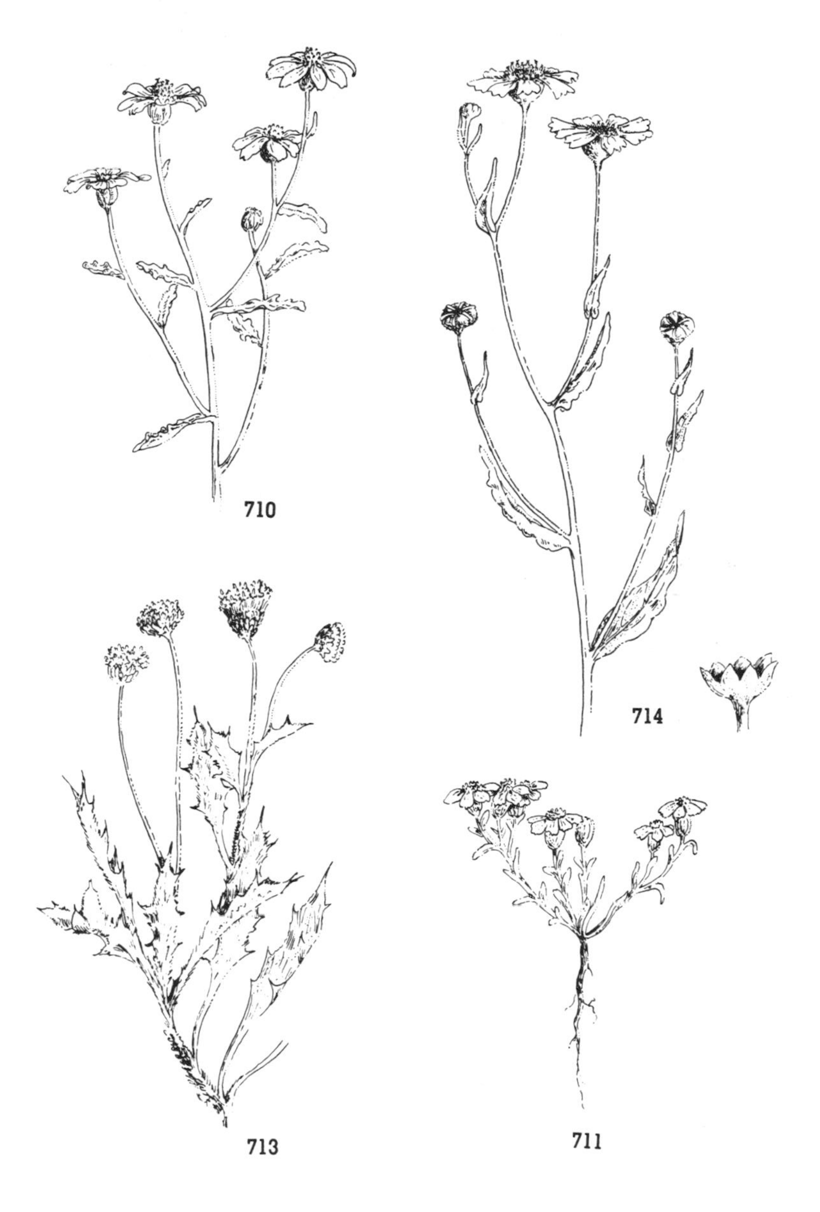

710. *Eriophyllum ambiguum*
711. *Eriophyllum lanosum*
713. *Trichoptilium incisum*
714.* *Monolopia major*

706. WALLACE ERIOPHYLLUM. *Eriophyllum Wallacei* (Wm. A. Wallace, who collected in the vicinity of Los Angeles in 1854 and later). **Fl.**: yellow. Handsome little annual, frequent on gravelly and sandy flats of both deserts; to Baja Calif. The variety *rubellum* (L., reddish), found along the western edge of the Colorado D., has white-to-reddish flowers.

707. LONG-STEMMED ERIOPHYLLUM. *Eriophyllum confertiflorum laxiflorum* (L., crowded-flowered; L., openly spreading or loose-flowered). **Fl.**: yellow. A several-stemmed, white-woolly perennial of the desert mountains, conspicuous in spring because of its bright clusters of flowers. Joshua Tree Nat. Mon. and western edge of Colorado D.; to Sonora, Baja Calif.

Fuzz from *Eriophyllum confertiflorum* was used by some of the southern California Indian tribes in the treatment of rheumatism. The woolly hairs of the stems were scraped off with the fingernail and rolled into small balls. Three of these balls were placed in succession on the affected parts and set on fire, with the result that blisters were raised. The counter-irritation doubtless somewhat relieved the pain.

708. LEMMON XERASID. *Syntrichopappus Lemmonii* (Gr., united-hair pappus; J. G. Lemmon—see **464**). **Fl.**: pink above, rose to purplish beneath, the margins white. Low annual. Distinguished from Frémont xerasid by the absence of a pappus. Cajon Pass, Antelope Valley. Called Xerasid from Gr., *xerasia*, "dryness," + *id* (contraction of *-ides*) "son of."

709. FRÉMONT XERASID. *Syntrichopappus Fremontii* (J. C. Frémont—see **315**). **Fl.**: golden yellow. Woolly annual, common to sandy places on the Mohave D. above 2,000 ft. First described by Asa Gray from a single specimen "gathered by Capt. Frémont in his journey across the continent in 1853–54, probably in the spring of '54 somewhere between the Rocky Mts. and the Sierra Nevada."

710. YELLOW-FROCKS. *Eriophyllum ambiguum* (Gr., woolly leaf; L., "uncertain," because its describer was for some time puzzled as to its taxonomic position). **Fl.**: yellow. Annual, 2–12 in. high, freely and widely branched from the base. Generally in soils about rocks. Palm Springs region, but mostly in the western and northern Mohave D.

711. WOOLLY ERIOPHYLLUM. *Eriophyllum lanosum* (L., woolly). **Fl.**: white. Low, loose-woolly annual, of the eastern Mohave D.; to Ariz., Utah, and Baja Calif.

712. MOHAVE ERIOPHYLLUM (not illustrated). *Eriophyllum mohavense* (L., of the Mohave). **Fl.**: white. Depressed annual, very rare on the central and western Mohave D. The only specimens known are, I believe, in the Gray Herbarium. A doubtful species.

713. YELLOW-HEAD. *Trichoptilium incisum* (Gr., "hairy feather," in reference to the dissected pappus bracts; L., "cut into," because the leaves are incisely dentate). **Fl.**: brilliant yellow. Low, floccose-woolly, fragrant annual, generally found growing in stony soil or rock crevices of the low, hot Colorado and southern Mohave deserts. The solitary heads are borne on slender, reddish peduncles.

714. GREATER MONOLOPIA. *Monolopia major* (Gr., "one husk," referring to the uniserial involucre; L., larger). **Fl.**: yellow. Erect, branched, white-woolly annual, occasionally seen on the western Mohave D. to central Calif.

715. PEBBLE PINCUSHION. *Chaenactis carphoclinia* (Gr., "gaping ray," the enlarging orifice and limb of the marginal corollas in most species emulating a kind of ray; Gr., "a small dry object" + "bed"). **Fl.**: whitish. Most common on dry patina flats or rock floors of the hot desert basins and low mountains, where we see the near absence of all plant life. Here the flat pebbles are so closely compacted as to leave scarcely any niche where seeds may strike root; the dazzling reflection is an added deterrent to the healthy growth of plants. A few of the chorizanthes also accommodate themselves to the same hot, barren surfaces. Both deserts; to N.Mex. and Mex.

716. ESTEVE PINCUSHION. *Chaenactis stevioides* (L., like *Stevia*, a genus of Compositae, named in honor of P. J. Esteve, sixteenth-century Spanish botanist). **Fl.**: white. A grayish-leaved annual frequent in the eastern Colorado D. and extending to the southern Mohave D., thence to Wyo. and Mex. In the western and northern Mohave D. it intergrades with the variety *brachypappa* (Gr., broad-pappused) with much-reduced pappus-palae. This is one of the favorite food plants of the large red-headed soldier beetle (*Lytta magister*), which in spring is found in great numbers on the stony flats.

717. FRÉMONT PINCUSHION. *Chaenactis Fremontii* (J. C. Frémont—see **315**). **Fl.**: white. A simple, erect or branching annual, 2–16 in. high. The marginal flowers of the head are often much enlarged and irregular. A very variable species, frequent about the bases of creosote bushes. Common on both deserts; to Ariz.

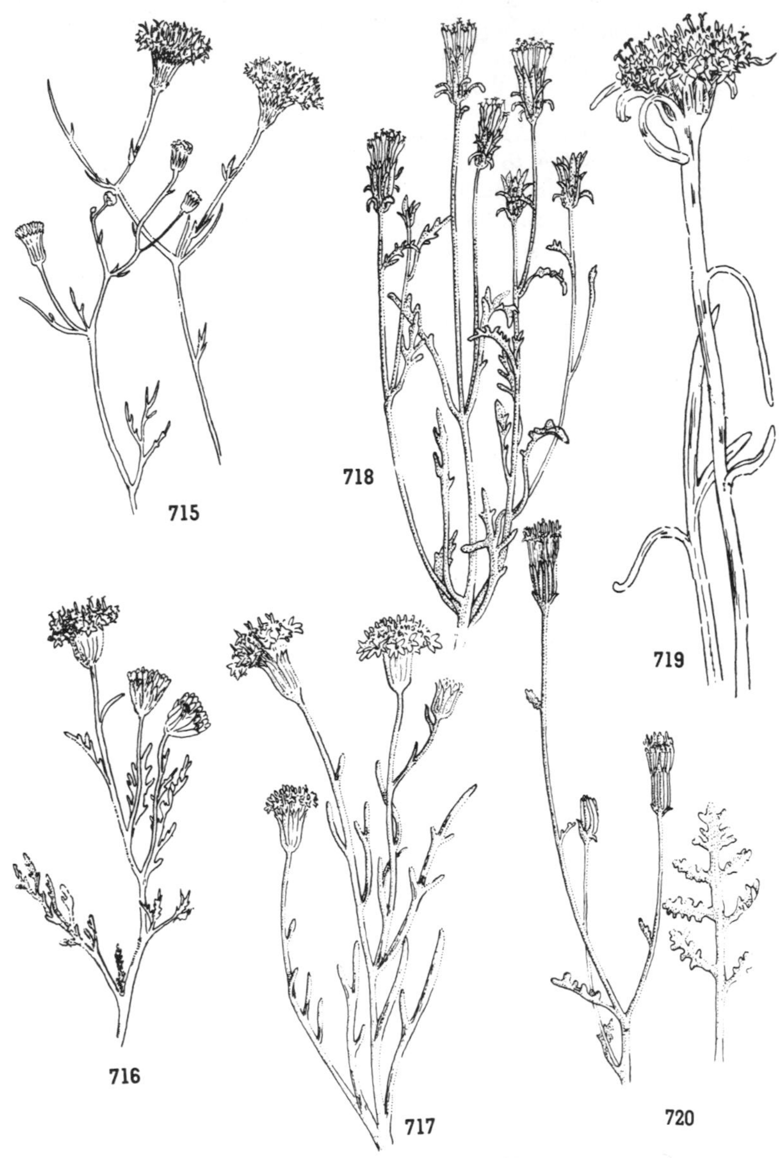

715. *Chaenactis carphoclinia*
716. *Chaenactis stevioides*
717. *Chaenactis Fremontii*
718. *Chaenactis macrantha*
719. *Chaenactis Xantiana*
720. *Chaenactis Douglasii*

721.* *Hymenoxys biennis* 722.* *Hymenoxys chrysanthemoides excurrens*
723.* *Actinea acaulis arizonica*

718. MOHAVE PINCUSHION. *Chaenactis macrantha* (Gr., large-flowered). **Fl.**: white to rose-tinted. A low annual, with reddish-brown stems, pink-tipped florets, and whitish bracts, which appear as if covered with flour. Occasional on stony slopes of lower elevations but most common in the piñon-juniper belt of mountains of the Mohave D.; to Ariz. and Utah.

719. XANTUS PINCUSHION. *Chaenactis Xantiana* (L. J. Xantus de Vesey, who collected at Fort Tejon for the Smithsonian Institution in 1857–1859). **Fl.**: white. A rather stout-stemmed annual, 4–16 in. high, with somewhat fleshy leaves. The marginal flowers are but little enlarged. Western Mohave D.; to eastern Ore. and Ariz.

720. DOUGLAS PINCUSHION. *Chaenactis Douglasii* (David Douglas—see **31**). Leafy-stemmed biennial or perennial, often 2 ft. high; found in the piñon-juniper belt of mountains of the northern Mohave D.; also in northern Calif.; to Wash. and Colo.

721. BIENNIAL GOLDFLOWER. *Hymenoxys biennis* (Gr., sharp or spiny membrane; L., biennial). **Fl.**: yellow. Biennial, sometimes perennial, with a single stem, which is branched above. Little San Bernardino Mts., Clark and Providence mountains; to Utah, Ariz.

722. BELL GOLDFLOWER. *Hymenoxys chrysanthemoides excurrens* (Gr., like chrysanthemum; L., running). **Fl.**: yellow. Annual, 1–2 ft. high, having flowers with somewhat bell-shaped involucres. To be seen in the lowlands bordering the Colorado R. from Blythe to Yuma; Ariz. and Mex.

723. ARIZONA ACTINEA. *Actinea acaulis arizonica* (Gr., ray; L., stemless; of Arizona). **Fl.**: bright yellow when young. A hairy, bitter-tasting perennial, without true stems. The several root crowns often protrude well above the surface of the dry, rocky soils where the plants grow. Mountain slopes of eastern and southern Mohave D.; to Nev. and Colo.

724. THURBER DYSSODIA. *Dyssodia Thurberi* (Gr., a disagreeable smell; G. Thurber—see **319**). **Fl.**: yellow. Low, strong-scented, slender-stemmed perennial. In crevices of limestone rocks. Providence and Clark mountains of eastern Mohave D.; to Tex. and Mex.

725. COOPER DYSSODIA. *Dyssodia Cooperi* (J. G. Cooper—see **132**). **Fl.**: orange. In this species of *Dyssodia* the stems are covered with fine hairs. The plants are from 12 to 20 in. high, with several erect stems. Eastern Mohave D., below 4,000 ft.; to Ariz. and Nev.

726. SAN FELIPE DYSSODIA. *Dyssodia porophylloides* (Gr., like *Porophyllum*). **Fl.**: orange. Strong-scented perennial, with several-to-many smooth, branching stems, 8–16 in. high. Southern Mohave and Colorado deserts; to Baja Calif. San Felipe, Calif., is the type locality.

727. HOLE-IN-THE-SAND PLANT. *Nicolletia occidentalis* (Joseph N. Nicollet, 1786–1843, American explorer, geologist, and astronomer, and John C. Frémont's first teacher in scientific studies, employed by the U.S. government as explorer for the country west of the Mississippi; L., Western). **Fl.**: yellow, later pink or purplish. Stout-stemmed, strong-odored perennial, 4–8 in. high. The stems are numerous and come up in early spring from tortuous, brittle, deep-seated roots. A striking feature of this *Nicolletia* is its habit of always growing in the bottom of shallow depressions which are about the size of hoof-prints and an inch or more deep. I feel certain the wind had not blown the sand from around the stem bases. Would that someone could explain this phenomenon! *Stillingia paucidentata* (see **268**), which has similar growth habits and which often grows beside it, does not exhibit this strange habit. Deep, coarse, sandy soil of washes of the middle and western Mohave D. This was one of the odd plants collected by Frémont on his 1844 journey on the Mohave.

728. ODORA. *Porophyllum gracile* (Gr., "passage" or "pore leaf," referring to the translucent oil glands, which make the herbage appear punctate; L., slender). **Fl.**: purplish-white. Many-branched, dark purplish-green shrub, with rank odor. Called *hierba del venado* (herb of the deer) by the people of Baja California, who make a tea from the bitter leaves for intestinal illness. Rocky soils of eastern San Diego Co. and the Colorado D.; eastern Mohave D.; to Ariz. and Tex. This is one of the plants taken by the naturalist aboard the good ship "Sulphur" on its voyage of discovery along the Mexican and California coast in the spring and summer of 1837.

729. SCALE-BROOM. *Lepidospartum squamatum* (Gr., "scale" + "spartum," from *Spartium*, generic name of the Spanish broom, *S. juncea;* L., scaly). **Fl.**: yellow. Rigid, green, broom-like shrub, with small, appressed, scale-like leaves. A common plant of the windy San Gorgonio Pass of the Colorado D., where it blooms more or less throughout the summer. The seedling plants and young shoots are woolly and have spatulate, entire leaves. Common also in cismontane southern Calif. and in a number of other places on the desert in washes and gravels where there are constant supplies of moisture; to Nev. and Ariz.

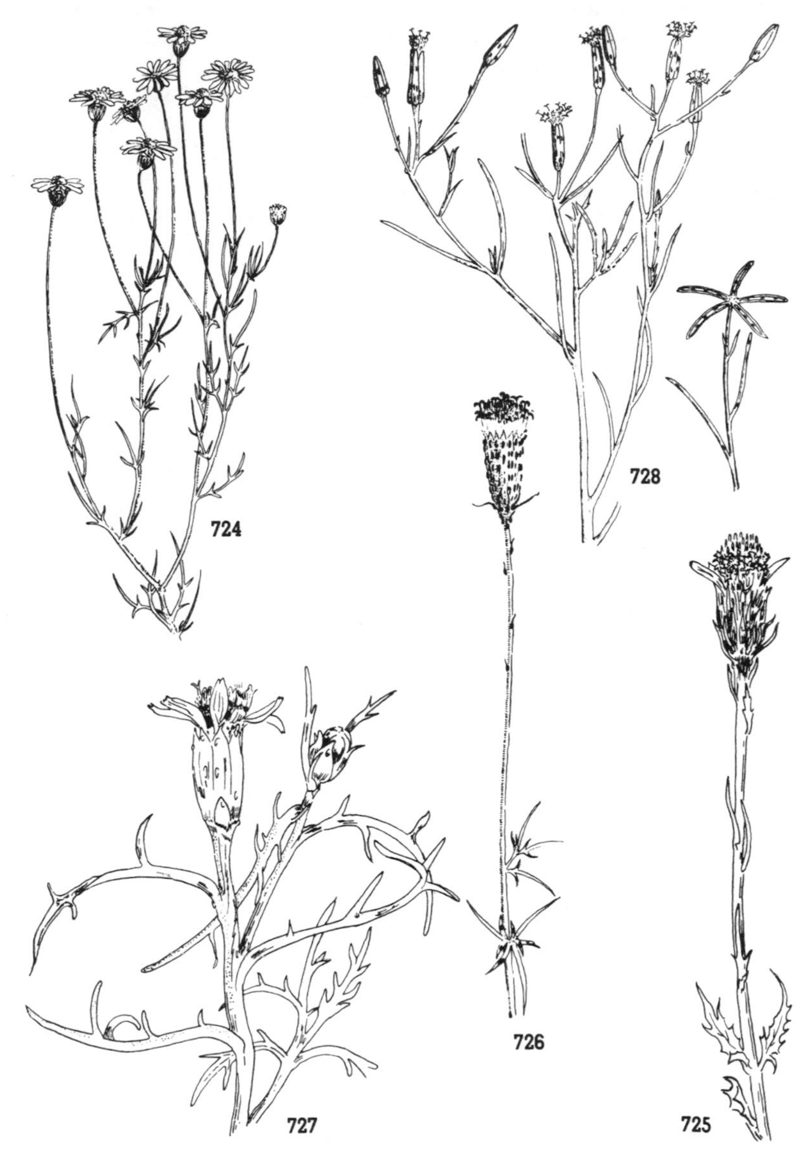

724. *Dyssodia Thurberi* 726. *Dyssodia porophyloides*
725. *Dyssodia Cooperi* 727. *Nicolletia occidentalis*
728. *Porophyllum gracile*

729. *Lepidospartum squamatum*
730. *Artemisia spinescens*
731. *Artemisia tridentata*
732. *Artemisia albula*
733. *Pectis papposa*
734. *Peucephyllum Schottii*

730. BUD SAGEBRUSH. *Artemisia spinescens* (Anc. name of the mugwort, given in memory of the sister and wife of Mausolus, king of Caria. It was she who built for him a magnificent tomb, called the Mausoleum, long considered one of the seven wonders of the world. L., spine-bearing). A compact little shrub, 4–8 in. high, with stocky, woody stems. It is the dominant plant over wide areas in many of the broad basins of southern Nevada. Almost pure stands occur in Perognathus Flat, aptly named with reference to the abundance of pocket mice of the genus *Perognathus* found there by members of the Death Valley Expedition. The short lateral branches, on which the flower heads are borne, later become hard, naked spines. Semi-alkaline soils of the middle and northern Mohave D.; to Ore., Wyo., N.Mex.

731. BIG SAGEBRUSH. *Artemisia tridentata* (L., three-toothed). This is the true aromatic sagebrush of the Great Basin. It reaches from the plains of Oregon and Idaho through the whole of Nevada and into southeastern California. In some places it forms pure sage plains. Sagebrush occupies the same dominant place among plants of the northern desert that creosote bush, together with burro bush, does in the deserts of the south. It is, according to W. A. Drayton, probably the most abundant shrub in western North America. The elevation of much of the Mohave Desert is too low for it, but in the higher desert ranges it is a familiar plant and there consorts with junipers, piñons, and tree yuccas. Its most distinguishing feature is its silvery leaf, generally three-toothed at the apex. In late summer sagebrush passes into a drought-rest state, most of the leaves are gradually dropped, and the plants look very scraggly. The wood and shaggy bark burn with an intense heat and an agreeable odor, and it is always with pleasure that we put our black kettles to boil over a sagebrush campfire. In September the ripening seeds are gathered and ground into meal by the Cahuilla Indians. Tea made from the bitter leaves has long been employed as a tonic, a hair- and eye-wash, an antiseptic for wounds, and a remedy for colds. The green herbage, rich in both fats and proteins, is valuable forage for sheep and goats. The little gray vireo often suspends its cup-shaped nest of grasses and shredded bark in the branches and decorates the exterior with the spatulate leaves. Sagebrush is also a favorite nest site for the black-throated sparrow, the California sage sparrow, the gnatcatcher, and the Costa hummingbird. The big, spongy, velvety-surfaced, purplish galls, so prevalent on the stems in midsummer, are caused by a small gall-midge bearing the scientific name of *Diarthronomyia artemisiae.*

732. PIÑON WORMWOOD. *Artemisia albula* (L., whitish). Low, gray-green, semi-woody plant, of the piñon-juniper woodlands, where it grows among rocks. The author, while working in the Joshua Tree National Monument during the summer of 1938, made the first collection of this plant known from the western part of the desert. Like many of the wormwoods it has a pleasing wild odor when crushed. It is known from the higher elevations of the Joshua Tree National Monument, where it blooms in midsummer. Otherwise known from Clark Mt.; north and east to Can. and Mo.

733. CHINCH-WEED. *Pectis papposa* (Gr., "to comb," the leaves in some being pectinate; L., with pappus). Very green, heavy but agreeably scented annual, generally coming up after summer rains on the sand and clay flats of both deserts; eastward to Utah and south to Mex. The Hopi Indians of Arizona sometimes used the ashes to mix with corn flour in making *peki,* while the Zuñis used it as a seasoning for meat stew or rubbed it on the body as a perfume.

734. DESERT-FIR. *Peucephyllum Schottii* (Gr., "fir leaf," because of its superficial resemblance to a fir tree; Arthur Schott, one of the naturalists of the Mexican Boundary Survey). This perennial green shrub, 1–1½ yds. high, is found on the rocky benches and in many of the dry washes of both deserts; it is quite common in the Panamint Mts. and the Muddy Mts. of Nev.; to Baja Calif.

735. COTTON-THORN. *Tetradymia spinosa longispina* (Gr., "four" + "together," some species having heads of four flowers; L., thorny; L., long-thorned). **Fl.**: yellow. A shrub with whitish stems and conspicuous, long, straw-yellow spines. "In early spring," says Dr. Coville, "when the foliage is freshest it is very handsome, and later in the season when in fruit and covered with its white woolly-tufts of soft feathery plumes it is still more beautiful." Quite common among tree yuccas. The plants frequently have fusiform swellings on the stems made by the Tetradymia gallfly. Southwestern Mohave D.; to Ore. and Utah.

736. BALD-LEAVED FELT-THORN. *Tetradymia glabrata* (L., bald). **Fl.**: yellow. The species conforms to the original conception of the genus by having its flowers occur in heads of four. Unlike the other two species here described, its primary leaves are not modified into spines. The secondary leaves are green, and in spring succulent. The persistent, secondary flowering branches arising from the woody, dark-barked main stems are from four to six inches long. Along borders of washes of alkaline hills north of Barstow to Red Rock Canyon; to Ore., Utah.

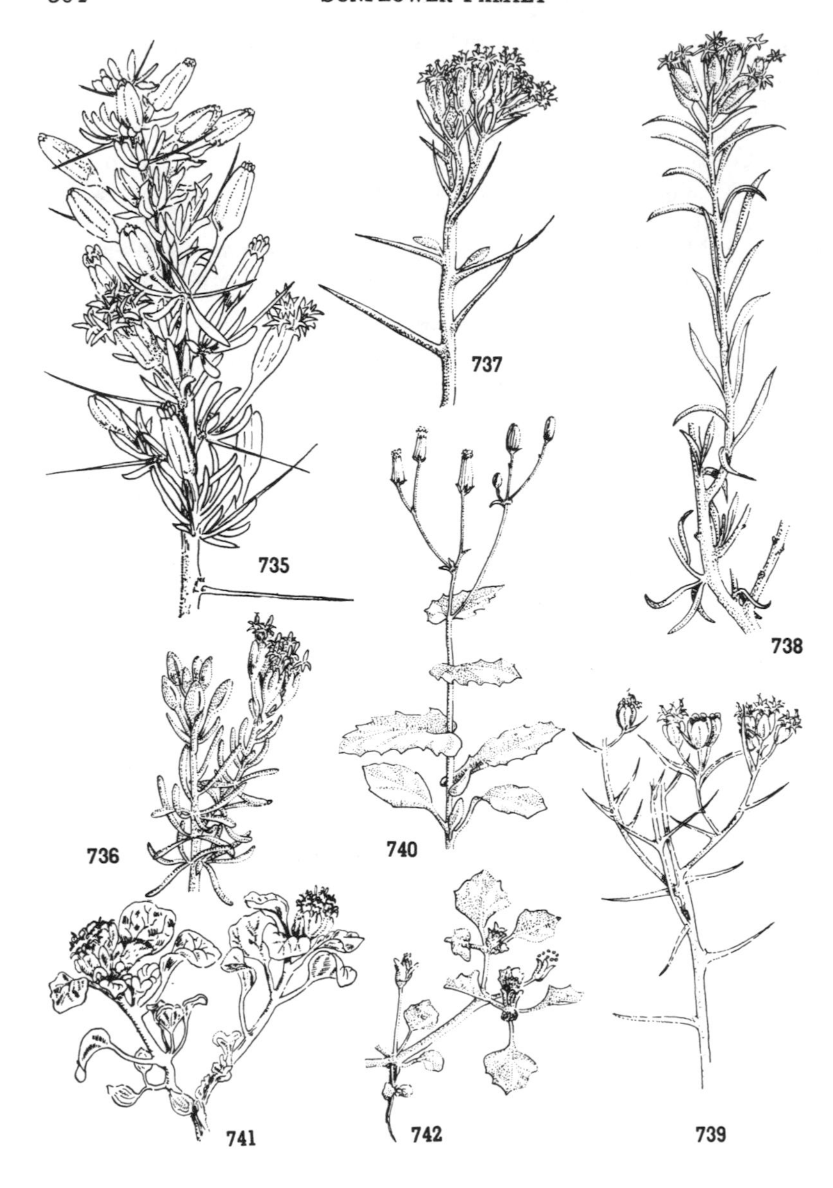

735. *Tetradymia spinosa longispina*
736. *Tetradymia glabrata*
737. *Tetradymia stenolepis*
738. *Tetradymia canescens*
739. *Tetradymia comosa*
740. *Senecio mohavensis*
741. *Psathyrotes ramosissima*
742. *Psathyrotes annua*

743. *Atrichoseris platyphylla*
744. *Senecio stygius*
745. *Senecio Douglasii*
746. *Cirsium mohavense*
747. *Trixis californica*

737. NARROW-SCALED FELT-THORN. *Tetradymia stenolepis* (Gr., narrow-scaled). **Fl.**: canary-yellow. Much-branched shrub, covered with a close-appressed, white wool. A handsome plant when in full bloom. The flowers are arranged in heads of five. A species of rather restricted distribution, occurring only in Calif. from the western arm of the Mohave D. to Inyo Co.

738. GRAY FELT-THORN. *Tetradymia canescens* (L., graying). **Fl.**: yellow. A low shrub, with freely branching stems, generally about 1 ft. tall. The pappus hairs are yellow to sordid. Dry slopes of piñon-covered mountains along the western Mohave D.; to B.C. and Utah.

739. WHITE FELT-THORN. *Tetradymia comosa* (L., hairy). **Fl.**: yellow. Erect shrub 1¼–4 ft. high, with many whitish stems and spines. A species quite common over coastal southern Calif. and occasional on our deserts; to Nev.

740. MOHAVE GROUNDSEL. *Senecio mohavensis* (L., *senex*, "old man," in allusion to the hoary pappus; of the Mohave). **Fl.**: yellow. Green annual, 8–15 in. high. The old English name, "groundsel," meaning a ground swallower, now given to several of our American *Senecios*, alludes to the luxuriant growth of some of the English species. Occasional in canyons and washes of both deserts; to Ariz.

741. VELVET ROSETTE. *Psathyrotes ramosissima* (Gr., "brittleness," because of the brittle stems and branches; L., most-branched). **Fl.**: yellow to purplish. A compact little plant, conspicuous along desert roadsides because of its velvety, gray-green leaves. The rounded "cushions," suggesting the common name, "turtle back," are sometimes almost a foot across. When crushed the leaves give off a strong, turpentine-like odor. In this species the outer involucral bracts have recurved or spreading tips. Rather common in open washes of eastern Mohave D.; Colorado D., Baja Calif., Ariz.

742. FAN–LEAF. *Psathyrotes annua* (L., annual). This rather rare species, with thinner, fan-shaped leaves, has its outer involucral bracts with erect tips. Southern Mohave D.; north to Inyo Co.; east to Nev., Utah, and Ariz.

743. PARACHUTE PLANT. *Atrichoseris platyphylla* (Gr., "without hair" + "*seris*," a cichoriaceous plant; Gr., wide-leaved). **Fl.**: white, some flowers with purple ends to rays. Sometimes called tobacco-weed because of the broad, brown-spotted, somewhat tobacco-like leaves, which lie flat to the earth. The name parachute plant was given because in large plants the inflorescence, with its canopy of white flowers and converging pedicels, appears like a parachute floating in mid-air, the lower stem being quite inconspicuous against its background of dark-surfaced soil. It is occasional to common about clay hills and on rocky soils of both deserts. The plants vary in height from 1 to 2½ ft. A rich, vanilla-like fragrance distils from the handsome flowers.

744. GROUNDSEL. *Senecio stygius* (see **740**; L., of the river Styx, hence infernal). **Fl.**: yellow. The herbage of this handsome *Senecio* is a lively green, and the lower leaves are up to 3 in. long. It is in full flower in late May in the limy soils of washes of the eastern Mohave D.

745. SAND-WASH GROUNDSEL. *Senecio Douglasii* (David Douglas—see **31**). **Fl.**: yellow. A long-blooming, almost shrubby *Senecio,* common to sandy washes of the western part of both deserts. In the mountains of the north and east Mohave D. it is replaced by *S. Monoensis,* a nearly related species, with bright green herbage and much less woody stems.

746. MOHAVE THISTLE. *Cirsium mohavense* (Gr., "a swelled vein," for which the thistle was a reputed remedy; of the Mohave). **Fl.**: phlox-pink to white. This tall biennial is our only widespread desert thistle. The lobes of the leaves are tipped with yellow spines. It occurs, sometimes in aggregations, in open, gravelly valleys, on rocky slopes, and about alkaline seeps of the Mohave D. In the far eastern Mohave D. the stout-stemmed, broad-headed New Mexico thistle, *C. neomexicani,* has been collected. Its flowers are white to pale pink. *C. californica,* with white blossoms, occurs in the Death Valley Nat. Mon. at elevations of 4,000–5,000 ft. The blossoms are larger than those of *C. mohavense,* and the stems are up to 7 ft. high.

747. TRIXIS. *Trixis californica* (Gr., "threefold," because of the trifid lower lip of the corolla; of California). **Fl.**: deep yellow. A low, rounded shrub, thickly beset with medium-sized, dark green leaves, which are somewhat inrolled at the edges and conspicuously glandular beneath. In late spring numerous heads of flowers appear at the ends of the short, tender branches, and if moisture is available blossoming continues more or less throughout summer. After the flowering season is over, the dry, fluted, involucral tubes are very conspicuous. The preferred habitat is crevices in rocks, beneath overhanging ledges, or in the shelter of bushes, where it is able to enjoy more or less shade; it is in this shaded situation that it makes its rankest growth. Colorado D.; to Tex. and Mex.

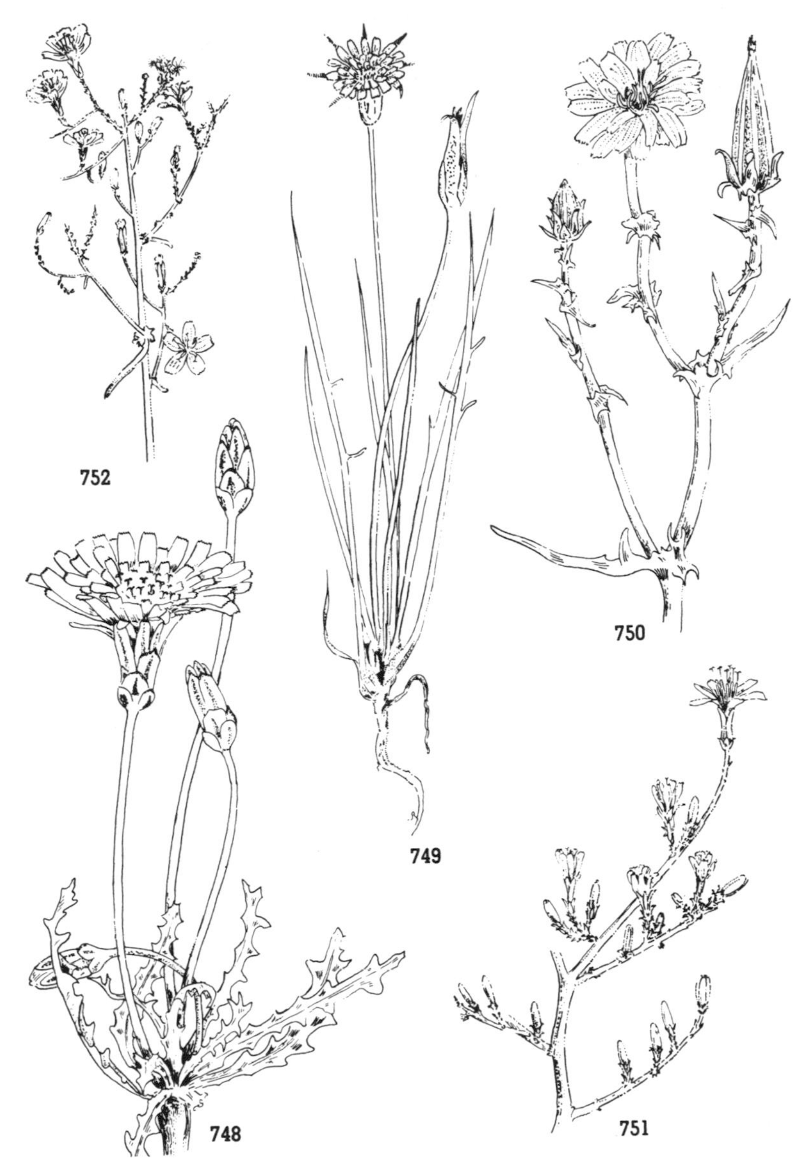

748. *Anisocoma acaulis*
749. *Microseris linearifolia*
750. *Rafinesquia californica*
751. *Stephanomeria pauciflora*
752. *Stephanomeria exigua*

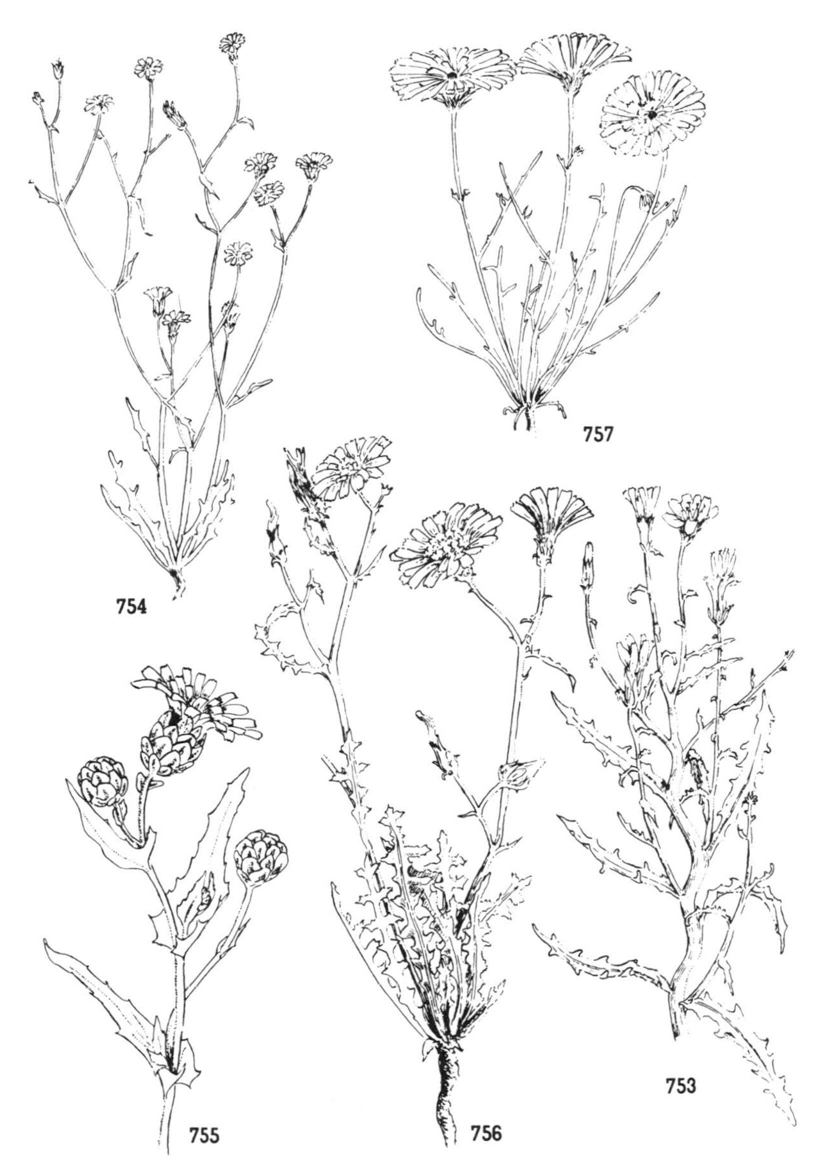

753. *Stephanomeria Parryi* 755. *Malacothrix Coulteri*
754. *Malacothrix Clevelandii* 756. *Malacothrix sonchoides*
757. *Malacothrix californica glabrata*

748. SCALE BUD. *Anisocoma acaulis* (Gr., "unequal tufts of hair," because the two sets of pappus bristles are unlike; L., stemless). **Fl.**: pale yellow. A free-flowering annual, of sandy flats and washes. The involucral bracts are often red-tipped. Both deserts, 2,000–7,500 ft.; to Nev. and Ariz.

749. SILVER-PUFFS. *Microseris linearifolia* (Gr., small + *Seris*, a name for several cichoriaceous plants, including endive and lettuce; L., linear-leaved). **Fl.**: yellow. Throughout desert and cismontane southern Calif., below 5,000 ft.; to Wash., Nev., N.Mex. Named silver-puffs because of the "puffs" which are made up of silvery-white pappi appearing after the flowers dry up.

750. CALIFORNIA CHICORY. *Rafinesquia californica* (Constantine S. Rafinesque, 1783–1840, scholarly recluse, eccentric American naturalist and traveler; of California). **Fl.**: white. Stout-branched annual, most common in cismontane southern Calif. but occasional on deserts; to Utah and Ariz. *R. neo-mexicana*, called desert chicory, is a closely related species also common on our deserts. It is a rather weak-stemmed plant, growing in the shade of shrubs and frequently climbing up through them. The outer, white ray flowers have rose-colored veins on the under side.

751. DESERT-STRAW. *Stephanomeria pauciflora* (Gr., "crown" + "division" or "part," the application being obscure; L., few-flowered). **Fl.**: pink to reddish. A woody-based, intricately branched, rounded perennial. When growing in the shelter of shrubs it sends up a long stem, then branches to form a compact, rounded mass superficially resembling, when dry, a tuft of hay or straw. Donkeys are passionately fond of it. Both deserts; to Kan. and Tex. The variety *myrioclada* (Gr., with numberless branches), having linear, almost thread-like, leaves, is known from the eastern Mohave D.

752. ANNUAL MITRA. *Stephanomeria exigua* (L., "scanty," with reference to the few leaves). **Fl.**: pink to reddish. A diffuse annual, 6 in. to 2 ft. high, found in open, sandy stretches of both deserts; to Wyo. and N.Mex. Minute, glandular hairs are found on the upper herbage. The variety *pentachaeta* (Gr., five-bristled) is a much more slender, open-growthed plant with no glands on the herbage.

753. PARRY ROCK-PINK. *Stephanomeria Parryi* (C. C. Parry—see **24**). **Fl.**: white to reddish. A rather coarse perennial, of the mountains bordering or in the desert (3,000–5,000 ft.) north to Inyo Co.; to Utah and Ariz.

754. CLEVELAND YELLOW-SAUCERS. *Malacothrix Clevelandii* (Gr., soft hair; Daniel Cleveland, 1838–1929, of San Diego, attorney and botanical collector, "whose name is well perpetuated in many specific names which give reference to his lifelong interest in the indigenous flora"). **Fl.**: bright yellow. Annual most often found in the disturbed soil along roadsides but frequent about creosote bushes. Both Calif. deserts; coastal southern Calif. to central Calif.

755. SNAKE'S-HEAD. *Malacothrix Coulteri* (Thomas Coulter, Irish botanist, who made a trip from San Gabriel to the Colorado River by way of the San Felipe Pass in the spring of 1831. He is credited with being the "botanical discoverer of the Colorado Desert." For him the Coulter pine was named). **Fl.**: light yellow. Common from March to May in sandy basins of both Calif. deserts; to San Joaquin Valley; and Utah.

756. YELLOW-SAUCERS. *Malocothrix sonchoides* (L., like *Sonchus,* the sow-thistle). **Fl.**: bright yellow. Annual, 4–11 in. high, with fragrant flowers. Sandy soils of the Mohave D., above 2,500 ft., to Inyo Co.; Ariz., Ida., and Neb.

757. DESERT DANDELION. *Malacothrix californica glabrata* (of California; L., "bald," in allusion to the hairless herbage). **Fl.**: canary-yellow, often with a small, bright red "button" in the center. An annual, common to the open, sandy basins and bordering valleys of both deserts; to Nev. and Ore. Often a single plant has a dozen flower heads all in blossom at the same time.

758. YELLOW TACK-STEM. *Calycoseris Parryi* (Gr., "cup" + "*Seris,*" a cichoriaceous genus; Dr. C. C. Parry—see **24**). **Fl.**: yellow. Called tack-stem because of the numerous dark-colored, tack-shaped glands present on the stems. The plants grow from 4 in. to 2 ft. high. The reverse side of each petal is brownish. Sandy soils of both deserts.

759. WHITE TACK-STEM. *Calycoseris Wrightii* (C. Wright—see **209**). **Fl.**: rays white, with rose or purplish dots or streaks on back. In this *Calycoseris* the tack-shaped glands are pale. Mohave and western Colorado deserts; to Texas.

760. THORNY SKELETON-PLANT. *Lygodesmia spinosa* (Gr., "a pliant twig" + "bundle," because the typical species has long, flexible, fasciculate stems; L., spiny). **Fl.**: rose to pink. A spiny perennial, with bright green herbage and queer, 3-rayed heads. At first sight the plant appears much like a *Stephanomeria* (see **752**). The woody

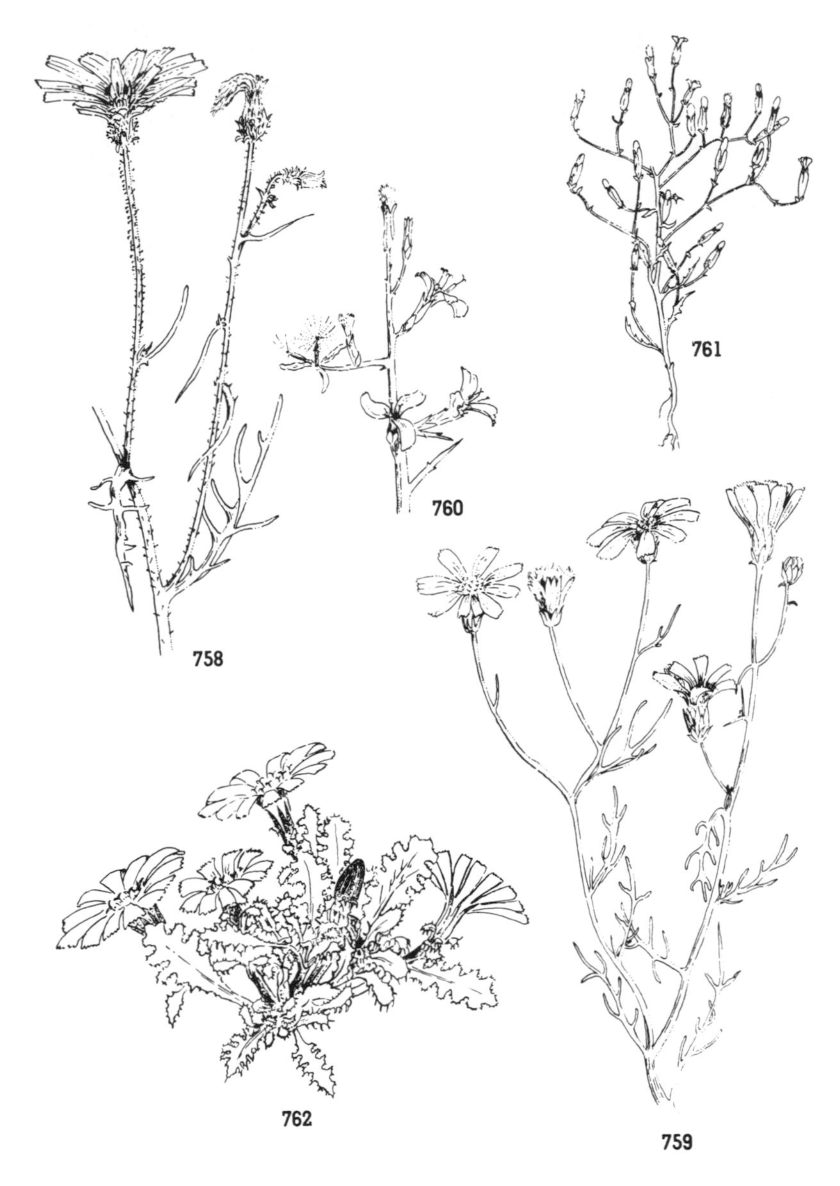

758. *Calycoseris Parryi*
759. *Calycoseris Wrightii*
760. *Lygodesmia spinosa*
761. *Lygodesmia exigua*
762. *Glyptopleura setulosa*

763. *Crepis occidentalis*

764. *Hymenopappus filifolius*

root crowns bear peculiar white, woolly tufts about the size of hazelnuts, said to be used by the Indians to stanch the flow of blood in wounds. Both deserts; to Colo. and Tex.

761. EGBERTIA. *Lygodesmia exigua* (L., scanty). **Fl.**: whitish. A low spring annual, 4–12 in. high, most attractive because of the numerous glistening, white pappus heads. It is especially conspicuous when growing on areas covered with blackish or red rock. Colorado and Mohave deserts; to Utah and N.Mex. Named Egbertia in honor of Adrian Egbert of Cave Springs, who for years maintained supplies of water along lonely desert roads in case of need by travelers. The water was kept in 5-gallon bottles under shelters of wood.

762. KEYSIA. *Glyptopleura setulosa* (Gr., "carved side," referring to the achenes; L., bristly). **Fl.**: creamy or yellowish. Very leafy, tufted, sweet-scented annual, of the Mohave sands. The light green leaves showing a narrow margin of bright white are almost smothered in blossoms during the flowering period. This attractive desert plant is named *Keysia* in honor of Mr. and Mrs. Wm. Keys, pioneers of the Little San Bernardino Mts., whose perennial kindness is known to so many desert travelers. In the mid-northern and eastern Mohave D. grows the closely related *G. marginata,* with white flowers (aging red) and more conspicuous, white leaf margins.

763. WESTERN CREPIS. *Crepis occidentalis* (Gr., *krepis,* "a sandal," also an ancient plant name; L., western). **Fl.**: yellow. Low, leafy perennial, with grayish-green herbage, occurring in the juniper and sagebrush belt and coming into flower late in May. Inyo and Coso mountains; north to Lake Co. and east to Colo.

764. YELLOW CUT-LEAF. *Hymenopappus filifolius* (Gr., membrane pappus; L., thread-leaved). **Fl.**: clear yellow. The several clustered stems arise from a perennial root. The stems are leafy, often very much so below but naked above. Both leaves and stems are white-woolly.

Anthony Zipprich

Index of Persons for Whom Desert Plants Have Been Named

[References are to numbered items instead of page numbers.]

General Index

[References are to numbered items unless otherwise stated.]

New Plant Names

As field and laboratory studies of plants are made by taxonomists and geneticists and the literature of botany is more carefully scrutinized, new scientific names are appropriately and necessarily given to some of our desert plants. Such new names as have been listed in Munz and Keck's *A California Flora* (University of California Press, Berkeley, 1959) are given here. The new names are listed in the left column; the old ones are printed opposite them on the right. This list will be helpful to those who wish to familiarize themselves with the more recent appellations.

Allium nevadense cristatum	*Allium cristatum*
Amelanchier utahensis Covillei	*Amelanchier alnifolia Covillei*
Ammoselinum giganteum	*Ammoselinum occidentalis*
Arceuthobium campylopodum divaricatum	*Arceuthobium divaricatum*
Argymone corymbosa	*Argemone intermedia corymbosa*
Artemisia ludoviciana albula	*Artemisia albula*
Asclepias asperula	*Asclepiodora decumbens*
Aster intricatus	*Aster carnosus*
Astragalus didymocarpus dispermus	*Astragalus dispermus*
Astragalus lentiginosus borregoensis	*Astragalus aginus*
Astragalus panamintensis	*Astragalus atratus panamintensis*
Bebbia juncea	*Bebbia juncea aspera*
Boerhaavia coccinea	*Boerhaavia caribaea*
Boerhaavia erecta intermedia	*Boerhaavia intermedia*
Brodiaea pulchella pauciflora	*Brodiaea capitata pauciflora*
Castilleja chromosa	*Castilleja angustifolia*
Caulanthus inflatus	*Streptanthus inflatus*
Ceanothus Greggii vestitus	*Ceanothus vestitus*
Chrysopsis villose hispida	*Chrysopsis hispida*
Condalia globosa pubescens	*Condalia spathulata*
Cordyllanthus ramosus eremicus	*Cordyanthus eremicus*
Cowania mexicana Stansburiana	*Cowania Stansburiana*
Cynanchium utahense	*Astephanus utahensis*
Dalea	*Parosela*
Dudleya saxosa	*Echeveria saxosa*
Echinocactus Johnsonii	*Echinocactus Johnsonii octocentrus*
Encelia frutescens actoni	*Encelia actoni*
Enceliopsis argophylla	*Enceliopsis Covillei*
Erigeron nevadicola	*Erigeron nevadensis*
Erigeron pumilus concinnoides	*Erigeron concinnus*
Eriogonum Baileyi brachyanthum	*Eriogonum brachyanthum*
Eriogonum deflexum Rixfordii	*Eriogonum Rixfordii*

Eriogonum latifolium nudum	*Eriogonum nudum*
Eriastrum densifolium	*Gilia densifolia*
Eriastrum eremicum	*Gilia eremica*
Erysimum capitatum	*Erysimum asperum*
Euphorbia incisa	*Euphorbia schizoloba*
Fagonia californica laevis	*Fagonia chilensis laevis*
Fendlerella utahensis	*Whipplea utahensis*
Forsellesia nevadensis	*Glossopetalon spinescens*
Forsellesia pungens	*Glossopetalon pungens*
Franseria confertiflora	*Franseria tenuifolia*
Fremontia californica	*Fremontodendron californicum*
Galium Wrightii Rothrockii	*Galium Rothrockii*
Gilia latiflora Davyi	*Gilia Davyi*
Gutierrezia microcephala	*Gutierrezia lucida*
Hamilolobus diffusa Jaegeri	*Sirymbrium diffusum*
Haplopappus Goodingii	*Haplopappus spinulosus Goodingii*
Hedeoma nana californica	*Hedeoma thymoides*
Helianthus petiolaris canescens	*Helianthus canus*
Hulsea calicarpha	*Hulsea vestita calicarpha*
Hymenoxys acaulis ludoviciana	*Actinea acaulis*
Hymenoxys Cooperi	*Hymenoxys biennis*
Hymenoxys odorata	*Hymenoxys chrysanthemoides*
Ipomopsis depressa	*Gilia depressa*
Ipomopsis polycladon	*Gilia polycladon*
Langloisia Matthewsii	*Gilia Matthewsii*
Langloisia setosissima	*Gilia setosissima*
Langloisia Shottii	*Gilia Shottii*
Leptodactylon pungens	*Gilia pungens*
Leucelene ericoides	*Aster Leucelene*
Linanthus aureus	*Gilia aurea*
Linanthus breviculus	*Gilia brevicula*
Linanthus dichotoma	*Gilia dichotoma*
Linanthus maculatus	*Gilia maculata*
Linanthus Parryae	*Gilia Parryae*
Linyum perenne Lewisii	*Linum Lewisii*
Lotus procumbens	*Lotus leucophyllus*
Lupinus arizonicus	*Lupinus sparsiflorus arizonicus*
Lupinus flavoculatus	*Lupinus rubens*
Machaeranthera cognata	*Aster cognatus*
Malacothrix glabrata	*Malacothrix californica glabrata*
Matelea parvifolia	*Vincetoxicum hastulatum*
Mentzelia leucophylla	*Mentzelia oreophila*
Mentzelia multiflora	*Mentzelia longiloba*
Mirabilus Bigelovii aspera	*Mirabilis aspera*
Muilla maritima serotina	*Allium serotina*
Nemadadus gracilis	*Nemadadus ramosissimus gracilis*
Opuntia erinacea ursina	*Opuntia ursina*
Oxybaphus comatus	*Allionia comata*
Penstemon antirhiniodes microphyllus	*Penstemon microphyllus*
Petrophytum caespitosum	*Spirea caespitosa*
Phacelia Fremontii	*Phacelia Hallii*
Phoradendron Bolleanum densum	*Phoradendron densum*
Phoradendron flavescens macrophyllum	*Phoradendron coloradense*
Phoradendron juniperinum	*Phoradendron ligatum*
Plantago Purshii picta	*Plantago spinulosa oblonga*
Proboscidia althaefolia	*Martynia althaefolia*
Proboscidia parviflora	*Martynia parviflora*
Rosa Woodsii glabrata	*Rosa mohavensis*
Salvia Dorrii	*Salvia carnosa pilosa*
Sarcostemma hirtellum	*Funastrum hirtellum*
Sesbania exaltata	*Sesbania macrocarpa*
Simmondsia chinensis	*Simmondsia californica*
Solidago pumila	*Solidago Petrodoria*
Sphaeralcea ambigua monticola	*Sphaeralcea pulchella*
Sphaeralcea ambigua rosacea	*Sphaeralcea rosacea*
Sphaeralcea Emoryi variabilis	*Sphaeralcea Fendleri californica*
Tragia stylaris	*Tragia ramosa*
Tetrococcus Hallii	*Halliophytum Hallii*
Tetradymie axillaris	*Tetradymia spinosa longispina*
Thelypodium integrifolium	*Thelypodium affine*
Thelypodium lasiophyllum	*Caulanthus lasiophyllus*
Trifolium Wormskioldii	*Trifolium involucratum*